A GUIDE TO THE MAMMALS OF OHIO

Meadow Jumping Mouse *(Zapus hudsonius)* — Watercolor by John A. Ruthven

Jack L. Gottschang

A GUIDE TO THE MAMMALS OF OHIO

With color and black-and-white photographs by
Karl Maslowski, Peter Maslowski, Stephen Maslowski,
Alvin Staffan, William Farr, and Ronald Austing

With original drawings by Elizabeth Dalvé

With a frontispiece by John A. Ruthven

Published by The Ohio State University Press
in cooperation with The Ohio Biological Survey

QL719
O3
G67

Copyright © 1981 by the Ohio State University Press
All Rights Reserved.

Library of Congress Cataloguing in Publication Data

Gottschang, Jack L   1923–
  A guide to the mammals of Ohio.
  Includes Index.
  1. Mammals—Ohio—Identification. I. Title.
QL719.03G67   599.09771   80-27661
ISBN 0-8142-0242-X

CONTENTS

Acknowledgments ix
Preface xi
Introduction 3
  General information about mammals 3
  Natural environment of Ohio 4
  Explanation of distribution maps 5
  Use of keys for identifying Ohio mammals 7
  Key to the orders of mammals in Ohio 9
  Checklist of the 54 living mammals of Ohio 11
ORDER MARSUPIALIA Pouched mammals 13
  Family Didelphidae 13
    Virginia opossum (*Didelphis virginiana*) 13
ORDER INSECTIVORA Moles and shrews 17
  Key to the families and species of Insectivora in Ohio 17
  Family Soricidae—Shrews 17
    Masked shrew (*Sorex cinereus*) 18
    Smoky shrew (*Sorex fumeus*) 21
    Pygmy shrew (*Microsorex hoyi*) 22
    Short-tailed shrew (*Blarina brevicauda*) 23
    Least shrew (*Cryptotis parva*) 26
  Family Talpidae—Moles 28
    Hairy-tailed mole (*Parascalops breweri*) 28
    Eastern mole (*Scalopus aquaticus*) 30
    Star-nosed mole (*Condylura cristata*) 32
ORDER CHIROPTERA Bats 35
  Family Vespertilionidae—Bats 36
    Key to the species of bats in Ohio 37
    Little brown myotis (*Myotis lucifugus*) 38
    Keen's myotis (*Myotis keenii*) 41
    Indiana myotis (*Myotis sodalis*) 41
    Small-footed myotis (*Myotis leibii*) 43
    Silver-haired bat (*Lasionycteris noctivagans*) 43
    Eastern pipistrelle (*Pipistrellus subflavus*) 44
    Big brown bat (*Eptesicus fuscus*) 46
    Red bat (*Lasiurus borealis*) 48
    Hoary bat (*Lasiurus cinereus*) 49
    Evening bat (*Nycticeius humeralis*) 50
    Rafinesque's big-eared bat (*Plecotus rafinesquii*) 51
ORDER LAGOMORPHA Rabbits and hares 53
  Eastern cottontail (*Sylvilagus floridanus*) 53
ORDER RODENTIA Rodents 59
  Key to the families of Rodentia in Ohio 59
  Family Sciuridae—Woodchucks, squirrels, chipmunks, etc. 60
    Key to the species of Sciruidae in Ohio 60
    Eastern chipmunk (*Tamias striatus*) 60
    Woodchuck (*Marmota monax*) 63
    Thirteen-lined ground squirrel (*Spermophilus tridecemlineatus*) 66
    Gray squirrel (*Sciurus carolinensis*) 69
    Fox squirrel (*Sciurus niger*) 73
    Red squirrel (*Tamiasciurus hudsonicus*) 74
    Southern flying squirrel (*Glaucomys volans*) 77
  Family Castoridae—Beaver 80
    Beaver (*Castor canadensis*) 80
  Family Cricetidae—Native rats and mice, voles, and lemmings 83
    Key to the species of Cricetidae in Ohio 83
    Eastern harvest mouse (*Reithrodontomys humulis*) 84
    Deer mouse (*Peromyscus maniculatus*) 86
    White-footed mouse (*Peromyscus leucopus*) 88
    Eastern woodrat (*Neotoma floridana*) 91
    Southern red-backed vole (*Clethrionomys gapperi*) 94
    Meadow vole (*Microtus*

CONTENTS

  pennsylvanicus) 95
  Prairie vole (*Microtus ochrogaster*) 98
  Woodland vole (*Microtus pinetorum*) 99
  Muskrat (*Ondatra zibethicus*) 102
  Southern bog lemming (*Synaptomys cooperi*) 105
 Family Muridae—Old World rats and mice 107
  Key to the species of Muridae in Ohio 107
  Norway rat (*Rattus norvegicus*) 108
  House mouse (*Mus musculus*) 110
 Family Zapodidae—Jumping mice 112
  Key to the species of Zapodidae in Ohio 112
  Meadow jumping mouse (*Zapus hudsonius*) 112
  Woodland jumping mouse (*Napaeozapus insignis*) 115
ORDER CARNIVORA Flesh eaters 119
 Key to the families of Carnivora in Ohio 119
 Family Canidae—Dogs, foxes, and wolves 119
  Key to the species of Canidae in Ohio 120
  Coyote (*Canis latrans*) 120
  Red fox (*Vulpes vulpes*) 123
  Gray fox (*Urocyon cinereoargenteus*) 126
 Family Procyonidae—Raccoons and allies 128
  Racoon (*Procyon lotor*) 128
 Family Mustelidae—Weasels, skunk, mink, otter, and badger 131
  Key to the species of Mustelidae in Ohio 131
  Ermine (*Mustela erminea*) 131
  Least weasel (*(Mustela nivalis*) 132
  Long-tailed weasel (*Mustela frenata*) 134
  Mink (*Mustela vison*) 137
  Badger (*Taxidea taxus*) 138
  Striped skunk (*Mephitis mephitis*) 140
ORDER ARTIODACTYLA Even-toed hoofed Ungulates 143
 Key to the skulls of Ungulates in Ohio 143
 Family Cervidae—Deer 143
  White-tailed deer (*Odocoileus virginianus*) 143
Extirpated species and those of incidental or doubtful occurrence 151
Key to the skulls of Ohio mammals 159
Appendix Table 1. Selected English-metric conversions 163
Appendix Table 2. Total pelts reported by Ohio fur buyers, 1951–52 through 1977–78, and monetary value of Ohio fur harvests, 1965–66 through 1977–78 164
Appendix Table 3. Summary of Ohio deer hunting seasons, 1943–78 165
Glossary 167
General References 171
Index 175

FIGURES

1. Landforms map of Ohio indicating limits of glaciation 5
2. Counties of Ohio 6
3. Common body measurements of mammals 8
4. Lateral view of mammal skull types 9
5. Mammal tooth types 10
6. Female Virginia opossum (*Didelphis virginiana*) with young 14
7. Contrasting head characteristics of selected groups of Insectivora 18
8. Adult masked shrew (*Sorex cinereus*) on a fallen sugar maple leaf 19
9. Short-tailed shrew (*Blarina brevicauda*) showing side and belly glands 25
10. Skeleton of a big brown bat (*Eptesicus fuscus*) 35
11. Characteristics used in identifying bats 37
12. Dorsal view of eastern cottontail (*Sylvilagus floridanus*) skull 54
13. Female eastern cottontail (*Sylvilagus floridanus*) nursing young 55
14. Tooth enamel patterns in rodents 59
15. Ventral view of skulls of fox squirrel (*Sciurus niger*) and gray squirrel (*Sciurus carolinensis*) 60
16. Red squirrel (*Tamiasciurus hudsonicus*) 75
17. Ventral view of skull of the deer mouse (*Peromyscus maniculatus*), comparing anterior palatine foramina with that of the white-footed mouse (*P. leucopus*) 85
18. Enamel patterns on third molar of upper jaw of *Microtus* 85
19. Eastern woodrat (*Neotoma floridana*) 92
20. Ventral view of skull of the house mouse (*Mus musculus*), showing the first molar larger than second and third combined 107
21. Lateral view of the house mouse (*Mus musculus*) upper incisor, showing subapical notch 111
22. Dorsal view of skulls and lateral view of mandibles of red fox (*Vulpes vulpes*) and gray fox (*Urocyon cinereoargenteus*), showing distinguishing characteristics 121
23. Contrasting front views of skulls of the domestic dog (*Canis familiaris*) and the coyote (*Canis latrans*) 122
24. Anterior end of the coyote (*Canis latrans*) skull, ventral view 122
25. Mature white-tailed buck (*Odocoileus virginianus*), with eight points 144
26. Growth of the Ohio deer herd and deer taken by hunters, 1965–1979 145

COLOR PLATES

Frontispiece
  Meadow jumping mouse (*Zapus hudsonius*)
Facing page 36
  Short-tailed shrew (*Blarina brevicauda*) (albino)
  Least shrew (*Cryptotis parva*)
  Eastern mole (*Scalopus aquaticus*)
Facing page 37
  Little brown myotis (*Myotis lucifugus*)
  Big brown bat (*Eptesicus fuscus*)
  Red bat (*Lasiurus borealis*) with young
Facing page 52
  Hoary bat (*Lasiurus cinereus*)
  Eastern cottontail (*Sylvilagus floridanus*)
  Eastern chipmunk (*Tamias striatus*)
Facing page 53
  Woodchuck (*Marmota monax*)
  Thirteen-lined ground squirrel (*Spermophilus tridecemlineatus*)
  Fox squirrel (*Sciurus niger*)
Facing page 100
  Southern flying squirrel (*Glaucomys volans*)
  Beaver (*Castor canadensis*)
  Eastern harvest mouse (*Reithrodontomys humulis*)

Facing page 101
  White-footed mouse (*Peromyscus leucopus*)
  Meadow vole (*Microtus pennsylvanicus*)
  Woodland vole (*Microtus pinetorum*)
Facing page 116
  Southern bog lemming (*Synaptomys cooperi*)
  House mouse (*Mus musculus*)
  Meadow jumping mouse (*Zapus hudsonius*)
Facing page 117
  Red fox (*Vulpes vulpes*)
  Least weasel (*Mustela nivalis*) (summer)
  Least weasel (*Mustela nivalis*) (winter)
Facing page 132
  Long-tailed weasel (*Mustela frenata*)
Facing page 133
  Mink (*Mustela vison*)
  Badger (*Taxidea taxus*)
Facing page 148
  Striped skunk (*Mephitis mephitis*)
  Raccoon (*Procyon lotor*)
Facing page 149
  White-tailed deer (*Odocoileus virginianus*)

ACKNOWLEDGMENTS

This book has been many years in the making, and any attempt to name every person who has helped during its long progress would only result in unfortunate and unintentional omissions. Rather, I will say thank you to everyone who has in any way helped with the book; your efforts are appreciated! The late Dr. Charles Dambach originally encouraged me in this effort, and since that time the Ohio Biological Survey has continued to aid with finances, advice, and encouragement. Dr. Charles King and Nicki King have been especially helpful over the past several years, and to them I owe a special debt of gratitude.

I certainly want to thank John Ruthven for painting the jumping mouse that appears as the frontispiece of the book; Karl Maslowski, Ron Austing, and Alvin Staffan for their excellent photographic contributions; and Mrs. Elizabeth Dalvé for her drawings and the original distribution maps. Mr. Chris Stotler has carefully checked the distribution maps, and many of the records were supplied by him.

Although I have received much help with the book, I am completely and totally responsible for any and all errors, omissions, and oversights. Please call them to my attention.

Finally, the book is not intended to be a finished product—indeed, my greatest hope is that it will stimulate others to add and report new findings about the wild mammals living in Ohio.

J.L.G.

PREFACE

This book is a guide to the wild mammals living in Ohio today. It is hoped that it will be useful to the layman who desires to know more about these mammals, to the professional biologist and mammalogist who desires more technical information about Ohio mammals, and to the student who is looking for guidelines and starting points in the study of mammalogy. For most species information is presented regarding distribution and status in Ohio, identifying characteristics including dental formula, brief descriptions of life history and habits, occasionally taxonomic variations, and selected references. It will soon be obvious to the reader that there is only a limited amount of information presented for certain species of mammals and that sizable gaps occur in our knowledge of them. The reasons for writing this book are to call attention to these gaps, to stimulate interest, and to invite further investigation into the lives of Ohio mammals.

The fossil record indicates that mammoths, mastodons, giant pigs, giant beavers, ground sloths, and camels, as well as a large assemblage of smaller mammals, once roamed over Ohio and were an important and impressive part of the state's prehistoric fauna. The records for these early mammals have not been brought together in a single publication, but they have been reported in numerous single papers scattered through the early literature. For a discussion of this group, see LaRocque and Marple (1966), Hibbard et al. (1965), and Martin and Wright (1967).

Jared P. Kirtland (1838), in *A Catalogue of the Mammalia, Birds, Reptiles, Fishes, Testacea, and Crustacea in Ohio,* listed 50 kinds of mammals for the state. Brayton (1882) presented a report on the mammals of Ohio, and Thomas and Bole (1938) compiled an even more complete report in *List of the Mammals of Ohio.* The most recent listing is that of Bole and Moulthrop (1942) entitled *The Ohio Recent Mammal Collection in the Cleveland Museum of Natural History,* a summary of the authors' previous ten years' of research. Considerable work has subsequently been done with Ohio mammals and forms the basis for this book.

SELECTED REFERENCES

Bole, B. P., Jr., and P. N. Moulthrop. 1942. The Ohio recent mammal collection in the Cleveland Museum of Natural History. Sci. Publ. Cleveland Mus. Nat. Hist. 5:83-181.

Brayton, A. W. 1882. Report on the mammals of Ohio. Rep. Ohio Geol. Surv. 4:1-185.

Hibbard, C. W., D. E. Ray, D. E. Savage, D. W. Taylor, and J. E. Guilday. 1965. Quaternary mammals of North America. Pp. 509-25 *in* H. E. Wright, Jr., and D. G. Frey, eds., The Quaternary of the United States. Princeton Univ. Press, Princeton, N.J.

Kirtland, J. P. 1838. Report on the zoology of Ohio. Second Annu. Rep. Ohio Geol. Surv. 2:157-200.

LaRocque, A., and M. F. Marple. 1966. Ohio fossils. Ohio Geol. Surv. Bull. 54. 152 pp.

Thomas, E. S., and B. P. Bole, Jr. 1938. List of the mammals of Ohio. Bull. Ohio State Mus. 1:1-2.

A GUIDE TO THE MAMMALS OF OHIO

INTRODUCTION

General Information about Mammals

Because all mammals possess hair and mammary glands, they are not apt to be confused with any other kind of vertebrate animal (fish, amphibians, reptiles, or birds). Most of them, however, are inconspicuously colored, secretive, or nocturnal in their habits and thus seldom seen. A line of small mouse traps baited with peanut butter and oats and placed in strategic locations along underground tunnels or in runs just beneath the grass cover will often reveal the presence of a surprising number of heretofore-unnoticed small mammals.

Hair is the one distinguishing feature of a mammal. Collectively, the hairs covering the body are referred to as the coat or pelage. The fur is that part of the coat that consists of short, soft, thick hairs that lie next to the skin. Guard hairs lie over, and stand above, the fur or underhair; they are longer, generally more coarse, and spaced farther apart. Guard hairs primarily protect the fur, and the fur prevents loss of body heat or external penetration of cold air. Hairs are primarily dead structures produced by the skin and are periodically replaced by new hairs. The pushing out or shedding of hairs is referred to as molting. Most, although certainly not all, of our local mammals molt twice each year, once at the beginning of the summer and again at the onset of winter. Since the new coat is usually brighter than the old, and since summer and winter coats may be quite different in color, molt patterns and stages of molt are mentioned in many of the species descriptions. Many immature mammals are colored very differently from the adults and may go through several color changes before reaching adulthood.

In the text a distinction is made between immature, subadult, and adult animals. By definition immature animals are those that are still dependent upon the parents and probably have not yet left the nest. Subadults are free of the parents but are reproductively inactive, whereas adults are reproductively active or at least are capable of reproducing at certain periods of the year. Sexual maturity is reached at different ages by different species of mammals; generally speaking, the smaller the animal, the earlier it reaches sexual maturity. Mice, shrews, and perhaps moles are capable of reproducing when two or three months old. Most squirrels, rabbits, and other medium-sized mammals do not mate until they are approximately one year old. Many larger forms such as deer and bear do not mate until the spring of their third or fourth year.

Once a mammal has reached sexual maturity, it usually follows a fairly predictable yearly breeding cycle. The smaller species become sexually active in early spring; they produce several litters before midsummer and then stop reproducing until fall, when one or possibly two more litters are born. Medium-sized mammals (raccoons, opossum, fox) have only one or two litters in the spring of the year, and the larger mammals produce a single litter per year or less (a female bear usually has only a single cub every other year). Both males and females of most species are reproductively inactive throughout the winter months (from about October through mid-February in Ohio). The reproductive tracts of both male and female wild mammals noticeably decrease in size during the winter; the testes of males of most small and medium-sized species migrate from the scrotum into the abdominal cavity. During the nonbreeding season it is often difficult, even for the expert, to distinguish between males and females of many Ohio mammals on the basis of external appearance alone.

Most species of mammals inhabit only a particular kind of habitat. The white-footed mouse most commonly occurs in wooded areas, whereas the closely related eastern harvest mouse is strictly a field inhabitant. Gray squirrels live only in areas where there are trees, and the eastern mole seldom ventures out of its subterranean tunnels. A few species are found in both open fields and woodlands. One such species, the gray fox, may utilize underground dens and may even ocasionally climb trees. Within these primary habitats, which may extend for many miles and include hundreds or even thousands of acres, each animal has its own home range. An animal gets its food, rears its young, and has its home within the home range. The home range may be smaller than one acre (as in the case of a meadow vole) or as large as several square miles (as in the case of a predator such as a fox). Depending upon food availability, weather conditions, population pressures, and a number of other inadequately understood factors, an animal may use the same home range for years or for life, or may change home ranges several times within a single year. The area within its home range that immediately surrounds the animal's home is called its territory. Whereas several home ranges within a given area may overlap, territorial boundaries almost never overlap. Ordinarily an individual vigorously defends its home territory against not only other kinds of animals but also any member of the same species. However, members of the opposite sex are often tolerated, especially if the resident animal is not in the process of raising a family.

Natural Environment of Ohio

The state of Ohio is located just west of the Appalachian Mountains and south of the Great Lakes in the eastern portion of North America. It is situated between 38°27′ and 41°57′ N latitude and 80°34′ and 84°49′ W longitude, and extends approximately 338 km (210 mi) from north to south and 360 km (225 mi) from east to west. Ohio is generally exposed to a continental temperate climate with abundant and evenly distributed precipitation (approx. 90 cm, or 35″, per year) and with occasional temperature extremes of below −26°C to 38°C (−15°F to 100°F). The state is underlaid with sedimentary rocks formed during the Paleozoic Era 600 million to 200 million years ago. The older rocks (Ordovician, Silurian, and Devonian), primarily limestones, dolomites, and calcareous shales, are generally alkaline and outcrop in the western half of the state; the younger rocks (Mississippian, Pennsylvanian, and Permian), primarily sandstones, shales, and coals, are generally acidic and outcrop in the eastern half of the state. Elevations range from a high of 473 m (1,550 ft) above sea level near Bellefontaine in Logan County to a low of 138 m (450 ft) above sea level in the Ohio River at the mouth of the Great Miami River in extreme southwestern Hamilton County. The western and central two-thirds of the state lie in those physiographic sections known as the Lake Plains and the Till Plains. This is relatively flat, rolling country with little or no relief. It is characterized by large unbroken farmlands and numerous scattered, deciduous woodlots. Some elm-ash and other swamp forest remnants are found primarily in the Lake Plains section, and remnants of beech, mixed-oak, and prairie types are found in the Till Plains section. The existing typography is a product of the Illinoian and Wisconsinan glaciers (fig. 1), the last of which disappeared barely 13,000 years ago.

The eastern one-third of the state occupies the Allegheny Plateau and can be further subdivided into two principal physiographic sections. The northern one-third of the plateau, like the Till Plains, was invaded by the glaciers and is known as the Glaciated Allegheny Plateau. This is hilly, rolling country covered with deciduous forests of mixed oak, mixed mesophytic, and beech types. The remaining southeastern section comprises the Unglaciated Allegheny Plateau. The greatest local relief within the state occurs here along the Ohio River, in some places up to 244 m (800 ft). Mixed mesophytic and mixed oak forest types are typical of the area.

For more detailed discussions of the characteristics of the natural environment of Ohio, the reader is referred to Braun (1950); Gordon (1969); Stout, VerSteeg, and Lamb (1943); and Trautman (1957, 1977).

REFERENCES CITED

Braun, E. L. 1950. Deciduous forests of eastern North America. Blakiston Co., Philadelphia, Pa. Reprinted 1964. Hafner Publ. Co., New York. 596 pp.

Gordon, R. B. 1969. The natural vegetation of Ohio in pioneer days. Ohio Biol. Surv. Bull. N.S. 3(2). 113 pp.

INTRODUCTION 5

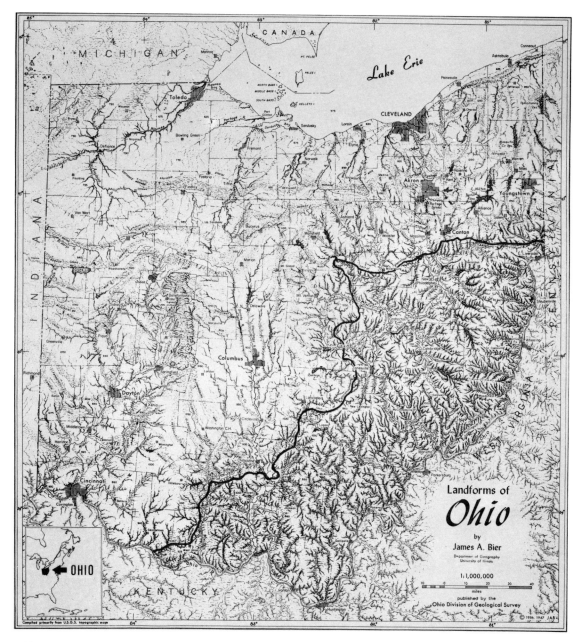

Fig. 1. Landforms map of Ohio indicating limits of glaciation. (Adapted from, and used by courtesy of, James A. Bier and the Ohio Division of Geological Survey.)

Stout, W. E., K. VerSteeg, and G. F. Lamb. 1943. Geology of water in Ohio. Ohio Geol. Surv. Bull. 44. 694 pp.

Trautman, M. B. 1957. The fishes of Ohio. Ohio State Univ. Press, Columbus, Ohio. 683 pp.

———. 1977. The Ohio country from 1750 to 1977—a naturalist's view. Ohio Biol. Surv. Biol. Notes No. 10. 25 pp.

Explanation of Distribution Maps

The characteristics of the natural environment have been important factors in determining the distribution of mammals in Ohio. However, the specific implications are imperfectly understood. To complicate the system further, man has so altered the environment that current distribution of certain species no longer reflects original distribution patterns.

Distribution maps are presented for most species in two forms:

1. Range maps. Range in Ohio is presented by a shaded pattern as determined in most instances from standard references such as Burt (1957), Hall and Kelson (1959), and Burt and

Grossenheider (1976). Ranges of a few species are not indicated because of sparse records. Much more trapping data and specifi information must be collected before the complete range for most mammals in Ohio can be mapped.

2. Dot maps. A large dot in the center of a county indicates that a known record exists for the given species in that county. Absence of a dot in any county does not necessarily mean that the species does not occur there but simply that no record for that species known to me exists for that county. Names of counties in Ohio are shown in figure 2.

The records as indicated by the dot maps are based on:

(1) my trapping results
(2) specimens that I have observed or records that I have received from the following museums and agencies:

Bowling Green State University, Bowling Green, Ohio

University of California, Museum of Vertebrate Zoology, Berkeley, California

Cincinnati (Society) Museum of Natural History, Cincinnati, Ohio

University of Cincinnati Museum, Cincinnati, Ohio

Fig. 2. Counties of Ohio.

Cincinnati Zoo, Cincinnati, Ohio
Cleveland Museum of Natural History, Cleveland, Ohio
Dayton Museum of Natural History, Dayton, Ohio
Denison University, Granville, Ohio
Earlham College, Richmond, Indiana
University of Illinois, Museum of Natural History, Urbana, Illinois
Indiana University, Bloomington, Indiana
University of Kentucky, Lexington, Kentucky
Miami University, Oxford, Ohio
University of Michigan, Museum of Zoology, Ann Arbor, Michigan
University of Minnesota, J. F. Bell Museum of Natural History, Minneapolis, Minnesota
Ohio Department of Natural Resources, Division of Wildlife, Columbus, Ohio
The Ohio State University, Museum of Zoology, Columbus, Ohio
Ohio University, Athens, Ohio
United States National Museum, Washington, D.C.
West Virginia University, Morgantown, West Virginia

(3) records that I believe to be valid that have appeared in earlier publications
(4) information supplied by Mr. G. Chris Stotler, who has examined additional specimens in the following museums:

University of Akron, Akron, Ohio
American Museum of Natural History, New York, New York
Bowling Green State University, Bowling Green, Ohio
Cincinnati Society (Museum) of Natural History, Cincinnati, Ohio
University of Cincinnati Museum, Cincinnati, Ohio
Cleveland Museum of Natural History, Cleveland, Ohio
Dayton Museum of Natural History, Dayton, Ohio
Defiance College Museum, Defiance, Ohio
University of Illinois, Museum of Natural History, Urbana, Illinois
Kent State University, Kent, Ohio
Marietta College, Marietta, Ohio
University of Michigan, Museum of Zoology, Ann Arbor, Michigan
Muskingum College, New Concord, Ohio
Ohio Historical Society, Ohio State Museum, Columbus, Ohio
Ohio State University, Department of Zoology, Mammals Teaching Collection, Columbus, Ohio
Ohio State University, Museum of Zoology, Columbus, Ohio
Ohio University, Athens, Ohio
Old Mill Museum, Mill Creek Park, Youngstown, Ohio
United States National Museum, Washington, D.C.

REFERENCES CITED

Burt, W. H. 1957. Mammals of the Great Lakes region. Univ. of Mich. Press, Ann Arbor. 246 pp.
Burt, W. H., and R. P. Grossenheider. 1976. A field guide to the mammals. 3d ed. Houghton Mifflin, Boston, Mass. 289 pp.
Hall, E. R., and K. R. Kelson. 1959. The mammals of North America. Ronald Press Co., New York. 2 vols., 1162 pp.

Use of Keys for Identifying Ohio Mammals

All mammals belong to the group (phylum) of animals known as Chordata. This is a large, diverse group (including some 50,000 members) that is subdivided into several classes, of which Mammalia is one; classes are in turn subdivided into orders, families, genera, species, and, in some cases, subspecies or varieties. The purpose of the keys in this book is to identify the specimen to species. The first key (p. 7) is used to identify the seven orders of wild mammals found living in Ohio today. The characteristics referred to should be familiar to almost everyone. Following the general discussion of each order, there is a key for identifying the families and genera of mammals in that particular order. For example, if it is determined that the specimen in question is in the order Insectivora, turn to page 17 and, by using the key there, further narrow the identification to the family Talpidae (moles) or Soricidae (shrews). Within each family there are usually several genera; these can be identified by further following the choices given in the key. Beyond the generic level the characteristics used to separate species are often minute or difficult to identify. In some cases there is only one Ohio species in a given genus, but each species is described in detail in the text. Taxonomists further subdivide species into subspecies, which normally represent geographical variations at the species level. Subspecies are only briefly discussed in this text.

The keys are dichotomous; i.e., the user is given only two choices at each stage. The specimen is either longer than a certain length or shorter, either gray or brown, either hairy or naked, and so on. Once a choice has been made, continue to the next set of alternatives until a choice finally leads to the identification of the animal. A word of caution: every key should work perfectly, but few do. Keys are created by humans and are based on personal judgment. (What one person sees as light gray in color may be dark brown to someone else.) Therefore, after keying out an animal, compare it with a picture (if available) and read the description in the text. If there seems to be a discrepancy, run through the key again, perhaps taking an alternate choice at a point where hesitation arose the first time. Also, note that it is risky to identify a mammal only from a picture. Always key out the animal first, and then compare your identification with a picture.

Measurements of length throughout the book are most often given in millimeters (mm), and weights are given in grams (g). It is well to remember that mm × 0.04 = inches (approximately). For convenience it is also good to know that a reasonably new twenty-five-cent piece weighs about 5.5 g. Other useful metric and English conversions are presented in Appendix Table 1. Common body measurements

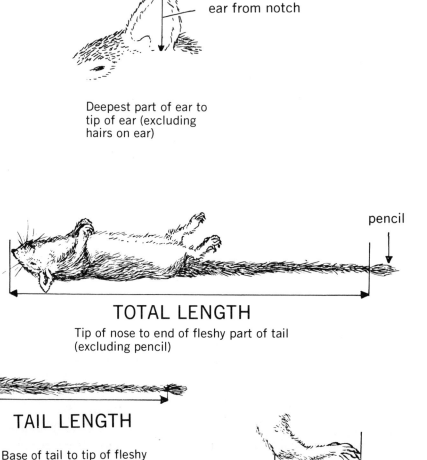

Fig. 3. Common body measurements of mammals. (Drawing by Elizabeth Dalvé.)

made on individual mammals are shown in figure 3.

Certain technical terms are useful in the keys and are essential for proper identification and accurate description of a mammal. (A glossary is provided for selected terms.) When needed, diagrams have been included showing structures that may not be familiar to the reader. For example, all skull bones and teeth referred to in the text are identified in figures 4 and 5.

Finally, the reader is reminded that collecting some wild mammals in Ohio requires a scientific collecting permit. Upon presentation of appropriate credentials, such a permit can be obtained from the local state game protector or the Ohio Division of Wildlife.

KEY TO THE ORDERS OF MAMMALS IN OHIO
1. Forelimbs modified into wings ....CHIROPTERA (Bats) p. 35
1'. Forelimbs not modified into wings 2
2. Diastema (gap) between incisors and molars (fig. 4) 3
2'. No diastema between incisors and molars ... 4

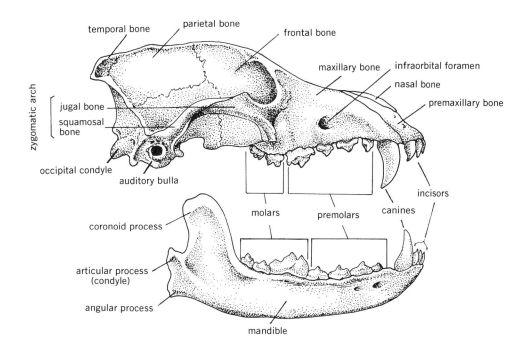

CARNIVORE SKULL

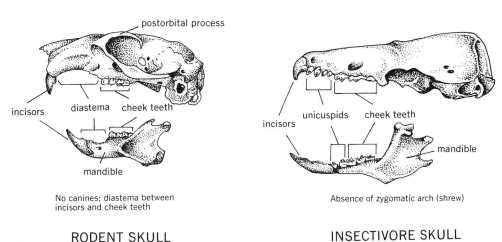

RODENT SKULL

No canines; diastema between incisors and cheek teeth

INSECTIVORE SKULL

Absence of zygomatic arch (shrew)

Fig. 4. Lateral views of mammal skull types.

LATERAL VIEW

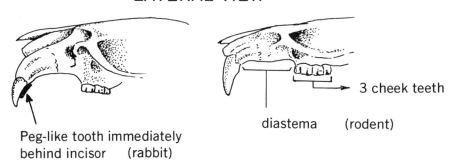

Peg-like tooth immediately behind incisor (rabbit)

diastema → 3 cheek teeth (rodent)

CHEEK TEETH (ventral view)

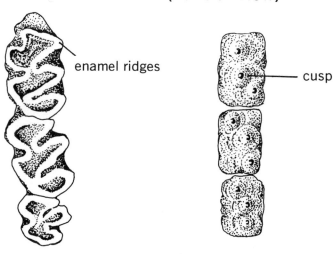

enamel ridges

cusp

Enamel pattern in "loops and whorls" (W pattern)

Teeth without "loops and whorls"

Fig. 5. Mammal tooth types. (Drawing by Elizabeth Dalvé.)

3. One pair of chisel-like incisors in upper jaw (fig. 5) .... RODENTIA (Mice, Squirrels, Beavers, etc.) p. 59
3'. Two pairs of chisel-like incisors in upper jaw; second pair immediately behind first, small and easily overlooked (fig. 5) .... LAGOMORPHA (Rabbits) p. 53
4. On the hind foot, inner toe is clawless, prehensile, and thumb-like; female with a marsupial pouch; male with scrotum in front of the penis ... MARSUPIALIA (Opossums) p. 13
4'. On the hind foot, inner toe has a claw and is not prehensile 5
5. Feet with hoofs ARTIODACTYLA (Deer) p. 143

5'. Feet with claws 6
6. Length of head and body usually less than 11.5 cm; eyes and ears hidden by fur; no ear pinnae present; canines about equal in size to other teeth; enamel on molars arranged in W patterns (fig. 5) .... INSECTIVORA (Moles, Shrews) p. 17
6'. Length of head and body more than 11.5 cm; eyes and ears prominent; ear pinnae present; canines larger than other teeth; molar teeth usually without distinct W enamel pattern (fig. 5) .... CARNIVORA (Foxes, Weasels, Raccoons, Skunks, etc.) p. 119

CHECKLIST OF THE 54 LIVING MAMMALS OF OHIO

Class Mammalia
  Order Marsupialia
    Family Didelphidae
      Didelphis virginiana Kerr* — Virginia Opossum (Possum)
  Order Insectivora
    Family Soricidae
      Sorex cinereus Kerr — Masked Shrew
      Sorex fumeus Miller — Smoky Shrew
      Microsorex hoyi (Baird) — Pygmy Shrew
      Blarina brevicauda (Say) — Short-tailed Shrew
      Cryptotis parva (Say) — Least Shrew
    Family Talpidae
      Parascalops breweri (Bachman) — Hairy-tailed Mole
      Scalopus aquaticus (Rafinesque) — Eastern Mole
      Condylura cristata (Linnaeus) — Star-nosed Mole
  Order Chiroptera
    Family Vespertilionidae
      Myotis lucifugus LeConte — Little Brown Myotis (Little Brown Bat)
      Myotis keenii Merriam — Keen's Myotis (Keen's Bat)
      Myotis sodalis Miller and Allen — Indiana Myotis (Indiana or Pink Bat)
      Myotis leibii (Audubon and Bachman) — Small-footed Myotis (Small-footed or Leib's Bat)
      Lasionycteris noctivagans (LeConte) — Silver-haired Bat
      Pipistrellus subflavus (F. Cuvier) — Eastern Pipistrelle (Georgian Bat)
      Eptesicus fuscus (Palisot de Beauvois) — Big Brown Bat
      Lasiurus borealis (Müller) — Red Bat
      Lasiurus cinereus Beauvois — Hoary Bat
      Nycticeius humeralis (Rafinesque) — Evening Bat
      Plecotus rafinesquii Lesson — Rafinesque's Big-eared Bat (Eastern Big-eared Bat)
  Order Lagomorpha
    Family Leporidae
      Sylvilagus floridanus (J.A. Allen) — Eastern Cottontail (Cottontail Rabbit)
  Order Rodentia
    Family Sciuridae
      Tamias striatus (Linnaeus) — Eastern Chipmunk
      Marmota monax Linnaeus — Woodchuck (Groundhog)
      Spermophilus tridecemlineatus (Mitchill) — Thirteen-lined Ground Squirrel (Striped Gopher)
      Sciurus carolinensis Gmelin — Gray Squirrel (Eastern Gray Squirrel)
      Sciurus niger Linnaeus — Fox Squirrel
      Tamiasciurus hudsonicus (Erxleben) — Red Squirrel (Chickaree)
      Glaucomys volans (Linnaeus) — Southern Flying Squirrel (Eastern Flying Squirrel)
    Family Castoridae
      Castor canadensis Kuhl — Beaver (American Beaver)
    Family Cricetidae
      Subfamily Cricetinae
        Reithrodontomys humulis (Audubon and Bachman) — Eastern Harvest Mouse
        Peromyscus maniculatus (Wagner) — Deer Mouse (Prairie Deer Mouse)
        Peromyscus leucopus (Rafinesque) — White-footed Mouse
        Neotoma floridana (Ord) — Eastern Woodrat (Allegheny Wood or Pack Rat)
      Subfamily Microtinae
        Clethrionomys gapperi (Vigors) — Southern Red-backed Vole
        Microtus pennsylvanicus (Ord) — Meadow Vole (Field Mouse)
        Microtus ochrogaster (Wagner) — Prairie Vole
        Microtus pinetorum (LeConte) — Woodland Vole (Pine Vole or Pine Mouse)
        Ondatra zibethicus (Linnaeus) — Muskrat
        Synaptomys cooperi Baird — Southern Bog Lemming (Cooper's Mouse)
    Family Muridae
      Rattus norvegicus (Berkenhout) — Norway Rat (Common or Brown Rat)
      Mus musculus Linnaeus — House Mouse

CHECKLISTS OF THE 54 LIVING MAMMALS OF OHIO (continued)

Family Zapodidae
   Zapus hudsonius (Zimmerman) — Meadow Jumping Mouse
   Napaeozapus insignis (Miller) — Woodland Jumping Mouse
Order Carnivora
  Family Canidae
   Canis latrans Say — Coyote (Brush Wolf)
   Vulpes vulpes Linnaeus — Red Fox
   Urocyon cinereoargenteus (Schreber) — Gray Fox
  Family Procyonidae
   Procyon lotor (Linnaeus) — Raccoon
  Family Mustelidae
   Mustela erminea Linnaeus — Ermine (Short-tailed Weasel)
   Mustela nivalis Linnaeus — Least Weasel
   Mustela frenata Lichtenstein — Long-tailed Weasel
   Mustela vison Schreber — Mink
   Taxidea taxus Schreber — Badger
   Mephitis mephitis Schreber — Striped Skunk
Order Artiodactyla
  Family Cervidae
   Odocoileus virginianus (Boddart) — White-tailed Deer

*Scientific and common names of species follow Jones et al. (1975) with commonly used synonyms also supplied in several instances.

SELECTED REFERENCES

Bier, J. A. 1967. Landforms of Ohio. Ohio Div. Geol. Surv. Map.

Bole, B. P., Jr., and P. N. Moulthrop. 1942. The Ohio recent mammal collection in the Cleveland Museum of Natural History. Sci. Publ. Cleveland Mus. Nat. Hist. 5:83-181.

Braun, E. L. 1950. Deciduous forests of eastern North America. Blakiston Press, Philadelphia. Reprinted 1964. Hafner Publ. Co., New York. 596 pp.

Brayton, A. W. 1882. Report on the mammals of Ohio. Rep. Ohio Geol. Surv. 4:1-185.

Burt, W. H. 1957. Mammals of the Great Lakes region. Univ. of Mich. Press, Ann Arbor. 246 pp.

Burt, W. H., and R. P. Grossenheider. 1976. A field guide to the mammals. 3d edition. Houghton Mifflin, Boston, Mass. 289 pp.

Claypole, E. W. 1891. *Megalonyx* in Holmes County, Ohio. Amer. Geol. 7:122-32.

Gordon, R. B. 1969. The natural vegetation of Ohio in pioneer days. Ohio Biol. Surv. Bull. N.S. 3(2). 113 pp.

Hall, E. R. 1936. Mustelid mammals from the Pleistocene of North America, with systematic notes on some recent members of the genera *Mustela, Taxidea,* and *Mephitis*. Publ. Carnegie Inst. Washington 473:41-120.

Hall, E. R., and K. R. Kelson. 1959. The mammals of North America. Ronald Press Co., New York. 2 vols., 1162 pp.

Hibbard, C. W., D. E. Ray, D. E. Savage, D. W. Taylor, and J. E. Guilday. 1965. Quaternary mammals of North America. Pp. 509-25 *in* H. E. Wright, Jr., and D. G. Frey, eds. The Quaternary of the United States. Princeton Univ. Press, Princeton, N.J. 922 pp.

Hoare, R. D., J. R. Coash, C. Innis, and T. Hole. 1964. Pleistocene peccary *Platygonus compressus* LeConte from Sandusky County, Ohio. Ohio J. Sci. 64(3):207-14.

Jones, J. K., Jr., D. C. Carter, and H. H. Genoways. 1975. Revised checklist of North American mammals north of Mexico. Occasional Papers, The Museum, Texas Tech. Univ., Lubbock. 79409. 24:1-14.

Kirtland, J. P. 1838. A catalogue of the Mammalia, birds, reptiles, fishes, Testacea and Crustacea in Ohio. Second Annu. Rep. Ohio Geol. Surv., pp. 157-200.

La Rocque, A., and M. F. Marple. 1966. Ohio fossils. Ohio Div. Geol. Surv. Bull. 54, pp. 134-36.

Martin, P. S., and H. E. Wright, Jr. 1967. Pleistocene extinctions: the search for a cause. Proc. VII Congress, Intern. Assoc. Quaternary Research. Yale Univ. Press, New Haven, Ct. 453 pp.

Mills, R. S. 1975. A ground sloth, *Megaloynx,* from a Pleistocene site in Darke Co., Ohio. Ohio J. Sci. 75(3):147-55.

Stout, W., K. VerSteeg, and G. F. Lamb. 1943. Geology of water in Ohio. Ohio Geol. Surv. Bull. 44. 694 pp.

Thomas, E. S. 1952. The Orleton Farms mastodon. Ohio J. Sci. 52(1):1-5.

Thomas, E. S., and B. P. Pole, Jr. 1938. List of the mammals of Ohio. Bull. Ohio State Mus. No. 1. 2 pp.

Trautman, M. B. 1957. The fishes of Ohio. Ohio State Univ. Press, Columbus. 683 pp.

_____. 1977. The Ohio Country from 1750 to 1977—a naturalist's view. Ohio Biol. Surv. Biol. Notes No. 10. 25 pp.

VerSteeg, K. 1938. Mastodon discovered in Ohio. Science 88(2291):498.

_____. 1944. Another mastodon found in Ohio. Science 100(2599):357.

Wood, A. E. 1952. Tooth-marks on bones of the Orleton Farms mastodon. Ohio J. Sci. 52(1):27-28.

ORDER MARSUPIALIA
Pouched Mammals

This order includes the pouched animals, which, at present, are found primarily in Australia and surrounding islands, although there are a number of forms in South and Central America. The more important characteristics of the marsupials are: (1) presence (usually) of a fur-lined pouch, or marsupium, in the female; (2) presence (usually) of epipubic bones; (3) absence of a true placenta, the young being born in an extremely immature state; (4) flaring out or widening posteriorly of the nasal bones; (5) inflection (turning in) of the angle of the mandible (lower jaw); (6) presence of no more than 3 premolars; (7) presence of 44 or more teeth; (8) absence of a true tympanic bulla; (9) brain of small size and not convoluted.

FAMILY DIDELPHIDAE
Essential characteristics of this family are found in the discussion of *Didelphis virginiana,* the only species of this family found in Ohio.

Virginia Opossum or Possum
Didelphis virginiana Kerr

The opossum is the only species of this unusual order found in North America. This animal has been characterized as being "stupid" (the size of its brain in proportion to the body size is relatively small, about the size of a walnut meat), primitive (its ancestry can be traced back some 100,000,000 years or more), and poorly adapted for the cold winters of the north (with a naked tail and naked ears). Regardless of these "handicaps," the opossum has been highly successful in increasing its distribution throughout the United States. In prehistoric times marsupials were found throughout North America, but for some unknown reason they completely disappeared. More recently, however, the opossum has slowly migrated back into the United States from South and Central America, and today it is found as far north as the New England States and northern Minnesota. It is found in all parts of Ohio. Although principally an animal of the eastern United States, the opossum was successfully introduced into California in the early 1900s.

Didelphis virginiana Kerr

Because of its secretive, nocturnal habits, the presence of opossums may not be suspected; but anyone who spends much time out of doors is certain to see this animal, and its unmistakable track can usually be found in the mud along the banks of creeks and rivers. Even larger cities have become havens for these enterprising mammals, which use sewers and storm wells for homes and garbage cans and dumps as sources of food. The University of Cincinnati is located in the heart of the city, yet I have seen opossums sleeping in the warm sunshine atop a brick wall outside my office window.

An adult opossum is somewhat larger than a house cat. The head, with its two prominent black, naked, leaf-like ears, is usually cream-colored or white and conical in shape. The tail is long, scaly, practically hairless, and prehensile. The long, white, black-tipped guard hairs overlying the shorter white underfur give the animal a grizzled-gray appearance, although there is considerable variation in color. Some animals are almost pure white, some are very dark, and, according to Hartman (1952), there is also a brown phase. However, most Ohio specimens are gray. The legs are short and black, and the feet have five white toes. The inside toe on each hind foot rests at right angles to the rest of the foot and is opposable. It is the only digit that does not bear a claw. The position of the "big toe," combined with the absence of the claw, makes the track of the opossum unmistakable. Females, of course, are immediately identified by the fur-lined pouch on the abdomen. Males have the scrotum located in front of, rather than behind, the penis.

The skull of the opossum is easily identified by (1) the inflection of the angular process of the lower jaw bone (dentary bone); (2) the small brain case; (3) the prominent sagittal crest; and (4) the large number of teeth (50 in the adult animal, more than in any other Ohio mammal). The dental formula is as follows:

$$i\frac{5-5}{4-4}, c\frac{1-1}{1-1}, pm\frac{3-3}{3-3}, m\frac{4-4}{4-4} = 50$$

Following are measurements of 31 adult opossums (*Didelphis marsupialis*) from Ohio:

	Total Length (mm)	Tail (mm)	Hind Foot (mm)
Extremes	635–869	239–352	57–74
Mean	739	305	65

The opossum is a prolific animal that may successfully raise 12 or 13 young each year (fig. 6). In Ohio the first (or only) litter is usually born in March or April after a gestation period of only 13 days. Some females in the southern part of the United States produce two litters each year. Petrides (1949) is of the opinion that two litters per female are born frequently in Ohio, but that apparently not all females produce two broods. Holmes

Fig. 6. Female Virginia opossum *(Didelphis virginiana)* with young. (Photograph by William Farr.)

and Sanderson (1965) estimate that 10 percent of the female opossums in Illinois produce two litters each year, but I have not found any pouch young in Ohio specimens taken in late summer or fall. More observations are needed to clarify this point. Opossums mate for the first time when they are one year old.

The opossum litter size ranges from 5 or 6 to 22, but the average size is 9 to 13. Newborn young are approximately the size of a honey bee, and the entire litter can be accommodated in an ordinary soup spoon. The newborn opossum has correctly been referred to as a "free-living embryo." This naked, blind, earless, pink, grub-like young is immediately faced with the problem of getting from the region of the vaginal orifice to the marsupium, or pouch, located several inches away on the belly of the mother. It was once believed that the mother aided in this task by lifting the embryo into the pouch, but it is now known that this is not so. Mr. Carl Hartman (1923, 1928), among others, has demonstrated many times that, except for wetting the fur on her abdomen with her tongue, the mother takes no active part in helping the embryo find the pouch. The front limbs of the young opossum are surprisingly well developed, with the fingers possessing tiny deciduous claws. The young literally climb hand over hand to the edge of the pouch and tumble in. Having successfully completed its journey, the voyager must find one of the pinhead-size mammary gland nipples among the hair lining the pouch. Finding the teat is accomplished by moving the head from side to side until contact is made with one of the nipples. The nipple is immediately seized in the sucker-like mouth, and the young begins to draw milk from the mother. Within a short time after nursing begins, the nipple enlarges and the young opossum is then firmly attached to the mother. Although the number may vary by 1 or 2, most female opossums have 13 teats arranged in an oval along the floor of the marsupium. When litters of more than 13 are born, it becomes a matter of "first come, first served." Once the available teats have been occupied, the late arrivals are doomed to starvation.

The young remain in the pouch from 50 to 70 days and are weaned when approximately 80 days of age. The family remains together from three to three and one-half months before finally dispersing. On several occasions I have seen a mother opossum scurrying along with eight or more quarter-grown young clinging desperately to the fur on her back and sides. Contrary to popular belief, the female does not loop her tail over her back, providing the young with a means of attachment by looping their tails around hers.

Although opossums are not well adapted to the freezing temperatures of winter, they do not hibernate. During the most severe weather, they take refuge under woodpiles or houses or in warm, leaf-lined or grass-lined nests prepared in hollow trees or logs. They may remain inactive for several days or for a week at a time.

Lay (1942) listed the home range for the opossum in Texas as 4.6–15.6 hectares (11.5–38.5 acres), but it is my impression, having done field work with the animal for over ten years, that in Ohio most individuals do not have a fixed home range. They are nomads, moving from place to place as food, weather, and environmental factors dictate.

The diet of the opossum is highly varied. In the few stomachs that I have examined, I have found fruit seeds, grass, mouse fur, insect remains, carrion, and crayfish remains. Hartman (1952), Lay (1942), Hamilton (1958), and others have found nuts, poultry, eggs, snakes, small birds, and garbage. Ordinary rat food pellets and vegetable scraps have kept numbers of opossums alive and healthy in our laboratory.

Opossums occasionally get into poultry yards, where they eat the eggs and kill the birds. They also destroy an undetermined number of game bird nests and young each year. Other than this, however, they probably do little damage. The opossum provides hours of relaxation and enjoyment for many hunters. Although not as popular in Ohio as in some other states, the meat of the opossum is baked, boiled, or fried for food.

Opossums are not aggressive animals. When approached, their first reaction is to run away. Occasionally, they take a threatening position and make a hissing noise, but I have never seen one actually attack. If there is no path to escape, the animal may "play possum." Falling on its side, the opossum lets its mouth hang open, with the tongue falling out and saliva dripping from the jaws. Opossums have a host of enemies; they fall prey to automobiles, man, dogs, foxes, hawks, and weasels—possibly in that order.

Opossum fur is used for trim on less expensive coats and hats. Rarely, pelts are sewn together to make an entire coat. Until the mid-1970s (at less than $.50 per pelt), many trappers considered opossums hardly worth "skinning out" (Appendix Table 2). Since 1977, however, a prime pelt may bring as much as $3.50.

SELECTED REFERENCES

Barr, T. R. B. 1963. Infectious diseases in the opossum: a review. J. Wildl. Manage. 27:53-71.

Coghill, G. E. 1939. Studies on rearing the opossum (*Didelphis virginiana*). Ohio J. Sci. 39:239-49.

Dexter, R. W. 1951. Earthworms in the winter diet of the opossum and raccoon. J. Mammal. 32:464.

Fitch, H. S., and L. L. Sandridge. 1953. Ecology of the opossum on a natural area in northeastern Kansas. Univ. of Kans. Publ. Mus. Nat. Hist. 7:305-38.

Gardner, A. 1973. The systematics of the genus *Didelphis* (Marsupialia: Didelphidae) in North and Middle America. Spec. Publ. Tex. Tech. Univ. 4:1-88.

Grote, J. C., and P. L. Dalby. 1973. An early litter for the opossum (*Didelphis marsupialis*) in Ohio. Ohio J. Sci. 73:240-41.

Hamilton, W. J., Jr. 1958. Life history and economic relations of the opossum (*Didelphis marsupialis virginiana*) in New York State. Cornell Agric. Exp. Stn., Mem. 354. 48 pp.

Hartman, C. G. 1923. Breeding habits, development, and birth of the opossum. Annu. Rep. 1921, Smithsonian Inst., Washington, D.C. Publ. 2689:347-63, 10 pls.

———. 1928. The breeding season of the opossum (*Didelphis virginiana*) and the rate of intra-uterine and postnatal development. J. Morphol. & Physiol. 46:143-215.

———. 1952. Possums. Univ. of Tex. Press, Austin. 174 pp.

Hazard, E. B. 1963. Records of the opossum in northern Minnesota. J. Mammal. 44:118.

Holmes, A. C. V., and G. C. Sanderson. 1965. Populations and movements of opossums in east-central Illinois. J. Wildl. Manage. 29:287-95.

Huggins, E. J., and G. E. Potter. 1959. Morphology of the urinogenital system of the opossum (*Didelphis virginiana*). Bios. 30:148-54.

Lay, D. W. 1942. Ecology of the opossum in eastern Texas. J. Mammal. 23:147-59.

Llewellyn, L. M., and F. H. Dale. 1964. Notes on the ecology of the opossum in Maryland. J. Mammal. 45:113-22.

McCrady, E., Jr. 1938. The embryology of the opossum. Wistar Inst. Anat. & Biol., Philadelphia. Amer. Anat. Mem. 16:5-233.

McKeever, S. 1958. Reproduction in the opossum in southwestern Georgia and northwestern Florida. J. Wildl. Manage. 22:303.

McManus, J. J. 1974. *Didelphis virginiana*. Amer. Soc. Mammal., Mammalian Species No. 40:1-6.

Ohio Department of Natural Resources. 1975. Fur harvest in Ohio, 1964-1974. Publ. No. 178, 2 pp.

Petrides, G. A. 1949. Sex and age determination in the opossum. J. Mammal. 30:364-78.

Reynolds, H. C. 1945. Some aspects of the life history and ecology of the opossum in central Missouri. J. Mammal. 26:361-79.

———. 1952. Studies on reproduction in the opossum (*Didelphis virginiana*). Univ. of Calif. Press, Berkeley. Univ. Calif. Publ. Zool. 52:223-84.

Sanderson, G. C. 1961. Estimating opossum populations by marking young. J. Wildl. Manage. 25:20-27.

Stoner, D. 1939. Remarks on abundance and range of the opossum. J. Mammal. 20:250-51.

Wiseman, G. L., and G. O. Hendrickson. 1950. Notes on the life history and ecology of the opossum in southeast Iowa. J. Mammal. 31:331-37.

ORDER INSECTIVORA
Moles and Shrews

The Insectivora are the most primitive of the placental mammals. These small, rat- or mouse-sized mammals present such a diversity of characteristics that it is difficult to define the group. In general, (1) there are five toes on each foot; (2) the snout is elongated, pointed, flexible, and tactile; (3) the eyes and ears are small, poorly developed, and usually completely or partially hidden beneath the fur; (4) the brain is small and poorly developed for a placental mammal; and (5) the teeth are simple, and the enamel on the crowns of the cheek teeth is arranged in W-patterns. Insectivores have a worldwide distribution with the exception of Australia and the southern two-thirds of South America. This order is represented in the United States and in Ohio by two families, Talpidae, the moles, and Soricidae, the shrews.

KEY TO THE FAMILIES AND SPECIES OF INSECTIVORA IN OHIO

1. Front feet broad and paddle-like; zygomatic arch present (fig. 7f)................
   .... TALPIDAE, Moles p. 28...2
1'. Front feet not broad and paddle-like; no zygomatic arch (fig. 7e)..................
   .... SORICIDAE, Shrews p. 17...4
2. Snout surrounded by 22 fleshy, finger-like projections (fig. 7b).....................
   .. *Condylura cristata,* Star-nosed Mole p. 32
2'. Snout without fleshy projections (fig. 7a) 3
3. Tail hairy; teeth 44 ....
   . *Parascalops breweri,* hairy-tailed Mole p. 28
3'. Tail essentially naked; teeth 36 ....
   ..... Scalopus aquaticus, Eastern mole p. 30
4. Tail short, equal to, or less than, 1/3 total body length 5
4'. Tail longer than 1/3 total body length 6
5. Fur gray; teeth 32 ....
   .*Blarina brevicauda,* Short-tailed Shrew p. 23
5'. Fur brown; teeth 30 ....
   .... *Cryptotis parva,* Least Shrew p. 26
6. At least four and usually five unicuspid teeth visible in upper jaw when viewed from the side (fig. 7c)......................*Sorex**...7
6'. Three unicuspid teeth visible in upper jaw when viewed from side (fig. 7d) ....
   ...... *Microsorex hoyi**, Pygmy Shrew p. 22
7. Length of tail usually more than 40 mm; fur color uniform over the body, dark brown in summer, gray in winter ....
   .... *Sorex fumeus,* Smoky Shrew p. 21
7'. Length of tail less than 40 mm; dorsal fur darker in color than belly fur, which has silver sheen all year ....
   .... *Sorex cinereus,* Masked Shrew p. 18

*Sorex and *Microsorex* are almost impossible to separate except by carefully examining the unicuspid teeth. If in doubt as to the identification of a specimen, send both skin and skull to a trained mammalogist.

FAMILY SORICIDAE
Shrews

The Family Soricidae includes the smallest mammals in the world. Shrews are present in Europe, Asia, Africa, North America, and extreme northern South America. There are 29 species of shrews in the United States of which 5 are found in Ohio.

All of the shrews in this country are mouse-sized or smaller. Shrews have short, fine-textured fur, dark brown, gray, or black in color. They have small beady eyes and long pointed noses. There are five toes on each foot. Although they may look like baby moles and are often identified as such, shrews can be separated from moles by the following characteristics: (1) the front feet are not paddle-like; (2) the teeth of all American forms are pigmented; (3) the zygomatic arch is incomplete; (4) minute, fur-covered external ears are present; and (5) auditory bullae are absent.

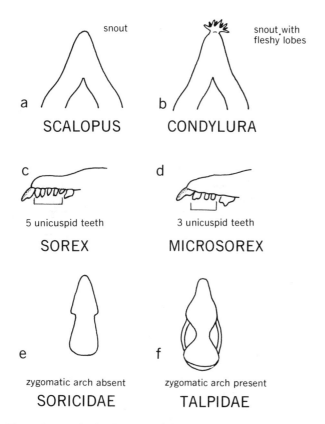

Fig. 7. Contrasting head characteristics of selected groups of Insectivora. (Drawing by Elizabeth Dalvé.)

The teeth of shrews are unusual and distinctive. In addition to being pigmented, the first pair of teeth, the incisors, project forward and form grasping pincers. Immediately behind the incisors, the unicuspids form a series of three to five simple, peg-like, sharp, pointed teeth. The last four teeth (the last premolar and the three molars) are multituberculate with sharp cusps arranged in distinct W-patterns (fig. 5). The different species of shrews have different numbers of unicuspid teeth, and frequently, one or more of the unicuspids is small and/or hidden from view. When counting the teeth, a hand lens or binocular microscope should be used, and the skull should be viewed from several different angles to make certain that none of the teeth are overlooked.

Masked Shrew
Sorex cinereus Kerr

The masked shrew (fig. 8) is the second smallest mammal in Ohio (see *Microsorex*). Most of the early records are from counties in the northeastern corner of the state (Hine, 1912; Bole and Moulthrop, 1942; Dexter, 1946, 1957). In 1962, however, I caught 15 masked shrews in Mercer County (Gottschang, 1965). Recent records also include specimens from Shelby, Franklin, Auglaize, and Paulding counties in western and central Ohio. In southern Ohio masked shrews have been trapped in Scioto, Jackson, Vinton, and Hocking counties. Extensive trapping efforts by the author and others in 15 or 16 counties in the southwestern corner and a dozen or so counties in the southeastern corner have failed to produce masked shrews. It is significant, I believe, that this shrew is present in the adjoining state of Pennsylvania but absent from those areas of Kentucky and West Virginia that are separated from Ohio by the Ohio River. The river and/or the Allegheny Plateau may serve as an effective geographical barrier to prevent emigration of the masked shrew from Ohio into these two neighboring states.

Sorex cinereus Kerr

The masked shrew is one of three species of long-tailed shrews in the state. Its tail is equal to, or longer than, one-half the body length. This tiny, long-nosed, mouse-like animal seldom exceeds 89 mm (3.5 in) in total length. The larger individuals may weigh as much as 5 g, but 3.5 or 4 g is the more common weight. Masked shrews in Ohio average somewhat smaller than those throughout the rest of their range. The eyes are small and bead-like. The ears are well developed but nearly concealed (masked) below the fur. In summer the fur on the upper parts is brown; the belly is lighter brown with a silver sheen. The summer coat is shed in November or December and replaced by dark gray or grayish-black

Fig. 8. Adult masked shrew *(Sorex cinereus)* on a fallen sugar maple leaf. (Photograph by Ronald A. Austing.)

hair. The belly still has a silver sheen. A spring molt occurs in March or early April. There is a definite line of demarcation between the darker dorsal fur and the lighter belly fur in both summer and winter. The skull is extremely small and delicate, and the complete absence of a zygomatic arch is typical of all shrews. The third and fourth unicuspid teeth in the upper jaw are smaller than the first and second. Frequently the third unicuspid is smaller than the fourth, a unique characteristic of the Ohio masked shrew. Usually all the teeth have mahogany-colored tips, although there is some variation; some old individuals have no pigment on the teeth, and others show very little coloration. The dental formula is as follows:

$$i\frac{3-3}{1-1}, c\frac{1-1}{1-1}, pm\frac{3-3}{1-1}, m\frac{3-3}{3-3} = 32$$

Following are measurements of 56 adult masked shrews *(Sorex cinereus)* from Ohio:

	Total Length (mm)	Tail (mm)	Hind Foot (mm)
Extremes	75–98	28–38	10–12
Mean	86	31	11

Because of their small size and secretive habits, masked shrews are seldom seen and are sometimes difficult to trap. They have been taken in a variety of habitats, but most commonly in hemlock-beech-maple woods, where there is a thick, soft leaf mulch covering the ground. Blossom (1932) and Hamilton (1930) have commented on the inability of these diminutive creatures to burrow in packed soil, but they can and do push their way through moss and leaf mold on the forest floor. I have also found them burrowing through the soft surface duff in old fields in Ohio. Masked shrews also make extensive use of the runways made by various other small mammals. Moisture is apparently important in the ecology of this shrew, for it is seldom encountered in excessively dry areas. Moist runways along streams or in meadows or woods are favorite haunts. Shrews do not hibernate, and they are active both day and night.

Merriam (1884) confined three masked shrews under a glass tumbler and observed two being killed and eaten by the third. Merriam concluded from his observation that shrews are vicious, cannibalistic little animals with almost insatiable appetites. Pearson (1947, 1948) and Morrison et al. (1957) have shown that shrews do have extremely high basal metabolic rates and that they require a large amount of food to satisfy their energy requirements. However, overemphasis may have been placed on the "vicious nature" of the shrew and on the amount of food necessary to keep one alive and healthy. Hamilton (1930) has made extensive studies on the food habits of North American shrews and has found that one-half the weight of the animal is a sufficient daily diet. On several occasions I have kept three or more shrews together in a five-gallon aquarium partly filled with soil, leaves and sticks. As long as sufficient food was available, they got along well together. It has been my observation that shrews become cannibalistic and "vicious" only when confined in too small a space or when deprived of food.

Blossom (1932) kept a female masked shrew in captivity for 77 days and observed that earthworms were consumed the most readily. Other foods eaten were salamanders, oats, cutworms, meadow mice, moths, sow bugs, insect larvae, grasshoppers, spiders, centipedes, hellgrammites, hamburger, liver, snails, and fish. During one seven-day period this 3.6-g shrew ate an average of 11.7 g of food each day, which is equal to 3.3 times its own body weight. Hamilton (1930) examined the stomachs of 62 shrews of this species and found the following percentages of food by bulk: insects, centipedes, arachnids, and sow bugs, 74 percent; vertebrates, 7 percent; worms and molluscs, 6 percent; vegetable and inorganic matter, 2 percent; and undetermined, 11 percent. I have also identified the remains of crickets, millipedes, and earthworms in the stomachs of masked shrews in Ohio. This species is obviously primarily carnivorous, but a certain amount of vegetable matter is regularly included in the diet. Masked shrews will attack mature, healthy mice placed in the same cage with them, but are unable to overcome and immediately kill them. Mouse remains reported from the stomachs of shrews caught in the field probably represent sick or dead animals that the shrews had found.

Quick, nervous, jerky movements are highly characteristic of these bundles of energy. As the shrew moves about, the end of its nose is constantly in motion. The eyes, although small and hidden, are important in helping the shrew find its way. Also its sense of hearing is keen. Shrews squeak and also make soft, twittering, bird-like sounds.

Few occupied nests of the masked shrew have been discovered, and there are only one or two reports describing the young. Indirect evidence suggests a gestation period of about three weeks. Litter size ranges from one to eight, and a single female may have more than one litter in a year. Blossom (1932) reports that the one young born in his laboratory was 19 mm long and weighed less than 0.1 g when still less than 1 day old. There was no trace of hair, the teeth had not erupted, and the eyelids were fused together. However, development is rapid, and the young leave the nest when 3 to 4 weeks old. It is generally believed that breeding activity extends from March into September, with an occasional litter born as late as October (Short, 1961; Buckner, 1966). Under normal conditions shrews born in the summer of one year become reproductively active during the spring and summer of the following year and die soon afterward. Buckner (1966) reported an exceptional population peak (23.3 shrews per hectare, or 9.3 per acre) for masked shrews in Manitoba in 1957. At that time juvenile and subadult shrews 4 months of age or younger were producing litters in the fall of the year. Normal life expectancy for the masked shrew is 12 to 15 months, but Buckner (1966) found in the litters he studied three individuals that had survived 17, 23, and 29 months, respectively. His work reminds us that most aspects of the population dynamics of this species are inadequately known.

Larger shrews, house cats, weasels, foxes, skunks, owls, and snakes are known to prey upon the masked shrew.

The masked shrew is considered beneficial since it, along with other shrews, feeds primarily upon small animals that might be injurious to crops. Presently two subspecies of *Sorex cinereus* are recognized in Ohio (Hall and Kelson, 1959, p. 26). *Sorex cinereus lesueurii* (Duvernoy) is considered to occupy the northwest corner of the state, extending as far south as Mercer County, and *Sorex cinereus ohionensis* (Bole and Moulthrop) occurs throughout the rest of Ohio. In 1965 I pointed out the lack of clarity in the taxonomy of *Sorex cinereus* in Ohio (Gottschang, 1965), which is still the case. More specimens are needed to determine (1) whether *S. c. lesueurii* actually occurs here, and (2) if both subspecies are present, the extent of their separate distributions.

SELECTED REFERENCES

Blossom, P. M. 1932. A pair of long-tailed shrews (*Sorex cinereus cinereus*) in captivity. J. Mammal. 13:136–43.

Bole, B. P., Jr. and P. N. Moulthrop. 1942. The Ohio recent mammal collection in the Cleveland museum of Natural History. Sci. Publ. Cleveland Mus. Nat. Hist. 5:83–181, esp. 89–95.

Buckner, C. H. 1966. Populations and ecological relationships of shrews in tamarack bogs of southeastern Manitoba. J. Mammal. 47:181–94.

Dexter, R. W. 1946. A new county record for the Ohio shrew. J. Mammal. 27:177.

———. 1957. Additional records of the Ohio shrew. J. Mammal. 38:513.

———. 1961. An albino shrew from Ohio. J. Mammal. 42:96.

Findley, J. S. 1955. Taxonomy and distribution of some American shrews. Univ. of Kans. Publ. Mus. Nat. Hist. 7(14):613–18.

Goodpaster, W. W., and D. F. Hoffmeister. 1968. Notes on Ohioan mammals. Ohio J. Sci. 68:116–17.

Gottschang, J. L. 1965. Range extension and notes on the shrew *Sorex cinereus* in Ohio. Ohio J. Sci. 65:46–47.

Hall, E. R., and K. R. Kelson. 1959. The mammals of North America. Ronald Press Co., New York. 2 vols., 1162 pp.

Hamilton, W. J., Jr. 1930. The food of the Soricidae. J. Mammal. 11:26–39.

Hine, J. S. 1912. Ohio moles and shrews. Ohio Nat. 12:494–96.

———. 1929. Distribution of Ohio mammals. Annu. Rep., Proc. Ohio Acad. Sci. 8(6):267.

Jackson, H. H. T. 1928. A taxonomic review of the American long-tailed shrews (Genera *Sorex* and *Microsorex*). North Amer. Fauna 51:1–238.

Merriam, C. H. 1882–84. The vertebrates of the Adirondack Region, northwestern New York. Mammalia. Trans. Linnaean Soc., New York. Vol. 1, 1882, pp. 27–106. Vol. 2, 1884, pp. 9–214.

Morrison. P. R., M. Pierce, and F. A. Ryser. 1957. Food consumption and body weight in the masked and short-tail shrews. Amer. Midl. Nat. 57:493–501.

Pearson, O. P. 1947. The rate of metabolism of some small mammals. Ecology 28:127–45.

———. 1948. Metabolism of small mammals, with remarks on the lower limit of mammalian size. Science 108:44.

Rudd, R. L. 1955. Age, sex, and weight comparisons in three species of shrews. J. Mammal. 36:323–39.

Short, H. L. 1961. Fall breeding activity of a young shrew. J. Mammal. 42:95.

Tuttle, M. D. 1964. Observation of *Sorex cinereus*. J. Mammal. 45:148.

Vickers, E. W. 1895. Comparative abundance of *Sorex personatus* Wog. and *Blarina brevicauda* Baird, in Ellsworth, Mahoning County. Third Annu. Rep., Ohio Acad. Sci. p. 15.

Smoky Shrew
Sorex fumeus Miller

The smoky shrew is principally a mammal of boreal regions, and in Ohio it is limited primarily to the Allegheny Plateau. Shady, damp, cool woods with soft leaf mold and lots of ground cover are the favorite haunts of this shrew. Unlike the smaller masked shrew, it seldom frequents open fields. Bole and Moulthrop (1942) state, "On the drainage area of the Chagrin River in Lake, Geauga, and eastern Cuyahoga Counties, *Sorex fumeus* is one of the most abundant mammals and occurs in almost every rich beech-hemlock woodland." Enders (1930) caught two smoky shrews near Chillicothe in Ross County, the closest this species has come to invading the Glaciated Till Plain of Ohio.

Sorex fumeus Miller

The three species of long-tailed shrews in Ohio look very much alike and are easily confused. The smoky shrew is the largest of the three, averaging about 117 mm (4.5 in) in total length. The tail is also longer, usually exceeding 40 mm. The fur, rather than being dark above and light below, is uniform in color over the entire body or, at most, only slightly lighter on the belly. Smoky shrews are dark brown in summer and gray in winter. The molt pattern is similar to that of the masked shrew. The spring molt occurs in March, and the summer coat is shed during August, September, or October. Most shrews have their winter coats by the third week in October. The skull is larger and heavier than that of the masked shrew. The number and relative size of the teeth is the same as for the masked shrew. Following are measurements of 96 adult smoky shrews (*Sorex fumeus*) from Ohio:

	Total Length (mm)	Tail (mm)	Hind Foot (mm)
Extremes	103–26	33–48	12–14
Mean	117	43	13

Some researchers believe that smoky shrews are colonial animals that tend to congregate in certain restricted areas. Bole (1939) estimated that in one study area there were 145 of these shrews per hectare (58 per acre), an unusually high figure. During a three-week period in 1937, Hamilton (1940) caught 70 smoky shrews on a 0.6 hectare (1.5 acre) plot in New York State, although the year before he estimated the population density in this same area to be 22.5 to 35 shrews per hectare (9 to 14 per acre).

The life history of this shrew is similar to that of the masked shrew. The gestation period is somewhat less than three weeks, and the young are born from April through August. Small invertebrates, mostly insects and insect larvae, comprise the bulk of the diet.

SELECTED REFERENCES

Bole, B. P., Jr. 1939. The quadrat method of studying small mammal populations. Sci. Publ. Cleveland Mus. Nat. Hist. 5:15–77.

Bole, B. P., Jr., and P. N. Moulthrop. 1942. The Ohio recent mammal collection in the Cleveland Museum of Natural History. Sci. Publ. Cleveland Mus. Nat. Hist. 5:83–181, esp. 96–97.

Enders, R. K. 1930. Some factors influencing the distribution of mammals in Ohio. Occ. Papers, Univ. of Mich. Mus. Zool. 212:1–27.

Hamilton, W. J., Jr. 1930. The food of the Soricidae. J. Mammal. 11:26–39.

———. 1940. The biology of the smoky shrew (*Sorex fumeus fumeus* Miller). Zoologica 25:473–92.

(See also Selected References for *Sorex cinereus*.)

Pygmy Shrew
Microsorex hoyi (Baird)

In the original description of this species, Baird (1858) listed and described a skin with skull from Zanesville, Ohio. This specimen was the only pygmy shrew recorded from Ohio until two adult females and one adult male were caught in Zaleski State Forest, Vinton County, on 11 December 1975 and 1 January 1976 (Svendsen, 1976). It is indeed a rare animal. A greater trapping effort, however, combined with careful identification of any long-tailed shrew caught in Ohio, may show that this tiny animal is better established in the state (especially in the Unglaciated Allegheny Plateau) than we have realized.

The pygmy shrew is probably the smallest animal in the world. Adults are 78 to 98 mm long (3 1/16 to 4 in). The tail is 27 to 35 mm long (1 1/16 to 1 2/3 in). Commonly adult shrews weigh 2 to 5 g, and the subspecies that occurs in Ohio is even smaller. The two females taken by Svendsen (1976) weighed 1.5 g each and the male 2.0 g.

Microsorex hoyi (Baird)

The pygmy shrew closely resembles the masked shrew, but can be separated from it by the teeth. Although the dental formulae are the same, the third and fifth unicuspid teeth in the upper jaw of the pygmy shrew are minute and cannot be seen from the side, whereas the five unicuspid teeth of the masked shrew are clearly visible from the side (fig. 7).

Pygmy shrews have been caught in woods, bogs, grassy fields, and brushy areas bordering forests. They utilize runways made by other small mammals as well as pencil-sized runs that they construct themselves. Saunders (1929) states that the holes he observed made by a captive pygmy shrew could very easily be mistaken for those made by large earthworms. The food habits, reproductive habits, and general life history of this shrew presumably are similar to those of other long-tailed shrews. Scott (1939) reported finding seven embryos in a female that he caught in Iowa. Raw fish, meat, and earthworms were accepted by a pygmy shrew kept in captivity for one month by Saunders (1929); and Hamilton (1930) reports grubs, earthworms, and insect fragments from the stomach of one specimen in Vermont.

Any small long-tailed shrew captured in Ohio should be examined carefully; it might be *Microsorex hoyi*!

SELECTED REFERENCES

Baird, S. F. 1858. Reports of explorations and surveys to ascertain the most practicable and economical route for a railroad from the Mississippi River to the Pacific Ocean. Vol. 8. General report upon the zoology of several Pacific Railroad routes. Part 1, Mammals:32.

Hamilton, W. J., Jr. 1930. The food of the Soricidae. J. Mammal. 11:26–39.

Hine, J. S. 1929. Distribution of Ohio mammals. Annu. Rep., Proc. Ohio Acad. Sci. 8(6):267–68.

Jackson, H. H. T. 1928. A taxonomic review of the American long-tailed shrews (Genera *Sorex* and *Microsorex*). North Amer. Fauna 51:1–238.

Long, C. A. 1974. *Microsorex hoyi* and *Microsorex thompsoni.* Amer. Soc. Mammal., Mammalian Species No. 33:1–4.

Prince, L. A. 1940. Notes on the habits of the pigmy shrew (*Microsorex hoyi*) in captivity. Can. Field Nat. 54:97–100.

Saunders, P. B. 1929. *Microsorex hoyi* in captivity. J. Mammal. 10:788–79.

Scott, T. G. 1939. Number of fetuses in the Hoy pigmy shrew. J. Mammal. 20:251.

Svendsen, G. E. 1976. *Microsorex hoyii* [*sic*] in southeastern Ohio. Ohio J. Sci. 76:102.

Short-tailed Shrew
Blarina brevicauda (Say)

Although most people have never seen the short-tailed shrew, it is quite possibly present in greater numbers and in a wider variety of habitats than any other species of wild mammal in Ohio. The short-tailed shrew is common statewide and has probably been taken in every country in which a mammalogist has trapped.

Blarina brevicauda (Say)

In contrast to the shrews discussed thus far, the short-tailed shrew has a dimunitive tail that does not exceed one-third of its body length. It is also the most mole-like of our shrews. The fur is short and soft and can be brushed or smoothed either forward or backward. In both summer and winter, it has a uniform gray coat, often with a silvery sheen when viewed from certain angles. The belly and underside of the tail are slightly lighter than the rest of the body. The tops of the feet are pink or white. Albinism has been reported by Svendsen and Svendsen (1975). Three albino specimens have been taken in Hamilton County, and I have several shrews in my collection with large white patches of fur. The short-tailed shrew is the largest shrew in Ohio. Adults may weigh more than 20 g and are commonly 101–14 mm (4.45 in) long. The skull of the short-tailed shrew is easily identified by its large size, angular outlines, and prominent crests and ridges. The number of teeth is 32, and the unicuspids are arranged in pairs. The first two unicuspids are large; the second two are about equal in size but smaller than the first two. The fifth unicuspid is so small that it is not visible from the side. All of the teeth are usually pigmented, but they may be practically colorless in some old adults. The dental formula is as follows:

$$i\frac{3-3}{1-1}, c\frac{1-1}{1-1}, pm\frac{3-3}{1-1}, m\frac{3-3}{3-3} = 32$$

Following are measurements of 50 adult short-tailed shrews (*Blarina brevicauda*) from Ohio:

	Total Length (mm)	Tail (mm)	Hind Foot (mm)
Extremes	101–23	18–28	11–15
Mean	112	22	14

This species lives in a wide variety of habitats, as reflected by the areas in which I have caught the short-tailed shrew: in freshly plowed fields, in grassy fields, in cornfields, in deciduous woods, along streams, on sandy beaches, in evergreen forests, in barns, and in cement basements of city homes. In fields overgrown with grass and/or weeds, runways may be found just above the surface of the ground. In woodlots that have a thick, soft mat of leaves on the ground, there is usually a honeycomb of runways just below the top layer of leaves. Unlike the smaller and weaker long-tailed shrews, the short-tailed shrew digs very well; virtually every woodlot and field has an elaborate system of tunnels located from several centimeters to half a meter (several inches to more than a foot) underground. Although most of their activity is confined to runways, short-tailed shrews are sometimes seen running about on the surface of the ground. Apparently when foraging for food, they do not hesitate to leave the security of their runways, and I have often caught them in traps removed some distance from any runways. My field notes show that these shrews have been caught during every hour of the day, but they are somewhat more active after dark.

I have not attempted a formal study of the food habits of the short-tailed shrew in Ohio, but I have examined many stomachs and have found the remains of insects, insect larvae, snails, mammals, birds, salamanders, worms, crayfish, vegetable matter, and so on. In 244 stomachs examined by Hamilton (1930), insects comprised 47.8 percent of the bulk, with plant materials constituting the second largest category. Surprisingly, during the winter months plant food made up as much as 25 percent of the bulk in the stomachs, perhaps reflecting the scarcity of insects, worms, and such during the colder months.

Captive animals have fed avidly on meadow mice, deer mice, house mice, liver, and hamburger. Whole rolled oats, peanut butter, sunflower seeds, dry bread, English walnut meats, and wild-bird seed have also been accepted. Phillips (1956) captured a specimen at Findlay, Ohio, and kept it alive for almost three months on a diet of English sparrows, house mice, raw hamburger, and chicken feed. During one three-week period the shrew was fed nothing but chicken feed. A dog chow-hamburger mixture was fed to shrews successfully over a period of many months by Pearson (1950).

Short-tailed shrews commonly store quantities of surplus food of all kinds. For example, Shull (1907) made the interesting discovery that these shrews collect and store snails. Depending upon the temperature, the snails either are placed in small piles aboveground or are stored in small rooms belowground. Short-tailed shrews even appear to respond to temperature changes by moving their caches from one location to another, discriminating between occupied and unoccupied shells.

Blarina brevicauda attacks and kills mice and other shrews placed in the same cage; however, it probably does not deliberately prey on these animals in the wild. A healthy mouse, because of its greater speed and agility, would surely be able to avoid the attack of a shrew in most instances.

The submaxillary glands in the mouth of the shrew have been demonstrated (Pearson, 1942) to contain a poison that flows with the saliva and enters the wound made by the shrew when it bites its prey. There is enough poison in the glands to kill an adult rabbit or 200 mice. The poison attacks the heart, respiratory system, and muscles, rendering the victim helpless within minutes. Men who have been bitten on the hand by the short-tailed shrew have complained of shooting pains in their hand and arm, and discomfort far greater than would normally be expected from the bite of an animal of this size. All shrews in Ohio are poisonous but to a lesser degree than the short-tailed shrew.

Short-tailed shrews are tremendously nervous, active animals that burn energy at a surprising rate. Even when resting, their breathing rate may be 130 to 140 breaths per minute, and the basal metabolic rate has been determined to be 130 kilocalories per hour. (An average man breathes 16 to 18 times per minute and has a basal metabolic rate of 78.) The amount of food eaten by the short-tailed shrew while satisfying its energy requirements has probably been overstated, but, nevertheless, it is a prodigious eater. A captive animal may eat more than its own weight in food every 24 hours; in the wild, however, it probably eats much less. Captive animals have been deprived of food for more than 36 hours without ill effects. They drink water readily in captivity.

My trapping records indicate that the short-tailed shrew breeds from the latter part of March through the second week of October in Ohio. Dusi (1951) uncovered a nest in Franklin County on 11 October 1949 that contained three almost fully-grown young. Although pregnant or nursing females may be found any time between March and October, there is a peak of reproductive activity in the spring and another in the fall. A gestation period of 21 or 22 days follows a rather long and involved courtship during which a pair of shrews may mate 20 or more times a day for several days in succession. In fact, Pearson (1944) has determined that several matings are necessary to induce ovulation in the female. The average litter size is five, although litters of one to ten have been reported. The young are born in nests made of any available materials, usually leaves and/or grass. These nests are globular in shape and are often located beneath logs or stones, or they may be in a specially prepared chamber some distance belowground. A nest discovered by Dusi (1951) was 23 cm (9 in) below the surface of the ground.

The young are born blind and naked, but they mature rapidly and are able to take care of themselves when one month old. Many females mate and produce a litter in their first season. It is quite likely that some males also mate during the summer and fall of their first year. Adult females may produce three or even four litters in a single year.

Short-tailed shrews have three prominent skin glands, one on either side midway between the fore and hind legs and another exactly down the middle of the belly (fig. 9). The belly gland is easily recognized in both sexes, but the side glands are much less developed in the females than in the males. During the breeding season the side glands of the male become especially noticeable and can be used as sure indicators of sexual activity. It is believed that the odor associated with the secretions from these glands could be a sexual attractant or an indicator to other male shrews that an area is already occupied. Many predators that catch and kill shrews will not eat them because of the overwhelming odor and taste of the secretions from the skin glands.

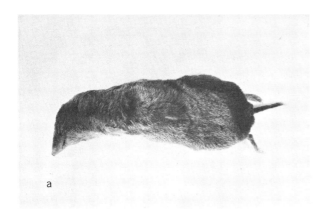

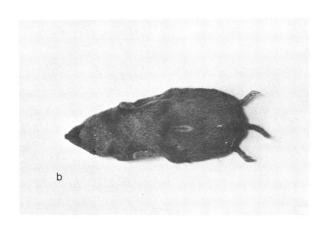

Fig. 9. Short-tailed shrew *(Blarina brevicauda),* showing side gland (9a) and belly gland (9b). (Photograph by the author.)

During some seasons the short-tailed shrew is exceedingly abundant, and at other times it is rather scarce. Some of the earliest workers cited populations of 125 or more short-tailed shrews per hectare (50 or more per acre). Shull (1907) estimated populations of 10 shrews per hectare (4 per acre). More recently, Blair (1940) estimated a normal population of 5.5 individuals per hectare (2.2 per acre), and Hamilton (1931) believes than a hectare of ground in a choice location can support about 10 pairs of shrews (4 pairs per acre). My trapping results under favorable conditions in southwestern Ohio indicate short-tailed shrew populations of 15 to 20 per hectare (6 to 8 per acre) in old fields and perhaps 20 to 24 per hectare (8 to 10 per acre) in woodlots.

Blaire (1940, 1941) live-trapped, marked, and released a number of shrews in Michigan. By retrapping the same animals over a period of several months, he was able to estimate the size of the area inhabited by any one shrew (its home range). One male ranged over 1.8 hectares (4.5 acres), but most of the shrews confined their activities to 0.4 hectare (1.0 acre) or less. Males moved about more than females. Several animals commonly shared the same home range; thus the short-tailed shrew probably does not actively defend a territory against outsiders.

Because of the great numbers of insects they destroy, short-tailed shrews are considered beneficial. Shull (1907) concluded that one shrew would eat the following items in one month's time: 8 voles, 90 insects, 78 insect larvae, 53 earthworms, and 18 snails. Short-tailed shrews serve as effective scavengers and thus are very important members of most ecosystems. I enjoy Arthur B. Williams's comments (1936, p. 365) concerning the short-tailed shrew and its position in a beech-maple climax community in northeastern Ohio:

Occupying the position of the most abundant mammal in the area, being the only one that approaches constance activity by day and by night, and at all seasons of the year, requiring an enormous amount of animal food at all times, and having no real check upon numbers except the barred owl, the short-tailed shrew appears as probably the most influential mammal in the area. He acts as an efficient check upon the numbers of mice and other species of shrew, and takes out [of] the humus and decaying logs and stumps an immense number of insects and their larvae and of other forms like millipedes, centipedes, sowbugs, and worms. It is not entirely fanciful to liken his network of under-cover tunnels and runways to a vast system of spider webs, laid down for the same purpose as the spider's web—the ensnaring of his prey.

SELECTED REFERENCES

Barrett, G. W., and K. L. Stueck, 1976. Caloric ingestion rate and assimilation efficiency of the short-tailed shrew, *Blarina brevicauda.* Ohio J. Sci. 76:25–26.

Blair, W. F. 1940. Notes on home ranges, and populations of the short-tailed shrew. Ecology 21:284–88.

———. 1941. Some data on the home ranges and general life history of the short-tailed shrew, red-backed vole, and woodland jumping mouse in northern Michigan. Amer. Midl. Nat. 25:681–85.

Buckner, C. H. 1966. Populations and ecological relationships of shrews in tamarack bogs of southeastern Manitoba. J. Mammal. 47:181–94.

Christian, J. J. 1950. Behavior of the mole (*Scalopus*) and the shrew (*Blarina*). J. Mammal. 31:281–87.

Dapson, R. W. 1968. Growth patterns in a post-juvenile population of short-tailed shrews (*Blarina brevicauda*) Amer. Midl. Nat. 79:118–29.

DeCapita, M. E., and T. A. Bookhout. 1975. Small mammal populations, vegetational cover, and hunting use of an Ohio strip-mined area. Ohio J. Sci. 75:305–13.

Doremus, H. M. 1964. Livetrapping the short-tailed shrew, *Blarina brevicauda*. J. Mammal. 45:144–46.

———. 1965. Heart rate, temperature, and respiration rate of the short-tailed shrew in captivity. J. Mammal. 46:424–25.

Dusi, J. L. 1951. The nest of a short-tailed shrew. J. Mammal. 32:115.

Findley, J. S., and J. K. Jones. 1956. Molt of the short-tailed shrew, *Blarina brevicauda*. Amer. Midl. Nat. 56:246–49.

Forsyth, D. J. 1972. Bioaccumulation of ^{36}Cl ring-labeled DDT by *Blarina brevicauda* and *Sorex cinereus* in an old-field ecosystem. M. S. Thesis, Ohio State Univ., Columbus. 37 pp.

Hamilton, W. J., Jr. 1929. Breeding habits of the short-tailed shrew, *Blarina brevicauda*. J. Mammal. 10:125–34.

———. 1930. The food of the Soricidae. J. Mammal. 11:26–39.

———. 1931. Habits of the short-tailed shrew, *Blarina brevicauda* (Say). Ohio J. Sci. 31:97–106.

Jameson, E. W., Jr. 1950. The external parasites of the short-tailed shrew, *Blarina brevicauda* (Say). J. Mammal. 31:138–45.

Merriam, C. H. 1895. Revision of the shrews of the American genera *Blarina* and *Notiosorex*. North Amer. Fauna 10:1–34.

Morrison, P. R., M. Pierce, and F. A. Ryser. 1957. Food consumption and body weight in the masked and short-tailed shrews. Amer. Midl. Nat. 57:493–501.

Moseley, E. L. 1930. Feeding a short tailed shrew. J. Mammal. 11:224–25.

Nader, I. A., and R. L. Martin. 1962. The shrew as prey of the domestic cat. J. Mammal. 43:417.

Oswald, V. H. 1958. Helminth parasites of the short-tailed shrew in central Ohio. Ohio J. Sci. 58:325–34.

Pearson, O. P. 1942. On the cause and nature of a poisonous action produced by the bite of a shrew *(Blarina brevicauda)*. J. Mammal. 23:159–66.

———. 1944. Reproduction in the shrew (*Blarina brevicauda* Say). Amer. J. Anat. 75:39–93.

———. 1945. Longevity of the short-tailed shrew. Amer. Midl. Nat. 34:531–46.

———. 1950. Keeping shrews in captivity. J. Mammal. 31:351–52.

Phillips, R. S. 1956. Notes on a captive short-tailed shrew. J Mammal. 37:543.

Pruitt, W. O., Jr. 1954. Notes on the shorttail shrew (*Blarina brevicauda kirtlandi*) in northern lower Michigan. Amer. Midl. Nat. 52:236–41.

Randolph, J. C. 1973. Ecological energetics of a homeothermic predator, the short-tailed shrew. Ecology 54:1166–87.

Rood, J. P. 1958. Habits of the short-tailed shrew in captivity. J. Mammal. 39:499–507.

Shull, A. F. 1907. Habits of the short-tailed shrew, *Blarina brevicauda* (Say). Amer. Nat. 41:495–522.

Svendsen, G. E., and M. G. Svendsen. 1975. An albino *Blarina brevicauda* from southeastern Ohio. Ohio J. Sci. 75:32.

Vickers, E. W. 1895. Comparative abundance of *Sorex personatus* Wog., and *Blarina brevicauda* Baird, in Ellsworth, Mahoning County. Third Annu. Rep., Ohio Acad. Sci. p. 15.

Whitaker, J. O., Jr., and M. G. Ferraro. 1963. Summer food of 220 short-tailed shrews from Ithaca, New York. J. Mammal. 44:419.

Williams, A. B. 1936. The composition and dynamics of a beech-maple climax community. Ecol. Monogr. 6:317–408.

Williams, M. W. 1962a. An albino short-tailed shrew from Vermont. J. Mammal. 43:424–25.

———. 1962b. Molt in *Blarina brevicauda*. J. Mammal. 43:423–24.

Least Shrew
Cryptotis parva (Say)

The least shrew, one of our smallest mammals, has been considered rare in Ohio, but Hall and Kelson (1959, p. 57) indicate that its occurrence is general throughout Ohio though rarely reported from the Unglaciated Allegheny Plateau. My trapping results during the past few years also indicate that the least shrew is a regular member of the old field habitat. The 45 specimens in my collection represent more than twice the number previously recorded from Ohio (Goodpaster, 1941; Bole and Moulthrop, 1942). Statements to the effect that least shrews do not readily enter traps, or that because of their small size and peculiar gait it is not possible to catch them in regular mousetraps, are not substantiated in my studies. All of the least shrews that I have captured have been caught just behind the head or across the

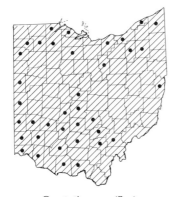

Cryptotis parva (Say)

shoulders as they were obviously reaching for the bait on the treadle of the trap. Skulls of the least shrew are commonly found in owl pellets, again suggesting that the scarcity of this shrew may actually be more apparent than real.

This tiny shrew is at once recognized as being different from the short-tailed shrew (*Blarina brevicauda*) by its much smaller size and dark brown coat, and its short tail immediately separates it from any of the other shrews in Ohio. Adults are dark reddish brown or mahogany above with light gray bellies and feet. There is a glistening silver sheen on the gray bellies of freshly molted specimens. In some the dark gray basal portion of each hair projects through the overlying guard hairs, giving the entire animal a light gray appearance. Weights of adults vary from 4.0 to 5.5 g. The least shrew is the only shrew in Ohio that has 30 rather than 32 teeth. There are 4 unicuspid teeth on either side of the upper jaw instead of the customary 5, and only 3 are visible from the side. The fourth or last unicuspid is so small that it can be easily overlooked. The dental formula is as follows:

$$i\frac{3-3}{1-1}, c\frac{1-1}{1-1}, pm\frac{2-2}{1-1}, m\frac{3-3}{3-3} = 30$$

Following are measurements of 57 adult least shrews (*Cryptotis parva*) from Ohio:

	Total Length (mm)	Tail (mm)	Hind Foot (mm)
Extremes	68–90	10–21	9–12
Mean	78	16	10

The least shrew probably molts twice each year. A series of 29 specimens that I caught in November 1962 shows that (1) there was a winter molt during the first two weeks of November; (2) all individuals of both sexes molted at about the same time; and (3) the entire molt was probably completed within a few days. I am fairly certain that there is a spring molt also, but I have no direct evidence to support my view. Summer specimens are lighter-colored and have thinner coats, but either of these could be caused by wear rather than molt.

The least shrew frequents weedy and brushy fields and hedgerows. In the southern part of the United States, it has been captured in salt marshes bordering the ocean, but in Ohio it usually avoids excessively wet areas. To my knowledge, Lyon's report (1936) is the only one of the least shrew occurring in a forest habitat (in "a damp woods not far from a damp meadow"). These shrews sometimes move about under and above the surface of the snow. Students in my mammalogy class once "shook loose" a pair of least shrews, male and female, from beneath the crust of snow covering a field on a cold day in March 1954. On only one or two occasions have I been successful in locating the runways made by these small shrews. These tiny runs have diameters no larger than that of a large lead pencil. Animals kept in captivity have been seen to construct a series of runways by using their feet, heads, and shoulders to scratch and push themselves through loose dirt and leaf mold. In the field they apparently also use the runways made by moles, chipmunks, field mice, and short-tailed shrews. They often leave their runways to forage for food.

The diet of this shrew is similar to that of *Blarina brevicauda*. The least shrew is, of course, much too small to overcome a mouse, but like other shrews it probably feasts on an occasional dead animal. We once kept six of these interesting animals alive and healthy on a diet of meadow mice and deer mice. The least shrew is also known to eat insects, millipedes, insect larvae, spiders, snails, and worms, as well as some vegetable matter.

The least shrew is not only tolerant of other members of the same species but at times seems even to be sociable. For example, both a male and a female have been found together in a nest containing young. One of my students once found 6 adult shrews occupying the same nest. Hamilton (1934) reported finding one nest containing 5 adult shrews and another containing 3. McCarley (1959) uncovered a nest near Nacogdoches, Texas, that had 31 least shrews in it, possibly due to the prevailing cool temperature.

Much has yet to be learned concerning the reproductive habits of the least shrew. Hamilton (1944) captured a female that later gave birth to six young; another female had five young with her when taken from a nest. The gestation period is 20 to 23 days. The young acquire a fine coat of hair when one week old. By the time they are three weeks old, they have the same appearance as the adults and are weaned. It is rather amazing that a female weighing 5 g can supply enough milk to care for a brood that collectively may equal three times her own weight! Some, or possibly most, females have more than one litter each year. The breeding season extends from March to November.

The importance to man of this small mammal in eating insect larvae should not be overlooked.

The taxonomy of this species in Ohio is not clear. At present three subspecies are recognized (Hall and Kelson, 1959, p. 57). *Cryptotis parva elasson* occupies most of the state; *C. p. harlani* is an Indiana form that barely reaches to the western border of Ohio; and *C. p. parva* is said to occupy the southern and eastern parts of Ohio. Many more specimens from throughout Ohio are needed before this interpretation can be verified.

SELECTED REFERENCES

Bole, B. P., Jr., and P. N. Moulthrop. 1942. The Ohio recent mammal collection in the Cleveland Museum of Natural History. Sci. Publ. Cleveland Mus. Nat. Hist. 5:83-181, esp. 97-99.

Broadbrook, H. E. 1952. Nest and behavior of a short-tailed shrew, *Cryptotis parva*. J. Mammal. 33:241-43.

Conaway, C. H. 1958. Maintenance, reproduction, and growth of the least shrew in captivity. J. Mammal. 39: 507-12.

Davis, W. B. 1938. A heavy concentration of *Cryptotis*. J. Mammal. 19:499-500.

Davis, W. B., and L. Joeris. 1945. Notes on the life-history of the little short-tailed shrew. J. Mammal. 26:136-38.

Getz, L. L. 1962. A local concentration of the least shrew J. Mammal. 43:281-82.

Goodpaster, W. W. 1941. A list of the birds and mammals of southwestern Ohio. J. Cincinnati Soc. Nat. Hist. 22:1-47.

Hall, E. R., and K. R. Kelson. 1959. The mammals of North America. Ronald Press Co., New York. 2 vols. 1162 pp.

Hamilton, W. J., Jr. 1934. Habits of *Cryptotis parva* in New York. J. Mammal. 15:154-55.

———. 1944. The biology of the little short-tailed shrew, *Cryptotis parva*. J. Mammal. 24:1-7.

Hatt, R. T. 1938. Feeding habits of the least shrew. J. Mammal. 19:247-48.

Hunt, T. 1951. Breeding of *Cryptotis parva* in Texas. J. Mammal. 32:115-16.

Lyon, M. W., Jr. 1936. Mammals of Indiana. Amer. Midl. Nat. 17:1-384.

McCarley, W. H. 1959. An unusually large nest of *Cryptotis parva*. J. Mammal. 40:243.

Mumford, R. E., and C. O. Handley, Jr. 1956. Notes on the mammals of Jackson County, Indiana. J. Mammal. 37:407-12.

Peterson, C. B. 1936. *Cryptotis parva* in western New York. J. Mammal. 17:284-85.

Pfeiffer, C. J., and G. H. Gass. 1963. Note on the longevity and habits of captive *Cryptotis parva*. J. Mammal. 44:427-28.

Springer, S. 1937. Observations on *Cryptotis floridana* in captivity. J. Mammal. 18:237-38.

Welter, W. A., and D. E. Sollberger. 1939. Notes on the mammals of Rowan and adjacent counties in eastern Kentucky. J. Mammal. 20:77-81.

Whitaker, J. O. 1974. *Cryptotis parva*. Amer. Soc. Mammal., Mammalian Species No. 43:1-8.

FAMILY TALPIDAE

Moles

Member of this family in Ohio are immediately recognized by the large paddle-like front feet that are turned outward from the body. The bones in the forelimbs have become shortened and fused together, an adaptation that enables the mole to push and dig through the soil. The powerful muscles required for this work are attached to a keel on the sternum, similar to the breastbone found in birds. The eyes and ears are small and hidden beneath the fur. The fur is short, soft, and easily smoothed in either direction. Members of this family are found in Europe and Asia as well as in North America. There are three species of moles in Ohio.

Hairy-tailed Mole
Parascalops breweri (Bachman)

The hairy-tailed mole is the smallest mole found in Ohio. It has been reported as occupying the eastern half of the state, but it may also exist farther west than we presently know. There are few records reported from Ohio, but it simply may have been overlooked, since, to most people, "a mole is a mole." Both Hine (1912) and Olive (1950) have reported hairy-tailed moles caught in Franklin

Parascalops breweri (Bachman)

$$i\frac{3-3}{3-3}, c\frac{1-1}{1-1}, pm\frac{4-4}{4-4}, m\frac{3-3}{3-3} = 44$$

Following are measurements of 28 adult hairy-tailed moles (*Parascalops breweri*) from Ohio:

	Total Length (mm)	Tail (mm)	Hind Foot (mm)
Extremes	140–80	21–40	17–24
Mean	155.6	29.1	19.1

County. These specimens are of interest because "it has been accepted that the range of *P. breweri* tends to be coincident with the counties of Ohio that lie within the Allegheny Plateau" (Olive, 1950). Most of these counties are unglaciated. Some plant communities associated with certain sandy and well-drained soils in northern Franklin County are somewhat similar to those on the Allegheny Plateau, and could possibly serve as an entry route for the hairy-tailed mole into the western glaciated counties. There is no apparent reason why this mole should not move west, since its requirements are essentially the same as those of the eastern mole. More ecological studies on the moles are needed to explain their distribution. The hairy-tailed mole inhabits meadows, open fields, gardens, and wooded areas wherever soil conditions are suitable for digging and there is sufficient food. Lyon (1936) states that the hairy-tailed mole is not known for certain from Indiana, although two questionable specimens reported to have been collected in Bloomington, Indiana, are in the Indiana University collection.

Adults of the hairy-tailed mole seldom exceed 165 mm (6.5 inches) in total length. The front feet, although paddle-like, are not as large as those of the eastern mole, and the palms are about as long as broad. The tail is proportionally longer and fatter than in the eastern mole and is covered with long, stiff hairs that form a noticeable "pencil" at the end. There is no webbing between the toes on the hind feet. The fur is somewhat darker than that of the eastern mole, but in other respects it is the same. The eyes and ears are small and hidden beneath the fur. The skull is similar to, but smaller than, that of the eastern mole, and there are 44 teeth instead of 36. The dental formula is as follows:

The diet of this mole is primarily the same as that of the eastern mole, although insects may be the preferred food with earthworms second. The hairy-tailed mole is active throughout the year, utilizing a series of shallow runways during the summer and moving to deeper quarters in the winter. Hamilton (1943) states that it often leaves its tunnels at night to forage for food on the forest floor. We know little about the reproductive habits of this animal. The single litter of four is born during April or May. The young mature rapidly and are ready to leave the nest and care for themselves when one month old. They mate for the first time when approximately ten months old.

This species, like the eastern mole, sometimes invades golf courses, home lawns, and gardens, but otherwise seldom interferes with man's activities. Its wholesale destruction of harmful insects makes it beneficial to man.

SELECTED REFERENCES

Eadie, W. R. 1939. A contribution to the biology of *Parascalops breweri*. J. Mammal. 20:150–73.

Gordon, R. E., and J. R. Bailey. 1963. The occurrence of *Parascalops breweri* on the highlands (North Carolina) plateau. J. Mammal. 44:580–81.

Hallett, J. G. 1978. *Parascalops breweri*. Amer. Soc. Mammal., Mammalian Species No. 98:1–4.

Hamilton, W. J., Jr. 1943. The mammals of eastern United States: an account of recent land mammals occurring east of the Mississippi. Comstock Publ. Co., Ithaca, N.Y. Vol. 2, 438 pp.

Hine, J. S. 1912. Ohio moles and shrews. Ohio Nat. 12:494–96.

Jackson, H. H. T. 1915. A review of the American moles. North Amer. Fauna 38:1–100.

Lyon, M. W., Jr. 1936. Mammals of Indiana. Amer. Midl. Nat. 17:1–384.

Olive, J. R. 1946. Economic importance of moles. Ohio Conserv. Bull. 10(12):4–5.

———. 1950. An extension of range of the hairy-tailed mole, *Parascalops breweri* (Bachman). J. Mammal. 31:458–59.

Scheffer, T. H. 1949. Ecological comparisons of three genera of moles. Trans. Kans. Acad. Sci. 52:30–37.

Thomas, E. S. 1951. Distribution of Ohio animals. Ohio J. Sci. 51:153–67.

Eastern Mole
Scalopus aquaticus (Rafinesque)

The eastern, common, or prairie mole bears the specific name *aquaticus* ("having to do with water"), a misnomer if intended to refer to the habitat of the mole. For although this animal is found in a wide variety of habitats throughout Ohio, it is not partial to water. It can swim, but does so only when forced.

Scalopus aquaticus (Rafinesque)

Concerning the distribution of the eastern mole in Ohio, Bole and Moulthrop (1942) concluded that it was definitely not present in the northeastern quarter of the state, although Jackson (1915) had reported a specimen from Salem in Columbiana County and another from Cleveland in Cuyahoga County. Thomas (1915), in discussing the distribution of Ohio mammals, says that "the Prairie Mole occupies all of the territory west of the plateau [Allegheny] and is apparently extending its range eastward." Guilday (1961) reported finding the bones of a prairie mole in an Indian mound located on a bluff on the Allegheny Plateau of western Pennsylvania. This is interesting, because "*Scalopus* is the common mole of western and central Ohio, but is replaced by the hairy-tailed mole in eastern Ohio throughout the Appalachian Plateau" (Guilday, 1961). Guilday suggests that the eastern mole may yet be found in the upper Ohio River Valley. The records that I have collected for this species suggest that in Ohio, as elsewhere, this mole is an animal of the flatlands and plains and is not found in the plateaus and mountains. It is very common in western and central Ohio, where the soils are soft and friable, less common eastward toward the Unglaciated Allegheny Plateau, and completely absent on the plateau.

The eastern mole is the largest mole in Ohio. The fur is gray, slightly lighter in summer than in winter. It is not unusual to find moles that have bright orange, yellow, or white patches of fur on the chest, wrist, or belly. Eadie (1954) has shown that these patches are the result of discoloration produced by the secretions of skin glands. The fur is soft, velvety, and thick; when viewed at different angles and under certain lighting conditions, it appears silvery or white. Albino moles are known from Ohio.

The skull is triangular in outline, broad in back and narrow in front. The individual bones of the skull are closely fused. The zygomatic arches are present and complete but delicate in structure. The cheek teeth are high-crowned and have many sharp cutting edges well adapted to crushing and grinding insects, snail shells, and other hard items. The tiny eyes, hidden beneath the fur, are permanently covered by the thin eyelids that have fused together. The eyes probably function only to distinguish between light and dark. The ears are well developed but hidden beneath the fur. They can be located by blowing the hair on the side of the head. The dental formula is as follows:

$$i\frac{3-3}{2-2},\ c\frac{1-1}{0-0},\ pm\frac{3-3}{3-3},\ m\frac{3-3}{3-3} = 36$$

Following are measurements of 16 adult eastern moles (*Scalopus aquaticus*) from Ohio:

	Total Length (mm)	Tail (mm)	Hind Foot (mm)
Extremes	147–92	21–36	20–27
Mean	172	27	23

Soil condition and food availability are apparently the two most important factors governing the local distribution of moles. If the ground is too hard or too wet and/or if adequate food is not available, the eastern mole will not invade an area. I have found them living along the sandy beaches of lakes, in woodlots, in pastures and meadows, and in parks, gardens, and golf courses. Moles are solitary creatures that spend their entire existence below the surface of the ground. An adventuresome (or confused?) individual occasionally leaves its burrow and is found on the surface of the ground, but this is most unusual.

A mole digs two rather extensive systems of tunnels, one used in summer, the other in winter. During spring and summer the mole burrows 2-5 cm (1-2 in) below the surface, producing the familiar unsightly ridges. As the mole moves along, its back pushes against the top of the tunnel, raising and cracking the ground. Many of these shallow tunnels, made as the mole forages for food, are used only once and then abandoned, whereas others serve as main thoroughfares and are kept in good repair and used over and over again. As winter approaches and temperatures drop, moles abandon their surface tunnels and descend to a second labyrinthian system of passageways located below the frost line where they are able to keep warm and find the food necessary to satisfy their voracious appetites. Moles do not hibernate but remain active throughout the year. Ordinarily, the only evidence that moles are "operating" in an area is the humpback ridges across gardens and lawns. However, when moles are excavating deep tunnels or building nests, the excess dirt, rather than being pushed aside, is brought to the surface and deposited in a "mole hill." These hills may be 12 to 15 cm (5 to 6 in) high and 30 to 60 cm (1 to 2 ft) in diameter. Since the dirt is added to the heap by pushing up from below, the opening to the tunnel is never exposed (as it is in a gopher mound). In Oregon, Kuhn et al. (1966) successfully used the hills made by Townsend's mole, *Scapanus townsendii*, as markers for locating their underground nests, a technique well worth trying with *Scalopus aquaticus*.

The principal foods eaten by the eastern mole, in order of their quantitative importance, are earthworms, insect grubs, insects, snails, crustaceans, and spiders. Some plant foods may be included in the diet; corn, wheat, oats, potatoes, grass, garden peas, tomatoes, and apples have been found in the stomach. Food habit studies conducted by Wilson (1898), West (1910), Dyche (1903), Scheffer (1917), and Hisaw (1923a) all indicate that vegetable matter makes up less than 20 percent of the total diet, and most agree that much of this is consumed accidentally.

Eastern moles capture prey in several ways. Inactive animals are simply grabbed with the mouth, chewed, and swallowed. A large and very active insect may be slammed against the bottom or side of the burrow with one of the big front feet and held there while the mole consumes it. Especially active victims are covered with dirt scraped from the sides of the burrow. As the prey struggles to free itself, the mole grabs and chews on the exposed parts. Earthworms are sometimes held between the two front feet and, like other items of food, are chopped into segments and thoroughly minced by the shearing action of the sharp teeth.

Eastern moles eat surprisingly large quantities of food. Christian (1950) kept an adult male in captivity for 25 days, feeding it freshly killed white mice. The amount of food consumed each day averaged 28.7 g, an amount between 50 and 55 percent of the total weight of the mole. Hisaw (1923a) noted: "The six captive moles under observation in this laboratory ate a total of 5,466.5 grams of food in 208 days, which is an individual average of 26.28 grams per day." This is a consumption of food equal to about 32 percent of the body weight of each animal. In comparison, a man weighing 73 kg (160 lb) would have to eat 22 kg (50 lb) of food a day to equal this gastronomic feat!

The reproductive habits of eastern moles living in Ohio have not been investigated, but it is assumed that they are the same here as elsewhere throughout their range. The gestation period is thought to be 25 to 30 days. One litter of two to five is produced in April of each year. The nursery is a globular nest of leaves or grass located off one of the main runways in the burrow system. The young are born blind, naked, and helpless. At one week they have a coat of fine, silky, gray hair, retained until they are approximately one-third grown. Young moles develop rapidly and probably are able to leave the nest and care for themselves by the age of one month. (This is a subject that needs further study.)

Eastern moles may be beneficial or detrimental, depending upon the situation. When lawns, golf courses, and gardens are disfigured by mole hills and ridges or when mole tunnels inadvertently provide passageways for the inroads of other small mammals (field mice, woodland mice, house mice), then the mole is obviously a detriment. However, moles destroy myriads of harmful insects and grubs; and in areas where they do not directly conflict with man's interests, they are undoubtedly beneficial. European mole pelts have been used for years in making moleskin coats and providing fur trim for coats and hats. The eastern mole, however, because of its small size and inferior grade of fur, has never been commercially important as a fur producer.

Eastern moles are difficult to control because of their subterranean habits and because they will not accept baits readily. The best means of control is the use of special traps, fumigation of burrows, and removal of their food. The last means, upsetting the moles' ecological food chain by eliminating the grubs and worms in the lawn, is the most effective control.

Because of the safety afforded by their underground retreats, Eastern moles are not often preyed upon by other animals. Occasionally, an individual may be caught and eaten by a hawk, owl, fox, domestic cat, weasel, or snake.

SELECTED REFERENCES

Arlton, A. V. 1936. An ecological study of the mole. J. Mammal. 17:349-71.
Bole, B. P., Jr., and P. N. Moulthrop. 1942. The Ohio recent mammal collection in the Cleveland Museum of Natural History. Sci. Publ. Cleveland Mus. Nat. Hist. 5:83-181, esp. 87-88.
Christian, J. J. 1950. Behavior of the mole (*Scalopus*) and the shrew (*Blarina*). J. Mammal. 31:281-87.
Conaway, C. H. 1959. The reproductive cycle of the eastern mole. J. Mammal. 40:180-94.
Davis, W. B. 1942. The moles (genus *Scalopus*) of Texas. Amer. Midl. Nat. 27:380-86.
Dyche, L. L. 1903. Food habits of the common garden mole. Trans. Kans. Acad. Sci. 18:183-86.
Eadie, W. R. 1954. Skin gland activity and pelage descriptions in moles. J. Mammal. 35:186-96.
Edwards, L. F. 1937. Morphology of the forelimb of the mole (*Scalops* [sic] *aquaticus* L.) in relation to its fossorial habits. Ohio J. Sci. 37:20-41.
Guilday, J. E. 1961. Prehistoric record of *Scalopus* from western Pennsylvania. J. Mammal. 42:117-18.
Hanawalt, F. A. 1922. Habits of the common mole, *Scalopus aquaticus machrinus* (Rafinesque). Ohio J. Sci. 22:164-69.
Henning, W. L. 1952. Studies in control of the prairie mole, *Scalopus aquaticus machrinus* (Rafinesque). J. Wildl. Manage. 16:419-24.
Hine, J. S. 1912. Ohio moles and shrews. Ohio Nat. 12:494-96.
Hisaw, F. L. 1923a. Feeding habits of moles. J. Mammal. 4:9-20.
_____. 1923b. Observations on the burrowing habits of moles (*Scalopus aquaticus machrinoides*). J. Mammal. 4:79-88.
Jackson, H. H. T. 1915. A review of the American moles. North Amer. Fauna 38:1-100.
Kuhn, L. W., W. Q. Wick, and R. J. Pedersen. 1966. Breeding nests of Townsend's mole in Oregon. J. Mammal. 47:239-49.
Olive, J. R. 1946. Economic importance of moles. Ohio Conserv. Bull. 10(12):4-5.
_____. 1950. Some parasites of the prairie mole, *Scalopus aquaticus machrinus* (Rafinesque). Ohio J. Sci. 50:263-66.
Scheffer, T. H. 1910. The common mole. Kans. State Agric. College Exp. Sta., Bull. 168. 19 pp.
_____. 1917. The common mole of eastern United States. U.S.D.A. Farmer's Bull. 583. 12 pp.
Silver, J., and A. W. Moore. 1941. Mole control. U.S. Fish & Wildlife Serv., Conserv. Bull. 16:1-17.
Thomas E. S. 1951. Distribution of Ohio animals. Ohio J. Sci. 51:153-67.
West, J. A. 1910. A study of the food of moles in Illinois. Bull. Ill. Lab. Nat. Hist. 9:14-22.
Wilson, H. 1898. The economic status of the mole (*Scalopus* and *Condylura*). Pa. Dept. Agric. Bull. 30. 26 pp.
Yates, T. L., and D. J. Schmidly. 1978. *Scalopus aquaticus*. Amer. Soc. Mammal., Mammalian Species No. 105:1-4.

Star-nosed Mole
Condylura cristata (Linnaeus)

The star-nosed mole, certainly the most unusual looking mammal in Ohio, is commonly recorded only in the northeastern part of the state. There are few records from other areas of the state, where its status is questionable. Because of the dependence of these moles upon a very specialized habitat, they are quite local in their distribution. They are semi-aquatic and most frequently occur in swamps, bogs, and wet meadows. Although tunnels are constructed, these moles are probably the least fossorial of all our North American Talpidae. Many of their underground runways have their openings below the water level in streams and ponds, and I once observed a star-nosed mole emerging from a culvert. Several observers (Tenney, 1871; Wood, 1922) have reported seeing these animals moving about on top of the snow during winter. Hamilton (1931) and others have also seen star-nosed moles in winter, swimming under the ice in frozen ponds. Hamilton (1931), who has captured individuals of this species in a minnow trap in a small, swift stream, finds the star-nosed mole an excellent swimmer. Most of the live specimens he examined were taken underwater. He also caught practically all his winter specimens in muskrat traps set at the entrance to muskrat dens. However, several Ohio muskrat trappers to whom I have talked were not familiar with this mole and had never caught one in their traps. I have taken star-nosed moles in traps set in surface runways winding through tall marsh grass.

Condylura cristata (Linnaeus)

The unusual appearance of this mole is due to the 22 fleshy, finger-like projections arranged symmetrically around the end of its nose. These nasal rays are constantly in motion and undoutedly serve as sensory organs. In one captive the lowest and most centrally located pair of rays moved around much more than the others, but I was not able to determine the significance of this.

The body of this mole is about the same length as that of the eastern mole but much thinner. The tail is sparsely haired and longer than that of either of the other Ohio moles, equaling one-half or more the body length; it also has a constriction at its base. During winter and spring the tail may become greatly swollen and enlarged, but since this occurs only in certain individuals, its significance is not clearly understood. Hamilton (1931) suggests that the fatty material stored in the tail may furnish additional energy during the breeding period. The thick fur is dark gray or more often black, darker than that of either of our other two moles.

The long, thin, delicately tapering skull is generally similar to that of the other moles. The second upper incisor tooth is small and easily overlooked, but there are 44 teeth in all. The dental formula is as follows:

$$i\frac{3-3}{3-3}, c\frac{1-1}{1-1}, pm\frac{4-4}{4-4}, m\frac{3-3}{3-3} = 44$$

Following are measurements of 11 adult star-nosed moles (*Condylura cristata*) from Ohio:

	Total Length (mm)	Tail (mm)	Hind Foot (mm)
Extremes	170-90	58-73	22-26
Mean	180	63	24

Examination of stomach contents suggests that the star-nosed mole in strictly carnivorous. Some plant remains are found in most stomachs, but these are probably eaten accidentally along with the main food of aquatic insects and worms rooted out of the mud on pond bottoms. Insect larvae, various kinds of crustaceans and molluscs, and an occasional fish complete the diet.

As in other moles, reproductive activities of the star-nosed mole take place in the spring, and the young are born in April, May, or June (Eadie and Hamilton, 1956). These same authors state, however, that young may be born from late March until early August. The gestation period is unknown; mating activities and birth of the young have not been observed. However, litters of two to seven young have been recorded. The several nests that have been uncovered and described were globular in form and constructed of leaves and/or grass. Manure piles, compost heaps, old stumps, or special rooms off the burrow system several inches below the surface of the ground may serve as nest-building sites.

The star-nosed mole is certainly one of the most peculiar and most interesting small mammals in North America. Although it is widely distributed over the eastern United States and is often abundant in many localities, it is curious that few mammologists have studied its habits in detail.

SELECTED REFERENCES

Ayers, H. 1885. On the structure and development of the nasal rays in *Condylura cristata*. Biol. Zentralbl. 4:358-59.

Eadie, W. R., and W. J. Hamilton, Jr. 1956. Notes on reproduction in the star-nosed mole. J. Mammal. 37:223-31.

Hamilton, W. J., Jr. 1930. The biology of the star-nosed mole, *Condylura cristata* Linnaeus. Ph.D. Dissertation, Cornell Univ., Ithaca, N.Y.

_____. 1931. Habits of the star-nosed mole, *Condylura cristata*. J. Mammal. 12:345-55.

Kennard, F. H. 1929. A star-nosed mole's nest at Newton Centre, Massachusetts. J. Mammal. 10:77-78.

Simpson, S. E. 1923. The nest and young of the star-nosed mole (*Condylura cristata*). J. Mammal. 4:167-71.

Tenney, S. 1871. The star-nosed mole. Amer. Nat. 5:314.

Vickers, E. W. 1894. Notes on *Zapus hudsonicus, Condylura cristata,* and *Sorex platyrhinus.* Second Annu. Rep., Ohio Acad. Sci., p. 15.

———. 1895. A contribution to the distribution of *Condylura cristata* (Linn.) Desmarest, starnosed mole in Ohio. Third Annu. Rep., Ohio Acad. Sci. p. 15.

Wilson, H. 1898. The economic status of the mole (*Scalopus* and *Condylura*). Pa. Dept. Agric. Bull. 30. 26 pp.

Wood, N. A. 1922. The mammals of Washtenaw Co., Michigan. Occ. Papers, Univ. of Mich. Mus. Zool. 123:1–23.

ORDER CHIROPTERA
Bats

The members of this large order of mammals possess wings and are able to fly. Among all the mammals of the world, only the bats have this unique ability. Those of the New World are typically small, mouse-sized animals with wingspreads that vary from 15 to 46 cm (6 to 18 in). In southern Asia and in the South Pacific, some bats attain a wingspread of 120 to 150 cm (4 to 5 ft) and weigh up to 1 kg (2 lb) or more. The bones in the forelimbs of these mammals are elongated and tubular (fig. 10). The "fingers," except for the thumb, are tremendously elongated and support a thin, double layer of skin that serves as the flight membrane. In most bats this thin layer of skin joins to the hind leg on each side and extends between them enclosing all or most of the tail (interfemoral or tail membrane). The thumb is the only digit on the forelimb possessing a claw (some of the Old World bats have a claw on the second finger as well).

Most species have a keel developed on the sternum where the powerful flight muscles are attached. Since flight is the principal means of locomotion, bats have lost the ability to walk, although some move remarkably well on the ground by hop-

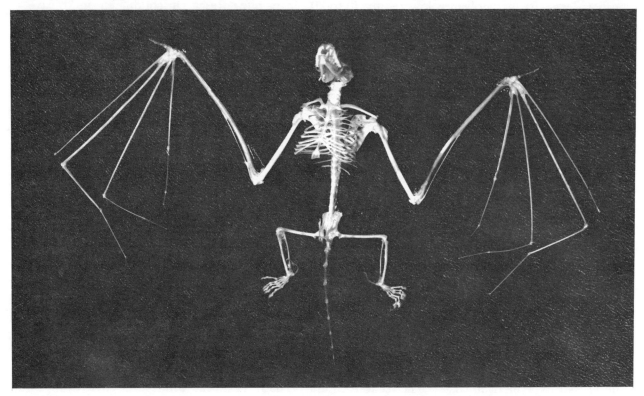

Fig. 10. Skeleton of a big brown bat *(Eptesicus fuscus)*. (Photograph by the author.)

ping or pulling themselves along with their wings. The claws on the thumbs are used effectively in climbing. The five toes on each hind foot are clawed and are used to anchor the bat as it sleeps or rests in its peculiar upside-down attitude. The knee joint is rotated as compared with that of other mammals; it is directed outward and posteriorly.

With few exceptions, North American bats are strictly insectivorous. Not all bats in the world, however, depend upon insects as their source of food. Many of the larger bats in the Old World and South America are fruit eaters, some collect nectar and pollen from night-blooming flowers, and still others are carnivorous. Almost everyone is familiar with the vampire bats of South America, Central America, and Mexico that feed upon nothing but blood.

All bats are crepuscular and/or nocturnal in habit. Usually they are active for about one hour in the early morning and again at dusk. Between flights bats occupy a dark, secluded retreat, hanging upside-down and in a deep sleep. A bat in flight expends a great amount of energy, but while asleep it uses almost no energy at all. While resting, even during the hottest time of year, bats enter a state of semi-hibernation in which all metabolic activity is drastically reduced. When we sleep, our metabolism is lowered approximately 10 percent; but when a bat sleeps, its metabolism is lowered by 70 to 80 percent. Some of the smallest bats live for more than 20 years. If man could follow the same daily routine as the bat, conserving his energy in the same way, imagine the possible impact on his longevity!

For many years biologists have wondered how bats, flying in half-light or total darkness, can find and catch flying insects, many as small as mosquitoes or minute gnats. We now know that, as the bat cruises through the air, it periodically sends out short bursts, or "blips," of ultrasonic sound (energy). Upon striking an object, the sound waves are immediately echoed back, thus allowing the bat to "zero in" on prey or to avoid an object in its path. The sounds are sent from either the mouth or the nose of the bat, depending upon the species, and echoes are received by the ears. The ultrasonic sounds used by bats are in the frequency range of 15,000 to 100,000 hertz (cycles per second) and are considerably beyond the audible range of man. Fifty thousand hertz seems to be the frequency commonly employed by several of our more common North American bats. The echo-locating system is so delicately tuned that a blindfolded bat with a 30 cm (12-inch) wingspread might thread its way through a darkened room laced with fine wires placed 25 cm (10 inches) apart and never touch a wire!

It has been demonstrated by Guthrie (1933) and Wimsatt (1960), as well as others, that the ovaries of female bats are quiescent during the fall and winter months and that ovulation does not occur until spring, usually in April. The sperm that are introduced into the reproductive tract of the female in the fall remain alive and active until the following spring when the eggs are released from the ovaries. At this time fertilization actually takes place. This is referred to as delayed fertilization and is characteristic of all North American bats.

Bats have been accused of being dirty, disease-carrying, parasite-ridden mammals. The truth is that they are no dirtier than any other kind of mammal, nor do they have any more parasites (although they do have some very interesting ones). Some bats, including species in Ohio (Krogwold, 1977), are rabies vectors, but then so are many other wild animals. Mammalogists agree, however, that live bats should not be handled without gloves and that people, especially children, should be warned not to touch or pick up any bat that they might find lying on the ground. These bats may be sick and are likely to bite. Research conducted in a cave densely populated with bats in Texas (Constantine, 1962) demonstrated that viable rabies virus can be transmitted through the air to susceptible mammal hosts.

The Order Chiroptera has a worldwide distribution, with the exception of the permanently cold polar regions. More than eight hundred species are recognized. They are divided into 17 families, 4 of which are represented in the United States, but only 1 is found in Ohio.

FAMILY VERSPERTILIONIDAE
Bats

All bats found in Ohio are vespertilionids, as are most of the bats in the United States. This family of "common bats" has a worldwide distribution with the exception of some oceanic islands. There are 30 species of vespertilionid bats throughout the United States, and in some areas they are extremely abundant. In the large limestone caves of the Southwest,

SHORT-TAILED SHREW
Blarina brevicauda

LEAST SHREW
Cryptotis parva

EASTERN MOLE
Scalopus aquaticus

RED BAT WITH YOUNG
Lasiurus borealis

LITTLE BROWN MYOTIS
Myotis lucifugus LeConte

BIG BROWN BAT
Eptisicus Fuscus

it is not unusual to find many thousands of these animals roosting together. No concentrations as large as these have been found in Ohio. Nevertheless, bats are common here, with possibly 11 different species represented in the state.

As a matter of convenience, the bats in Ohio can arbitrarily be divided into two groups: (1) the migratory bats, including *Lasiurus, Lasionycteris,* and *Plecotus;* and (2) the nonmigratory bats, including *Eptesicus, Nycticeius, Pipistrellus,* and *Myotis.* The migratory bats spend the warmer months of the year in Ohio, but with the approach of winter and the disappearance of their insect food, they migrate south into the warmer latitudes. Exactly where in the South these bats spend their winters is not yet clear. Since all the migratory species have fur on the dorsal surface of the interfemoral membrane, they are also known as furry-tailed bats. These solitary bats are primarily nongregarious, hunting and roosting alone or in small groups of three or four.

All our smaller bats are nonmigratory, the majority remaining in Ohio or at least in the northern latitudes throughout the year. They winter in caves, tunnels, hollow trees, or buildings in a deep state of hibernation. Nonmigratory bats are gregarious animals that gather by the hundreds or thousands at communal hibernation and nursery sites. The bats in a hibernating colony may be so tightly packed together that only the tips of the noses and the wrists on the wings of the individuals can be seen protruding from the furry mass of bodies. Several different species of bats often use the same general roosting area, each species occupying a specific location within the general area. The microclimate preference of bats are so pronounced that it is often possible to predict what species of bat will be present in a particular region of a cave.

KEY TO THE SPECIES OF BATS IN OHIO

1. Ears extremely large, much larger than head................*Plecotus rafinesquii,* Rafinesque's Big-eared Bat p. 51

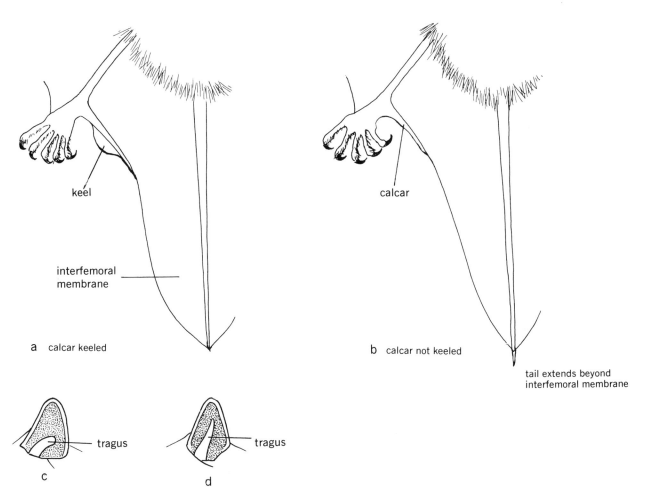

Fig. 11. Characteristics used in identifying bats. (Drawing by Elizabeth Dalvé.)

1'. Ears not larger than head 2
2. Dorsal surface of interfemoral membrane completely or partially covered with dense growth of fur; hairs on body may be tipped with silver 3
2'. Dorsal surface of interfemoral membrane naked or not densely furred; hairs on body not tipped with silver 5
3. Only basal half of interfemoral membrane furred; hairs on body tipped with silver white; first premolar in upper jaw not minute .... *Lasionycteris noctivagans,* Silver-haired Bat p. 43
3'. Interfemoral membrane completely furred; over-all color not silver, but hairs on body may have silver tips; first premolar in upper jaw minute and easily overlooked .... *Lasiurus* 4
4. Forearm less than 44 mm long; general coloration reddish or yellowish red .... *L. borealis,* Red bat p. 48
4'. Forearm more than 44 mm long; general coloration heavily frosted dark brown or gray . ... *L. cinereus,* Hoary Bat p. 49
5. Premolars 1 above and 2 below on each side . 6
5'. Premolars not 1 above and 2 below on each side 7
6. Incisors 2 above and 3 below on each side; total number of teeth 32; interfemoral membrane scantily furred near base; tail not extending beyond edge of interfemoral membrane (fig. 11a); color chocolate brown with long silky hair .... *Eptesicus fuscus,* Big Brown Bat p. 46
6'. Incisors 1 above and 3 below on each side; total number of teeth 30; interfemoral membrane without hair except at base; tail definitely extending beyond interfemoral membrane (fig. 11b); color brown with hair not long and silky .... *Nycticeius humeralis,* Evening Bat p. 50
7. Premolars 2 above and 2 below on each side; total number of teeth 34; color light brown to yellow; hairs distinctly tricolored, gray at the base, yellow in the middle, and tipped with brown .... *Pipistrellus subflavus,* Eastern Pipistrelle p. 44
7'. Premolars 3 above and 3 below on each side; total number of teeth 38; color brown or gray; hairs usually not distinctly tricolored, or, if tricolored, lacking a definite yellow band .... *Myotis* 8
8. Calcar with a small but usually evident keel (fig. 11a); sagittal crest on skull (fig. 22) evident 9
8'. Calcar not keeled (fig. 11b); sagittal crest on skull (fig. 22) not evident 10
9. Fur pinkish-gray in color, on back obviously tricolored; no black facial mask .... *M. sodalis,* Indiana Myotis p. 41
9'. Fur brown with golden sheen; distinct black facial mask .... *M. leibii,* Small-footed Myotis p. 43
10. Ear when laid forward extends beyond tip of muzzle; tragus long (equal to or more than one-half the length of ear), narrow and pointed (fig. 11d); hairs on toes do not extend beyond tips of claws .... *M. keenii,* Keen's Myotis p. 41
10'. Ear when laid forward does not extend beyond tip of muzzle; tragus short and rounded (fig. 11c); some hairs on the toes extend to or beyond tips of claws .... *M. lucifugus,* Little Brown Myotis p. 38

SELECTED REFERENCES

Allen, G. M. 1940. Bats. Harvard Univ. Press, Cambridge, Mass. 368 pp.

Barbour, R. W., and W. H. Davis. 1969. Bats of America. Univ. Press of Ky., Lexington. 286 pp.

Cockrum, E. L. 1955. Reproduction in North American bats. Trans. Kans. Acad. Sci. 58:487–511.

Constantine, D. G. 1962. Rabies transmission by nonbite route. Public Health Report 77:287–89.

Davis, W. H., and O. B. Reite. 1967. Response of bats from temperate regions to change in ambient temperature. Biol. Bull. 132:320–28.

Guthrie, M. J. 1933. The reproductive cycles of some cave bats. J. Mammal. 14:199–216.

Hartmann, C. G. 1933. On the survival of spermatozoa in the female genital tract of the bat. Quart. Rev. Biol. 8:185–93.

Krogwold, R. A. 1977. A study of bat rabies in Ohio. M. S. Thesis, Ohio State Univ., Columbus. 58 pp.

Leen, N., and A. Novick. 1969. The world of bats. Holt, Rinehart & Winston, New York. 171 pp.

Mills, R. S., G. W. Barrett, and J. B. Cope. 1977. Bat species diversity patterns in east central Indiana. Ohio J. Sci. 77:191–92.

Mohr, C. E. 1933. Observations on the young of cave-dwelling bats. J. Mammal. 14:49–53.

Rausch, R. 1946. Collecting bats in Ohio. J. Mammal. 27:275.

Tjalma, R. A., and B. B. Wentworth. 1957. Bat rabies report of an isolation of rabies virus from native Ohio bats. J. Amer. Vet. Med. Assoc. 130:68–70.

Wimsatt, W. A. 1960. An analysis of parturition in Chiroptera, including new observations on *Myotis l. lucifugus.* J. Mammal. 41:183–200.

Little Brown Myotis or Little Brown Bat
Myotis lucifugus LeConte

This small brown bat is a miniature edition of the big brown bat (see p. 46) and also closely resembles the other species of *Myotis* found in Ohio. Although we do not have records for this bat from every county, it is almost certainly found throughout the state. It probably is the most common bat of the genus *Myotis*, and in some areas of Ohio it is even more abundant than the big brown bat.

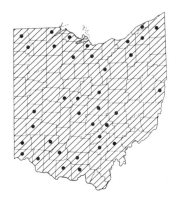

Myotis lucifugus LeConte

For someone not familiar with bats and not experienced in the use of keys, separating the little brown myotis from other bats in Ohio may prove difficult. The fur is similar to that of the big brown bat, but is not as long and silky. The color is olive brown or yellowish brown. The individual hairs on the back are dark at the base and brown at the ends. There is a dark brown or blackish spot of fur on each shoulder. The calcar is not keeled, and the wing membrane reaches to the base of the foot. The ears do not extend past the tip of the nose when laid forward, and the tragus, which is approximately one-half as long as the ear, is straight and has a slightly rounded tip. The skull characteristics of the little brown myotis are similar to those described for the big brown bat. The normal dentition for all Ohio species of *Myotis* is 38 teeth. The dental formula for all *Myotis* is as follows:

$$i\frac{2-2}{3-3}, c\frac{1-1}{1-1}, pm\frac{3-3}{3-3}, m\frac{3-3}{3-3} = 38$$

Following are measurements of 24 adult little brown myotis (*Myotis lucifugus*) from Ohio:

	Total Length (mm)	Tail (mm)	Hind Foot (mm)
Extremes	81–92	33–43	6–11
Mean	86	38	9

In the late evening as little brown myotis pursue their insect prey over clearings and in woods, they exhibit a quick, nervous erratic, flight pattern. Fishermen often see the shadowy outlines of these bats skimming along the surface of the water in pursuit of insects, or they may notice little furrows on the water surface as the bats scoop up water with their lower lips. Occasionally a bat will even attack the artificial fly being cast by the fisherman. The little brown myotis is highly efficient at catching small flying insects. Gould (1955) concluded after much study that the average weight of insects caught by one bat in one hour was 1 g. Further calculations led to the conclusion that this figure represents about 500 insects caught per hour, or one catch every seven seconds!

During the winter most little brown myotis remain in the north, where they establish hibernating colonies. They use totally dark areas with high humidity as hibernation sites. Caves are most often chosen, although hollow trees and buildings are also used. Usually the bats hibernate in tightly packed clusters of five to fifty individuals. Weather conditions determine the length of the hibernation period, which in Ohio is from November to April.

Emerging from hibernation in the spring, the little brown myotis at first finds little food and must rely upon what little stored fat remains for its source of energy. At this time the sexes separate, the males becoming solitary while the females form nursery colonies. Some of the bats may utilize the hibernation site as a summer roost, but most of them disperse to new summer quarters that may be attics, hollow trees, caves, barns, window shutters, or loose siding.

Some mating takes place in the spring, but most pregnancies probably result from matings of the previous fall. One naked, blind young bat is born per year normally, in May, June, or July. Twinning may occur, but all pregnant females from Ohio that I have examined have contained a single embryo. The gestation period is 50 to 60 days. The females carry their young with them for 1 or 2 days only; after that, the young are left hanging in the roost while the mothers hunt. To humans each young bat looks exactly like the other, but whether the female bat is able to recognize her own offspring and consistently return to it is still under investigation. The young bats are weaned and are able to fly when 3 to 4 weeks old, and they reach adult size when 8 weeks old. Females are capable of fertile mating in their first year, and males mate in their second year. Little brown myotis continue to reproduce for 12 years or more, and certain individuals have been known to live as long as 20 years.

Information gained from little brown myotis that have been banded and then recaptured shows that individuals have a tendency to return to the same roosts year after year, although some shifting between roosts as much as 240 km (150 mi) apart may occur. These bats often show marked homing behavior. Dr. Elizabeth Smith and K. Hale (1953) recovered two bats in Wilmington, Ohio, that had moved 365 km (228 mi). One of the bats was recovered after one year, the other after two years. There are many other reports confirming the homing behavior of the little brown myotis as well as other species of bats.

Little brown myotis have few enemies. Occasionally an irresponsible person will discover a roost or hibernation site and destroy the bats in it. Snakes have been seen picking bats from cave roofs, and weasels sometimes use bats as food during the winter. Hawks and owls catch a few. The little brown myotis has been successfully inoculated with rabies in the laboratory (Reagan and Brueckner, 1951), and the Ohio Department of Health has recorded rabies-positive specimens as reported by Krogwold (1977).

SELECTED REFERENCES

Brenner, F. J. 1974a. A five-year study of hibernating colony of *Myotis lucifugus*. Ohio J. Sci. 74:239-44.

———. 1974b. Body temperature and arousal rates of two species of bats. Ohio J. Sci. 74:296-300.

Cagle, F. R., and E. L. Cockrum. 1943. Notes on a summer colony of *Myotis lucifugus lucifugus*. J. Mammal. 24:474-92.

Cockrum, E. L. 1949. Longevity in the little brown bat, *Myotis lucifugus lucifugus*. J. Mammal. 30:433-34.

Craft, T. J., and R. W. Dexter. 1955. Swimming ability of the little brown bat. J. Mammal. 36:452-53.

Davis, W. H., M. J. Cawein, M. D. Hassell, and E. J. Lappat. 1967. Winter and summer circulatory changes in refrigerated and active bats. *Myotis lucifugus*. J. Mammal. 48:132-34.

Davis, W. H., and H. B. Hitchcock. 1965. Biology and migration of the bat, *Myotis lucifugus*, in New England. J. Mammal. 46:296-313.

Gould, E. 1955. The feeding efficiency of insectivorous bats. J. Mammal. 36:399-407.

Guilday, J. E. 1948. Little brown bats copulating in winter. J. Mammal. 29:416-17.

Guthrie, M. J., and K. R. Jeffers. 1938. Growth of follicles in the ovaries of the bat *Myotis lucifugus lucifugus*. Anat. Rec. 71:477-96.

Hitchcock, H. B., and K. Reynolds. 1942. Homing experiments with the little brown bat, *Myotis lucifugus lucifugus* (Le Conte). J. Mammal. 23:258-67.

Humphrey, S. R. 1966. Flight behavior of *Myotis lucifugus* at nursery colonies. J. Mammal. 47:323.

———. 1971. Population ecology of the little brown bat, *Myotis lucifugus* in Indiana and north-central Kentucky. Ph.D. Dissertation, Okla. State Univ., Stillwater. 149 pp.

Hurst, R. N., and J. E. Wiebers. 1968. Thermogenetic patterns in the little brown bat, *Myotis lucifugus* (Le Conte). J. Mammal. 49:791-94.

Krogwold, R. A. 1977. A study of bat rabies in Ohio. M. S. Thesis, Ohio State Univ., Columbus, Ohio. 58 pp.

Kunz, T. H., E. L. P. Anthony, and W. T. Rumage III. 1977. Mortality of little brown bats following multiple pesticide applications. J. Wildl. Manage. 41:476-83.

Miller, G. S., Jr., and G. M. Allen. 1928. The American bats of the genera *Myotis* and *Pisonyx*. Bull. U. S. Natl. Mus. 144:1-218.

Pittman, H. H. 1924. Notes on the feeding habits of the little brown bat (*Myotis lucifugus*). J. Mammal. 5:231-32.

Reagan, R. L., and A. L. Brueckner. 1951. Transmission of a strain of rabies virus to the large brown bat (*Eptesicus fuscus*) and to the cave bat (*Myotis lucifugus*). Cornell Vet. 41:295-98.

Smith, E. 1956. Pregnancy in the little brown bat. Amer. J. Physiol. 185:61-64.

———. 1957. Experimental study of factors affecting sex ratios in the little brown bat. J. Mammal. 38:32-39.

Smith, E., and K. Hale. 1953. A homing record in the bat, *Myotis lucifugus lucifugus*. J. Mammal. 34:122.

Stegeman, L. C. 1954a. Variation in a colony of little brown bats. J. Mammal. 35:111-13.

———. 1954b. Notes on the development of the little brown bat, *Myotis lucifugus lucifugus*. J. Mammal. 35:432-33.

———. 1956. Tooth development and wear in *Myotis*. J. Mammal. 37:58-63.

Stones, R. C., and T. Oldenburg. 1968. Occurrence of torpid *Myotis lucifugus* in a cold mine in summer. J. Mammal. 49:123.

Stones, R. C., and J. E. Wiebers. 1965. Seasonal changes in food consumption of little brown bats held in captivity at a "neutral" temperature of 92°F. J. Mammal. 46:18-122.

———. 1966. Body weight and temperature regulation of *Myotis lucifugus* at a low temterature, 10°C. J. Mammal. 47:520-21.

Wimsatt, W. A. 1960. An analysis of parturition in Chiroptera, including new observations on *Myotis l. lucifugus*. J. Mammal. 41:183-200.

Keen's Myotis or Keen's Bat
Myotis keenii Merriam

Although not often reported from Ohio, Keen's myotis is probably more common than suspected. A woodland species, it is less often encountered in towns and around human habitations than the other mouse-eared bats. Bole and Moulthrop (1942) believed it to be the commonest species of woodland *Myotis* in northeastern Ohio.

Myotis keenii Merriam

This bat is similar in coloration and size to the little brown myotis but the ears are noticeably larger; when laid forward they extend as much as three or four millimeters beyond the end of the nose. The tragus is long and narrow, and the calcar lacks a keel.

Following are measurements of 12 adult Keen's myotis (*Myotis keenii*) from Ohio:

	Total Length (mm)	Tail (mm)	Hind Foot (mm)
Extremes	80–90	35–40	7–9
Mean	84	37	8

Keen's myotis, usually a solitary animal, often flies late into the night. When sleeping, a single individual will hang behind a shutter or under a piece of loose bark, but little is known about its summer activities in Ohio. Brandon (1961) reported finding Keen's myotis in Scioto County on 12 June 1960. He writes: "One adult female and two young bats were collected and kept in captivity for three days in a small cardboard box. The manner in which the bats were collected was unusual. The two young bats fell from the ceiling onto the writer's sleeping bag, where they were found squeaking loudly. For safekeeping until morning, they were placed in a shoe. In the morning, the shoe revealed not only the two young bats, but also the adult female." Since the female nursed only one of the young and since females of this species are known to have but one young each season, it was concluded that the second young belonged to a different female. Hamilton (1943) has suggested that Keen's myotis gives birth to its young later in the summer than do other species of *Myotis*, but Brandon's observation implies that young are born in early June as in other *Myotis*. Mills (1971) states that young bats taken in early August were well developed and that later attempts to distinguish them from adults were unreliable.

Keen's myotis frequent the deeper recesses of caves as hibernation sites. They hang alone or in small groups of five or six in areas where the humidity is high and the temperature a little warmer than throughout the rest of the cave. Rysgaard (1942) found them utilizing crevices and drill holes as hibernation sites in Minnesota caves.

SELECTED REFERENCES

Bole, B. P., Jr., and P. N. Moulthrop. 1942. The Ohio recent mammal collection in the Cleveland Museum of Natural History. Sci. Publ. Cleveland Mus. Nat. Hist. 5:83–181, esp. 114.

Brandon, R. A. 1961. Observations of young Keen bats. J. Mammal. 42:400–401.

Easterla, D. A. 1968. Parturition of Keen's myotis in southwestern Missouri. J. Mammal. 49:770.

Hamilton, W. J., Jr. 1943. The mammals of eastern United States: an account of recent land mammals occurring east of the Mississippi. Comstock Publ. Co., Ithaca, N. Y. 438 pp.

Hine, J. S. 1929. Distribution of Ohio mammals. Annu. Rep., Proc. Ohio Acad. Sci. 8(6):267.

Mills, R. S. 1971. A concentration of *Myotis keenii* at caves in Ohio. J. Mammal. 52:625.

Rysgaard, G. N. 1942. A study of the cave bats of Minnesota with especial reference to the large brown bat, *Eptesicus fuscus fuscus* (Beauvois). Amer. Midl. Nat. 28:245–67.

Indiana Myotis or Indiana Bat or Pink Bat
Myotis sodalis Miller and Allen

Prior to 1928 this bat was undescribed. Few specimens have been reported from Ohio, although there are large hibernating colonies in Indiana and Kentucky. Barbour and Davis (1969, p. 90) published a number of locations in Ohio from which they have obtained recoveries of individuals banded at two caves in Kentucky.

Myotis sodalis Miller and Allen

It is likely that some specimens of the Indiana myotis have been mistaken for the little brown myotis, which it closely resembles. However, the keel on the calcar and the purplish-brown (pinkish-gray) fur should facilitate correct identification of this bat. Also, the individual hairs on the back are distinctly tricolored, having a black base, gray midline, and brown tip; this is not true of *M. lucifugus*. Some individuals do not have the keel on the calcar; the fur alone must be used for identification.

Following are measurements of 21 adult Indiana myotis (*Myotis sodalis*) from Ohio:

	Total Length (mm)	Tail (mm)	Hind Foot (mm)
Extremes	77–88	34–40	6–9
Mean	82	37	7

Indiana myotis are gregarious animals that gather in winter and form great hibernating colonies in caves to which they return year after year. Whole areas of the ceiling in a cave may be covered with a solid carpet of these bats. They pack themselves so closely together that several thousand individuals may occupy only half a square meter of space. In one cave that I visit each winter, the greatest concentrations of Indiana myotis are always found in a room where the ceiling is only a few feet from the surface of a flowing stream. Almost entire populations of hibernating bats have been destroyed because of flooding in areas like this (De Blase et al., 1965). Since the cave (technically a cavern) is open at both ends, there are noticeable air currents that continually bathe the hibernation area.

Humphrey et al. (1977) discovered and studied for two years a nursery population in Indiana. The nursery roost was located under the loose bark of a dead tree, and on occasion the bats moved their young to a nearby living shagbark hickory. Each female bore a single young. These authors believe that previous references to *M. sodalis* as strictly a cave bat year-round are incorrect. They state: "Tree roosts like the one we studied evidently are traditional summer homes for *M. sodalis*." The bats foraged over riparian and floodplain trees. Similar habitat was used by two other populations in Ohio.

Destruction and commercialization of caves plus increased activity by spelunkers have limited the distribution of wintering grounds for this species. It is estimated that over 90 percent of the hibernating population is restricted to only five caves in the Ozark and Cumberland Plateau regions in Indiana, and in West Virginia. Because of its limited winter range and the wanton destruction of hibernating colonies, the U.S. Fish and Wildlife Service (1973) has classified *Myotis sodalis* an endangered species. The Ohio Division of Wildlife (1976) has followed suit, and measures are being taken to protect the hibernation caves and to prevent further destruction of these bats.

SELECTED REFERENCES

Barbour, R. W., and W. H. Davis. 1969. Bats of America. Univ. Press of Ky., Lexington. 286 pp.

Barbour, R. W., W. H. Davis, and M. D. Hassell. 1966. The need of vision in homing by *Myotis sodalis*. J. Mammal. 47:356–57.

Brenner, F. J. 1974. Body temperature and arousal rates of two species of bats. Ohio J. Sci. 74:296–300.

Cope, J. B., and S. R. Humphrey. 1977. Spring and autumn swarming behavior in the Indiana bat, *Myotis sodalis*. J. Mammal. 58:93–95.

Craft, T. J., M. I. Edmondson, and R. Agee. 1958. A comparative study of the mechanics of flying and swimming in some common brown bats. Ohio J. Sci. 58:245–49.

DeBlase, A. F., S. R. Humphrey, and K. S. Drury. 1965. Cave flooding and mortality in bats in Wind Cave, Kentucky. J. Mammal. 46:96.

Fenton, M. B. 1966. *Myotis sodalis* in caves near Watertown, New York. J. Mammal. 47:526.

Guthrie, M. J. 1933. The reproductive cycles of some cave bats. J. Mammal. 14:199–216.

Hall, J. S. 1962. A life history and taxonomic study of the Indiana bats, *Myotis sodalis*. Reading Public Museum & Art Gallery, Reading, Pa. Sci. Publ. No. 12. 68 pp.

Humphrey, S. R., and J. B. Cope. 1977. Survival rates of the endangered Indiana bat, *Myotis sodalis*. J. Mammal. 58:32–36.

Humphrey, S. R., A. R. Richter, and J. B. Cope. 1977. Summer habitat and ecology of the endangered Indiana bat, *Myotis sodalis*. J. Mammal. 58:334–46.

LaVal, R. K.., R. L. Clawson, M. L. LaVal, and W. Caire. 1977. Foraging behavior and nocturnal activity patterns of Missouri bats, with emphasis on the endangered species *Myotis grisescens* and *Myotis sodalis*. J. Mammal. 58:592-99.

Mumford, R. E., and J. B. Cope. 1958. Summer records of *Myotis sodalis* in Indiana. J. Mammal. 39:586-87.

Ohio Division of Wildlife. 1976. Endangered wild animals in Ohio. Ohio Dept. Nat. Resources Publ. 316 (R576). 3 pp.

Richter, A. R., D. A. Seerley, J. B. Cope, and J. H. Keith. 1978. A newly discovered concentration of hibernating Indiana bats, *Myotis sodalis,* in southern Indiana. J. Mammal. 59:191.

U. S. Fish and Wildlife Service. 1973. Threatened wildlife of the United States. U. S. Dept. Interior, Bureau of Sport Fisheries and Wildlife. Resource Publ. 114. 289 pp. (esp. p. 209).

Small-footed Myotis or Small-footed Bat or Leib's Bat

Myotis leibii (Audubon and Bachman)

The only occurrence of this bat in Ohio was reported in 1842 by Audubon and Bachman (1851). Curiously enough, however, this is the type specimen for the species; that is, it was the first one of its kind ever taken, and the one on which the description of the species is based. The locality is given as Erie County, Ohio. The small-footed myotis is rare throughout its range in the eastern United States, although Hall and Kelson (1959) state that it is relatively common in the western United States.

Myotis leibii (Audubon and Bachman)

The small-footed myotis is the smallest *Myotis* in Ohio, approaching *Pipistrellus subflavus* in size. Its small size (34.4–48.0 mm) plus the following combination of characteristics will help identify this rare bat. The calcar is keeled; the ears, tragus, and flight membranes are black; there is a conspicuous black mask across the face; and the fur over most of the body is golden brown.

Mohr (1933), who has made the only detailed studies of this species, has caught and marked a number of them in caves in the mountains of Pennsylvania. Only a few individuals are usually found together, hanging singly in a cave, mine shaft, or tunnel. Mohr states: "The attitude of the bat as it hangs on the walls or ceilings of the caves is so characteristic that it offers a very definite means of identification at a distance. The arms instead of hanging practically parallel are extended about 30° from the vertical, and almost every [small-footed] bat was found in this position, others bats seldom." Martin et al. (1966) and other workers have found the small-footed myotis in hibernation, clinging horizontally to the undersurfaces of some of the rocks on the floor of caves and on hillsides in New York State. We should keep this habit in mind while searching for this species in Ohio. It is unlikely that it will be found in Ohio during the winter, although it probably wanders into the state during the summer.

The feeding habits, reproductive habits, and life history of the small-footed myotis are presumably similar to those of the little brown myotis.

SELECTED REFERENCES

Audubon, J. J., and J. Bachman. 1851. Viviparous quadrupeds of North America. V. G. Audubon, New York. P. 334.

Hall, E. R., and K. R. Kelson. 1959. The mammals of North America. Ronald Press Co., New York. 2 vols., 1162 pp.

Hine, J. S. 1929. Distribution of Ohio mammals. Annu. Rep., Proc. Ohio Acad. Sci. 8(6):268.

Martin, R. L., J. T. Pawluk, and T. B. Clancy. 1966. Observations on hibernation of *Myotis subulatus*. J. Mammal. 47:348-49.

Mohr, C. O. 1933. Pennsylvania bats of the genus *Myotis*. Proc. Pa. Acad. Sci. 7:39-43.

Schwager, H., and A. H. Benton. 1956. *Myotis subulatus leibii* in eastern New York. J. Mammal. 37:441.

Tuttle, M. D. 1964. *Myotis subulatus* in Tennessee. J. Mammal. 45:148-49.

Silver-haired Bat

Lasionycteris noctivagans (LeConte)

Perhaps a greater collecting effort will bear out the statement by Bole and Moulthrop (1942) that the silver-haired bat is not uncommon in northeastern Ohio, but in my experience this handsome bat is uncommon throughout Ohio. Only a few specimens (less than 35) have been taken from only 13 counties in the state.

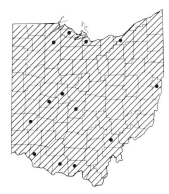

Lasionycteris noctivagans (LeConte)

Certainly, this is one species that should not be confused with any other. The long, silky fur is dark brown or black, and almost every hair has a prominent silver-white tip, which gives the animal its distinctive frosted-black color. The interfemoral membrane is heavily furred on its upper surface, but only on the basal half. The ears are naked, short, and rounded, and the tragus is short, broad, and broadly rounded. No other bat in Ohio has a total of 36 teeth. The dental formula is:

$$i\frac{2-2}{3-3}, c\frac{1-1}{1-1}, pm\frac{2-2}{3-3}, m\frac{3-3}{3-3} = 36$$

Following are measurements of 9 adult silver-haired bats (*Lasionycteris noctivagans*) from Ohio:

	Total Length (mm)	Tail (mm)	Hind Foot (mm)
Extremes	91–112	38–45	7–9
Mean	101	41	8

Silver-haired bats are solitary creatures that most often roost and hunt alone. They are woodland bats that hang in trees during the day and hunt for their insect food above the treetops and along streams and lakes within wooded areas in late evening. They probably migrate south in the winter; however, records from other states indicate that some may hibernate in the North. The dates for leaving and returning are not known for certain. The latest winter date that I have for Ohio is 5 November, and the earliest spring date, 28 April. Hamilton (1943) says: "These bats have been found hibernating in sky-scrapers, churches, wharf-houses and hulls of ships in New York City during the months between December and March, and a specimen has been found hibernating beneath the loose bark of a tree in British Columbia. Specimens have been found in hollow trees in North Carolina during the winter months, and there are records for Bermuda." Silver-haired bats breed in the North, where their one or two young are born during June or July.

No one has studied in detail the life history and habits of the silver-haired bat, although it certainly merits close study.

SELECTED REFERENCES

Bole, B. P., Jr., and P. N. Moulthrop. 1942. The Ohio recent mammal collection in the Cleveland Museum of Natural History. Sci. Publ. Cleveland Mus. Nat. Hist. 5:83–181, esp. 117.

Davis, W. H., and J. W. Hardin. 1967. Homing in *Lasionycteris noctivagans*. J. Mammal. 48:323.

Easterla, D. A., and L. C. Watkins. 1967. Silver-haired bat in southwestern Iowa. J. Mammal. 48:327.

Frum, W. G. 1953. Silver haired bat, *Lasionycteris noctivagans*, in West Virginia. J. Mammal. 34:499–500.

Gosling, N. M. 1977. Winter record of the silver-haired bat, *Lasionycteris noctivagans* LeConte, in Michigan. J. Mammal. 58:657.

Hamilton, W. J., Jr. 1943. The mammals of eastern United States: an account of recent land mammals occurring east of the Mississippi. Comstock Publ. Co., Ithaca, N. Y. 438 pp.

Eastern Pipistrelle or Georgian Bat
Pipistrellus subflavus (F. Cuvier)

This little butterfly-like mammal is the smallest bat in Ohio. It has been recorded more often from the eastern than the western part of Ohio, but it certainly has a statewide distribution. It is found throughout the eastern United States, from as far north as southern Canada to all but the southern tip of Florida.

Because of their small size, eastern pipistrelles are easily overlooked. I have seen them in winter hiber-

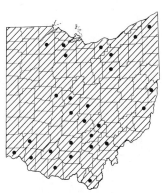

Pipistrellus subflavus (F. Cuvier)

nating in caves where they appeared to be nothing more than small patches of glistening fur stuck to the ceiling. In summer when hunting for insects in the late evening, they can easily be mistaken for large moths because of their weak and erratic flight. In addition to their size, there are several other distinguishing features useful in identification, particularly in separating *Pipistrellus* from the *Myotis* bats. The fur is decidedly tricolored. When separated by blowing into the fur, the main hairs are visible as gray at the base, then banded with yellowish brown, followed by a dark brown tip. The overall appearance is decidedly yellowish, yellowish-brown, or golden brown. (These pipistrelles, or "pips," are sometimes called yellow bats.) The ear is decidedly longer than it is wide, with a moderately rounded tip; when laid forward on the head it extends just beyond the tip of the nose. The tragus is straight and slightly less than one-half the total ear length. Fur extends onto the first one-third of the dorsal surface of the interfemoral membrane. The calcar is not keeled. The eastern pipistrelle is the only bat in Ohio with a total of 34 teeth. The dental formula is as follows:

$$i\frac{2-2}{3-3}, c\frac{1-1}{1-1}, pm\frac{2-2}{2-2}, m\frac{3-3}{3-3} = 34$$

Following are measurements of 16 adult eastern pipistrelles (*Pipistrellus subflavus*) from Ohio:

	Total Length (mm)	Tail (mm)	Hind Foot (mm)
Extremes	82-95	35-45	3-9
Mean	87	38	5

Although it is speculated that the eastern pipistrelle may remain active throughout the year in its southern range, in our latitude it hibernates in caves from October till April or May. The few times that I have observed these bats in hibernation, they were hanging either singly or in pairs in areas away from other bats; they did not group together or form large clusters. Whereas other species of bats tend to shift position or even to fly about periodically during the hibernation period, eastern pipistrelles remain relatively inactive throughout the entire winter. Indications are that these mammals exhibit strong homing behavior, for given individuals use the same hibernation caves year after year.

Most females probably mate in late fall and give birth to one, two, or three young the following summer between the first of June and mid-July. The young are able to fly when three to four weeks old, at which time they become independent of the mother. A longevity record of ten years has been noted for this bat, but Davis (1966) makes the observation that a relatively large percentage of individuals die between their first and second hibernation seasons. He cites inability in catching insects and insufficient fat storage as the two most probable causes of this early mortality.

The eastern pipistrelle feeds primarily on small flies, beetles, moths, bees, and bugs (Hamilton, 1943; Sherman, 1939). Gould (1955) recorded a 3.3 gram-per-hour insect catch rate for this species, a figure considerably higher that that for the big brown bat or the little brown bat. At least in the early evening, the eastern pipistrelle is most likely to be seen hunting at treetop level or higher.

SELECTED REFERENCES

Cockrum, E. L. 1952. Longevity in the pipistrelle, *Pipistrellus subflavus subflavus*. J. Mammal. 33:491-92.

Davis, W. H. 1959. Taxonomy of the eastern pipistrel. J. Mammal. 40:521-31.

_____. 1966. Population dynamics of the bat *Pipistrellus subflavus*. J. Mammal 47:383-96.

Davis, W. H., and R. E. Mumford. 1962. Ecological notes on the bat *Pipistrellus subflavus*. Amer. Midl. Nat. 68:294-98.

Goslin, R. 1947. A bat with white wing tips. J. Mammal. 28:62.

Gould, E. 1955. The feeding efficiency of insectivorous bats. J. Mammal. 36:399-407.

Hall, E. R., and W. W. Dalquest. 1950. A synopsis of the American bats of the genus *Pipistrellus*. Univ. of Kans. Publ. Mus. Nat. Hist. 1(26):591-602.

Hamilton, W. J., Jr. 1943. The mammals of eastern United States: an account of recent land mammals occurring east of the Mississippi. Comstock Publ. Co., Ithaca, N.Y. 438 pp.

Hine, J. S. 1929. Distribution of Ohio mammals. Annu. Rep., Proc. Ohio Acad. Sci. 8(6):267.

Jones, C., and J. Pagels. 1968. Notes on a population of *Pipistrellus subflavus* in southern Louisiana. J. Mammal. 49:134-39.

Lane, H. K. 1946. Notes on *Pipistrellus subflavus subflavus* (F. Cuvier) during the season of parturition. Proc. Pa. Acad. Sci. 20:57-61.

Poole, E. L. 1938. Notes on the breeding of *Lasiurus* and *Pipistrellus* in Pennsylvania. J. Mammal. 19:249.

Sherman, H. B. 1939. Notes on the food of some Florida bats. J. Mammal. 20:103-4.

Venables, L. S. V. 1943. Observations at a pipistrelle bat roost. J. Anim. Ecol. 12:19-26.

Big Brown Bat
Eptesicus fuscus (Palisot de Beauvois)

The big brown bat, a nonmigratory species, is one of the largest bats in Ohio; it may also be the most common. Because of its size and erratic, low-altitude flight pattern, and because it is usually the first bat out each evening, the big brown bat is probably seen more often than any other bat in Ohio. Certainly it is encountered in southwestern Ohio more often than any other species. This may be misleading, however, because most other bats are smaller and attract less attention. Also, many of them are not active until it is almost too dark to see them.

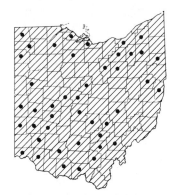

Eptesicus fuscus (Palisot de Beauvois)

Smith (1954) found the little brown myotis, *Myotis lucifugus,* more common than the big brown bat in northeastern Ohio. The big brown bat remains active for a greater period of time through the year than any of our other species of bats. I have seen big brown bats flying about the lights in Cincinnati in late October and early November; in 1969 one was flying around the lights outside the University of Cincinnati's biology building during the Christmas holiday. These bats are also the first to be seen in spring.

The big brown bat can be recognized by its relatively large size and long, silky hair that is a uniform rich chocolate-brown color over the entire body. The wing membranes and ears are decidedly dark or black. The interfemoral membrane is essentially naked, although there is usually a growth of fur at its base. The ears are short and broad. The tragus is straight, directed forward, wider in the middle than at either end, and tapered to a somewhat rounded tip.

The skull of this species is large and angular with a prominent sagittal crest. The skull is best identified by noting the number and arrangement of the teeth. The dental formula is as follows:

$$i\frac{2-2}{3-3}, c\frac{1-1}{1-1}, pm\frac{1-1}{2-2}, m\frac{3-3}{3-3} = 32$$

Following are measurements of 25 adult big brown bats (*Eptesicus fuscus*) from Ohio:

	Total Length (mm)	Tail (mm)	Hind Foot (mm)
Extremes	105-25	35-50	7.9-10
Mean	118	42	8.5

Soon after emerging from hibernation in the spring, female big brown bats congregate in maternity colonies while the males disperse and live a solitary existence. The males are occasionally found hanging singly or in groups of two or three behind window shutters, in attics, under the eaves of houses, in garages, under porches, and in other protected areas. In regions where there are no buildings, males find shelter in hollow trees, in crevices in rocks, or under loose bark on trees. The females form colonies in old barns, attics, hollow trees, and culverts. One colony that I visited on 18 May 1951 contained two to three hundred individuals squeezed between the rafters and shingles of an old barn. All the bats that I examined in this colony were females, and each of the ten collected contained two well-developed embryos, one in each uterine horn. When this colony was revisited on 6 July 1951, all the females were nursing young. Three immature males were also collected at this time. Presumably, female big brown bats in Ohio produce a single litter of two young in late May or early June each year. Christian (1956) has reported that 83 percent of the female big brown bats in Maryland produce two young per litter. Interestingly, a survey of the literature reveals that these bats in the far western states produced only one young a year.

The young bat is naked, blind, and helpless at birth. The mean weight, according to Christian (1956), is 2.5 g. The teeth of the newborn bat are slightly recurved, enabling it to hang onto the teats or fur of the mother as she moves about the roost or when she takes it along on her morning or evening flights. When the young are three to four days old, the mother no longer carries them on the flights, but "hangs them up" in the roost before leaving. Young big brown bats develop rapidly and gain adult size in three to four weeks.

By mid-July females have stopped lactating, and the adult males have joined the adult females and juveniles in the colony. Males reach their reproductive potential during their first summer, but females may not mate until the following spring or fall. Since mating occurs in August or September and the young are born in May or June, a gestation period of at least 8 months seems indicated. However, the actual gestation period after delayed fertilization is 50 to 60 days.

Most bats, including the big brown bat, have been observed mating while in their hibernation quarters during the winter; they also perform the mating act in the spring after emerging from hibernation. These observations have shed some doubt on the theory of delayed fertilization, but Wimsatt (1960) and Folk (1940) have successfully isolated females after fall matings and had them produce young the following spring without subsequent matings.

The life expectancy of the big brown bat is not known, but Banfield (1948) has reported one banded specimen in Quebec known to have survived at least nine years, three months. Christian (1956) has estimated the mean life expectancy at six years.

Although I have designated *Eptesicus fuscus* a nonmigratory species, this may not be entirely correct. Many big brown bats do hibernate in Ohio during the winter, for we have found them behind steam pipes in buildings, in attics, in hollow trees, and in caves. Not nearly enough are found, however, to account for all those that are seen during the summer months. Indications are strong that some, perhaps even most, of the big brown bats migrate southward in the winter. Each winter I visit a cave in Kentucky where there are 60,000 to 70,000 hibernating bats; included in this group, however, are never more than about 100 big brown bats, and this is the largest winter concentration that I have found. Goodpaster and Smith placed identification bands on Ohio bats for some years; it is hoped that their studies will eventually give us some clue as to where the big brown bat spends its winters.

The few big brown bats that are found hibernating in caves most frequently occupy a position near the entrance. Actually they may be in only semidarkness during the daylight hours, and certainly they are subject to greater environmental changes (light, wind, moisture, temperature) than other bats that are found in deeper recesses during hibernation. Most often individuals of this species hang alone or in small groups of five or six, not in great clusters as do so many of the smaller species of bats.

The big brown bat, of course, feeds on insects, although captive specimens eat milk-soaked bread, small pieces of meat, worms, and many other foods. Determining which insects are most commonly eaten is difficult because the bat chews its food so thoroughly that the stomach and feces contain only finely ground pieces. Probably any insect is acceptable. Flying insects are most often grabbed in the mouth, chewed, and devoured while the bat is in flight. A large insect may be held in the interfemoral membrane while it is devoured piecemeal. An especially large, active morsel such as a moth, butterfly, or June beetle may be held against a wall or limb with the wings while the bat bites it to death and then eats it. The big brown bat has been observed landing on the ground and capturing grasshoppers. It secures water by skimming low over a stream or pond and scooping up the water with its lower lip.

Because astronomical numbers of insects are consumed by these bats every day, they must be considered beneficial. However, summer colonies are sometimes bothersome when they roost in the attic or in a space between the walls of a home. Their droppings have the disagreeable odor of ammonia, and the scratching noise made by their claws as they move about is disturbing. The most effective defense against these bats is to cover the entrance of their roost with a board or piece of screening, making certain that all possible openings are tightly sealed. This is best done after the bats have left the roost for their evening flight. The use of repellents and toxic chemicals for bat control should be handled by pest control operators. Rabies virus has been isolated from several apparently normal, hibernating big brown bats in central Ohio (Tjalma and Wentworth, 1957).

SELECTED REFERENCES

Banfield, A. W. F. 1948. Longevity of the big brown bat. J. Mammal. 29:418.

———. 1950. A further note on the longevity of the big brown bat. J. Mammal. 31.455.

Beer, J. R. 1955. Survival and movements of banded big brown bats. J. Mammal. 36:242-48.

Beer, J. R., and A. G. Richards. 1956. Hibernation of the big brown bat. J. Mammal. 37:31-41.

Black, H. L. 1972. Differential exploitation of moths by the bats *Eptesicus fuscus* and *Lasiurus cinereus*. J. Mammal. 53:598-601.

Breener, F. J. 1968. A three-year study of two breeding colonies of the big brown bat, *Eptesicus fuscus*. J. Mammal. 49:775-78.

Cahalane, V. H. 1932. Large brown bat in Michigan. J. Mammal. 13:70-71.

Christian, J. J. 1956. The natural history of a summer aggregation of the big brown bat, *Eptesicus fuscus fuscus*. Amer. Midl. Nat. 55:66-95.

Clark, D. R., Jr. and T. G. Lamont. 1976. Organochlorine residues and reproduction in the big brown bat. J. Wildl. Manage. 40:249-54.

Craft, T. J., M. I. Edmondson, and R. Agee. 1958. A comparative study of the mechanics of flying and swimming in some common brown bats. Ohio J. Sci. 58:245-49.

Davis, W. H., R. W. Barbour, and M. D. Hassell. 1968. Colonial behavior of *Eptesicus fuscus*. J. Mammal. 49:44-50.

Folk, G. E., Jr. 1940. The longevity of sperm in the female bat. Anat. Rec. 76:103-9.

Goodpaster, W. W., and D. F. Hoffmeister. 1968. Notes on Ohioan mammals. Ohio J. Sci. 68:116-17.

Guthrie, M. J., and K. R. Jeffers. 1938. Growth of follicles in the ovaries of the bat *Myotis lucifugus lucifugus*. Anat. Rec. 71:477-96.

Hamilton, W. J., Jr. 1933. The insect food of the big brown bat. J. Mammal. 14:155-56.

Mills, R. S., G. W. Barrett, and J. B. Cope. 1977 Bat species diversity patterns in east central Indiana. Ohio J. Sci. 77:191-92.

Phillips, G. L. 1966. Ecology of the big brown bat (Chiroptera: Vespertilionidae) in northeastern Kansas. Amer. Midl. Nat. 75:168-98.

Reynolds, K. 1941. Notes on homing and hibernation in *Eptesicus fuscus*. Can. Field Nat. 55:132.

Rysgaard, G. N. 1942. A study of the cave bats of Minnesota, with special reference to the large brown bat, *Eptesicus fuscus fuscus* (Beauvois). Amer. Midl. Nat. 28:245-67.

Smith, E. 1954. Studies on the life history of noncave-dwelling bats in northeastern Ohio. Ohio J. Sci. 54:1-12.

Smith, E., and W. W. Goodpaster. 1956. Adjacent roosts of *Eptesicus* and *Myotis*. J. Mammal. 37:441-42.

Tjalma, R. A., and B. B. Wentworth. 1957. Bat rabies report of an isolation of rabies virus from native Ohio bats. J. Amer. Vet. Med. Assoc. 130(2):68-70.

Wimsatt, W. A. 1960. An analysis of parturition in Chiroptera, including new observations on *Myotis l. lucifugus*. J. Mammal. 41:183-200.

Red Bat
Lasiurus borealis (Müller)

The generic name of this bat is composed of two Greek words that literally mean "hairy tail"; certainly this is a characteristic feature of the red bat, which occurs throughout Ohio. The entire dorsal surface of the interfemoral membrane is heavily furred. (In Ohio the larger, much darker, and more frosted hoary bat, *Lasiurus cinereus*, shows this same characteristic.) The body fur, in contrast to the *Myotis* bats, is soft and fluffy, and in the males is always bright red, rusty, or reddish brown. The

Lasiurus borealis Müller)

individual hairs, when parted, have a dark gray or black base and a broad yellowish central band, followed by a reddish tip. Many of the dorsal hairs are tipped with silver, giving the animal an attractive frosted-red coloring. Dimorphism is common in this species; the females are usually yellow (blonde) or yellowish red. In both sexes there is a conspicuous patch of white fur on each shoulder just in front of the wing. The ears are short, broad, and rounded, and the outer surface is well furred. The tragus is short, rounded, and slightly curved forward. There are 32 teeth. The dental formula is as follows:

$$i\frac{1-1}{3-3}, c\frac{1-1}{1-1}, pm\frac{2-2}{2-2}, m\frac{3-3}{3-3} = 32$$

The first upper premolar is tiny and is tucked between the canine tooth and the large second premolar. It can be seen only from directly below or from inside the tooth row. Following are measurements of 21 adult red bats (*Lasiurus borealis*) from Ohio:

	Total Length (mm)	Tail (mm)	Hind Foot (mm)
Extremes	95–120	41–54	7–10
Mean	106	47	8

Red bats are found throughout the state but are sometimes overlooked because they are solitary rather than gregarious animals. Instead of forming colonies in caves and attics, they most often spend the day alone sleeping in a tree, bush, or cluster of weeds. I once found one attached to a corn stalk; another (a female with two young attached), given to me by a student, had been found hanging in a bush several feet above the ground. Because of this habit of roosting in vegetation, they are sometimes placed in the general category of "tree bats." I suspect that a given bat does not roost in the same tree or bush each day.

Some red bats remain in Ohio throughout the year, but the majority migrate south in winter. Davis and Lidicker (1956), however, state that this bat "can and normally does remain in regions where temperatures frequently drop below freezing. It winters regularly in West Virginia and probably also in Missouri, southern Illinois, and souther Indiana." They further state that "unseasonably high temperatures during winter months allow these bats to come out and feed. At such times the bats appear in late afternoon and return to their resting places long before dark." If indeed some individuals do remain in Ohio over winter, they would, of course, have to move to more protected sites, as do other species of bats. The red bat probably starts its migration south in September or October. The latest record I have for Ohio is 15 November. In some areas outside Ohio, they have been seen migrating in bird-like flocks.

Female red bats are somewhat unique in that they possess four teats and thus can nurse four young at a time. Packard (1956) states that in Wisconsin, on 26 June 1955, he observed a red bat lying on the ground; when he turned it over, he found four young clinging to the underside. Litter sizes range from one to four; Cockrum (1955) found the average number of embryos per female to be three. Mating occurs in the fall before the bats start their southward migration, and the young are born in May and June of the following year.

The food and feeding habits of this bat are similar to those of other bats in Ohio. Red bats are stronger fliers and exhibit less erratic patterns than the smaller *Myotis* species. Rabies virus has been demonstrated in the red bat (Wiseman et al., 1962).

SELECTED REFERENCES

Cockrum, E. L. 1955. Reproduction in North American bats. Trans. Kans. Acad. Sci. 58:487–511.

Davis, W. H., and W. Z. Lidicker, Jr. 1956. Winter range of the red bat, *Lasiurus borealis*. J. Mammal. 37:280–81.

Downes, W. L., Jr. 1964. Unusual roosting behavior in red bats. J. Mammal. 45:143–44.

McClure, H. E. 1939. Red bats at Lewis, Iowa. J. Mammal. 20:501–2.

———. 1942. Summer activities of bats (Genus *Lasiurus*) in Iowa. J. Mammal. 23:430–34.

Metzger, B. 1955. Notes on mammals of Perry County, Ohio. J. Mammal. 36:101–5.

Moseley, E. L. 1928a. The numbers of young red bats in one litter. J. Mammal. 9:249.

———. 1928b. Red bat as a mother. J. Mammal. 9:248–49.

Packard, R. L. 1956. An observation on quadruplets in the red bat. J. Mammal. 37:279–80.

Terres, J. K. 1956. Migration records of the red bat, *Lasiurus borealis*. J. Mammal. 37:442.

Wiseman, J. S., B. L. Davis, and J. E. Grimes. 1962. Rabies infection in the red bat, *Lasiurus borealis borealis* (Muller), in Texas. J. Mammal. 43:279–80.

Hoary Bat
Lasiurus cinereus Beauvois

The hoary bat is the largest bat in Ohio. It has been recorded in less than a dozen counties in the state (Gottschange, 1966), but it probably has a statewide distribution. These solitary bats are found generally throughout North America, but are seldom seen because they hunt primarily in forested areas late at night. In the western states, however, Mumford (1963) and Vaughan (1953) report outstanding success at times in netting hoary bats over streams in the late evening. In summer they commonly migrate as far north as central Canada, and there is at least one record (Hayman, 1959) of their having reached Iceland. Indications are that hoary bats first arrive in our latitude during the first week of May (Findley and Jones, 1964; Poole, 1932; Provost and Kirkpat-

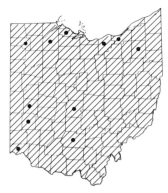
Lasiurus cinereus Beauvois

rick, 1952). The probably start south for the winter in October.

The size of the hoary bat is the most significant characteristic in identification. Its wing span may reach 304 to 330 mm (12 to 13 in). Like the red bat it has a heavily furred interfemoral membrane; the ears are short and rounded; the tragus is short and blunt; and the calcar is keeled. It also roosts in trees and is migratory. The hoary bat differs from the red bat principally in size, color, and length of fur. The individual hairs of the hoary bat are not as long and silky as those of the red bat and are much more heavily frosted. The general color is heavily frosted dark brown or gray (hoary). There is a small cream-colored spot of fur on the forearm just behind the thumb and another immediately in front of the elbow. The skull, although larger, is like that of the red bat, and the dental formula is the same.

Following are measurements of 6 adult hoary bats (*Lasiurus cinereus*) from Ohio:

	Total Length (mm)	Tail (mm)	Hind Foot (mm)
Extremes	111–33	45–59	10–13
Mean	125	54	12

Although females have two pairs of teats, most reports state that hoary bats give birth to two young. At one time it was generally believed that in spring adult females in the eastern United States moved farther north than Ohio before giving birth to their young (Findley and Jones, 1964). However, in early May 1965, a newborn hoary bat was found in Hamilton County, and other births have been recorded from this latitude in eastern Pennsylvania in mid-May and in central Indiana in June (Poole, 1932; Lyon, 1936; Provost and Kirkpatrick, 1952). Hoary bats exhibit delayed fertilization.

Hoary bats have been known to transmit rabies virus, but otherwise they conflict little with man's activities.

SELECTED REFERENCES

Black, H. L. 1972. Differential exploitation of moths by the bats *Eptesicus fuscus* and Lasiurus cinereus. J. Mammal. 53:598–601.

Bogan, M. A. 1972. Observations on parturition and development in the hoary bat, *Lasiurus cinereus*. J. Mammal. 53:611–14.

Brisbin, I. L., Jr. 1966. Energy-utilization in a captive hoary bat. J. Mammal. 47:719–20.

Findley, J. S., and C. Jones. 1964. Seasonal distribution of the hoary bat. J. Mammal. 45:461–70.

Gottschang, J. L. 1966. Occurrence of the hoary bat (*Lasiurus cinereus*) in Ohio. Ohio J. Sci. 66:527–29.

Hall, E. R. 1923. Occurrence of the hoary bat at Lawrence, Kansas. J. Mammal. 4:192–93.

Hayman, R. W. 1959. American bats reported in Iceland. J. Mammal. 40:245–46.

Hopkins, L. S. 1919. The hoary bat in Ohio. Ohio J. Sci. 20(2):35–37.

Lyon, M. W., Jr. 1936. Mammals of Indiana. Amer. Midl. Nat. 17:1–384.

Mumford, R. E. 1963. A concentration of hoary bats in Arizona. J. Mammal. 44:272.

Poole, E. L. 1932. Breeding of the hoary bat in Pennsylvania. J. Mammal. 13:365–67.

Preble, N. A. 1956. The hoary bat in eastern Kansas. J. Mammal. 37:111.

Provost, E. E., and C. M. Kirkpatrick. 1952. Observations on the hoary bat in Indiana and Illinois. J. Mammal. 33:110–13.

Tenaza, R. R. 1966. Migration of hoary bats on South Farallon Island, California. J. Mammal. 47:533–35.

Vaughan, T. A. 1953. Unusual concentration of hoary bats. J. Mammal. 34:256.

Vaughan, T. A., and P. H. Krutzsch. 1954. Seasonal distribution of the hoary bat in southern California. J. Mammal. 35:431–32.

Evening Bat
Nycticeius humeralis (Rafinesque)

The evening bat is a southern species that barely gets into Canada and is a summer resident, probably in moderate numbers, in Ohio and bordering states. It probably has been overlooked on occasion because it closely resembles the larger big brown bat and the slightly smaller little brown myotis. The fur is dull brown as in these two, but is generally shorter and less dense. The interfemoral membrane is not furred, and the end of the tail obviously extends several millimeters beyond it. The ears are short, thick, and rounded; the tragus is barely one-fourth as long as the ear, and it is rounded, rather than being pointed as in the little brown myotis. The calcar is not much help in identification since it may or may not be slightly keeled. There are only 30 teeth in the adult skull. This is the only bat that has a diastema between the one incisor and the canine tooth in the upper jaw. The dental formula is as follows.

$$i\frac{1-1}{3-3}, c\frac{1-1}{1-1}, pm\frac{1-1}{2-2}, m\frac{3-3}{3-3} = 30$$

I have no measurements for this bat from Ohio. However, in Kentucky, Barbour and Davis (1974, p. 106) report: "Total length 94–105 mm; tail 36–42 mm; foot 9 mm." Burt (1957, p. 223) reports total length 95 mm, tail length 40 mm, and hind foot 9 mm.

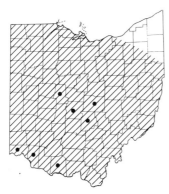

Nycticeius humeralis (Rafinesque)

Although the evening bat is said to roost in trees, most of the published works indicate that it is a colonial form that uses attics, old barns, and church steeples as communal sleeping areas. Most of the colonies investigated, however, were nursing colonies consisting of adult females and their young (Hooper, 1939; Gates, 1941; Humphrey and Cope, 1968). The adult males may roost alone and in trees during the period when the females are rearing their young. Females give birth to two young in May or June. When about three weeks old, the young first show interest in insect food and water and are first able to fly (Jones, 1967). Gates (1941) observed that when the young became big enough to move freely about the colony, they nursed from various lactating females. Conversely, in the excellent laboratory study by Jones (1967), it was found that a female would not nurse any young other than her own. The author states, however, that this may be a reflection of the manner in which the animals were maintained in relative isolation from other young and adults. He also found that it took between 45 and 50 days for the young to reach adult size.

Cope and Humphrey (1967) recorded evening bats arriving in the spring in Indiana during the first two weeks in May. In the fall one colony started to leave in mid-August, and 89 percent of the bats had departed by 21 September. Migratory dates for this bat in Ohio probably closely parallel those in Indiana.

SELECTED REFERENCES

Barbour, R. W., and W. H. Davis. 1974. Mammals of Kentucky. Univ. Press of Ky., Lexington. 322 pp.

Burt, W. H. 1957. Mammals of the Great Lakes region. Univ. of Mich. Press, Ann Arbor. 246 pp.

Cope, J. B., and S. R. Humphrey. 1967. Homing experiments with the evening bat, *Nycticeius humeralis*. J. Mammal. 48:136.

Gates, W. H. 1941. A few notes on the evening bat, *Nycticeius humeralis* (Rafinesque). J. Mammal. 22:53–56.

Goodpaster, W. W. 1941. A list of the birds and mammals of southwestern Ohio. J. Cincinnati Soc. Nat. Hist. 22:1–47.

Hine, J. S. 1929. Distribution of Ohio mammals. Annu. Rep., Proc. Ohio Acad. Sci. 8(6):267.

Hooper, E. T. 1939. Notes on the sex ratio in *Nycticeius humeralis*. J. Mammal. 20:369–70.

Humphrey, S. R., and J. B. Cope. 1968. Records of migration of the evening bat, *Nycticeius humeralis*. J. Mammal. 49:329.

Jones, C. 1967. Growth, development, and wing loading in the evening bat, *Nycticeius humeralis* (Rafinesque). J. Mammal. 48:1–19.

Mumford, R. E. 1953. Status of *Nycticeius humeralis* in Indiana. J. Mammal. 34:121–22.

Watkins, L. C. 1972. *Nycticeius humeralis*. Amer. Soc. Mammal., Mammalian Species No. 23:1–4.

Rafinesque's Big-eared Bat or Eastern Big-eared Bat
Plecotus rafinesquii Lesson

This southern species, formerly *Corynorhinus rafinesquii* (Lesson), is probably not a regular member of the Ohio mammalian fauna. There are only two specimens in the Ohio State University Museum of Zoology collection: an adult male collected in

December 1960 and an adult female collected in March 1953 from separate caves in Green Township, Adams County. These specimens at least suggest that big-eared bats pass through southern Ohio in their annual migratory flights. The greatly enlarged ears and rather grotesque facial mask render the bat unmistakable, but since it rarely takes wing until well after dark, it is seldom seen.

Plecotus rafinesquii Lesson

Measurements for the two specimens reported here are as follows:

	Total Length (mm)	Tail (mm)	Ear (mm)
Male	95	46	33
Female	97	44	32

SELECTED REFERENCES

Frum, W. G. 1948. *Corynorhinus macrotis,* big-eared bat, in West Virginia. J. Mammal. 29:418.

Goslin, R. M. 1954. Eastern big-eared bat in Ohio. J. Mammal. 35:430–31.

Hamilton, W. J., Jr. 1930. Notes on the mammals of Breathitt County, Kentucky. J. Mammal. 11:306–11.

Handley, C. O., Jr. 1959. A revision of American bats of the genera *Euderma* and *Plecotus.* Smithsonian Inst., Proc. U.S. Natl. Mus. 110:95–246.

Jones, Clyde. 1977. *Plecotus rafinesquii.* Amer. Soc. Mammal., Mammalian Species No. 69:1–4.

Maly, R. G. 1962. Second record of eastern big-eared bat in Ohio. J. Mammal. 43:108.

HOARY BAT
Lasiurus cinereus

EASTERN COTTONTAIL
Sylvilagus floridanus

EASTERN CHIPMUNK
Tamias striatus

Fox Squirrel
Sciurus niger

Woodchuck
Marmota monax

Thirteen-lined Ground Squirrel
Spermophilus tridecemlineatus

ORDER LAGOMORPHA
Rabbits and Hares

Except for the pika, animals in this order have long ears, hind legs that are longer than the front legs, and short tails. The tibia and fibula are joined, and articulate with the calcaneum. There are four upper incisors, the second pair being peg-like and inconspicuous. There is a groove on the anterior face of each of the first upper incisors. The five digits on each foot are clawed; the soles of the feet are completely furred. Lagomorphs are herbivorous, and coprophagy is common.

Lagomorphs are present on all continents (except Antarctica) and most large islands. They have been introduced successfully into Australia and New Zealand, where they constitute serious economic pests. The family Leporidae occurs in Ohio, and its characteristics are essentially those given for the order. The only member presently living in Ohio is *Sylvilagus floridanus,* the eastern cottontail.

Eastern Cottontail or Cottontail rabbit
Sylvilagus floridanus (J. A. Allen)

Twenty-four subspecies of *Sylvilagus floridanus* range from Costa Rica north to southern Canada and throughout the central and eastern United States. In Ohio eastern cottontails are present in every county in medium to high numbers.

The Ohio cottontail is a small rabbit, seldom weighing more than 1.6 kg (3.5 lb). The upper parts are predominantly gray or dark brown (agouti). A bright rufous patch on the nape is characteristic of this rabbit. There is a less conspicuous rufous patch on each shoulder. The very short tail is brown above and white below. The throat is brown, but the rest

Sylvilagus floridanus (J. A. Allen)

of the underside is white; the gray bases of the individual hairs may show through, imparting a gray tinge to the belly. The ears are moderately long, 63–76 mm (2.5–3 in). The skull is recognized by the presence of two pairs of upper incisors (fig. 5), by the notch in front of the supraorbital processes, and by the fusion of the postorbital processes with the frontal bones, sometimes leaving a slit-like foramen between the process and the cranium (fig. 12). The sides of the cranium in front of the orbits are thin and porous. The cheek teeth are high-crowned with transverse ridges (lophodont). The dental formula is as follows:

$$i\frac{2-2}{1-1}, c\frac{0-0}{0-0}, pm\frac{3-3}{2-2}, m\frac{3-3}{3-3} = 28$$

Following are measurements of 20 adult eastern cottontails (*Sylvilagus floridanus*) from Ohio:

	Total Length (mm)	Tail (mm)	Hind Foot (mm)
Extremes	350–580	37–64	82–107
Mean	425	47	95

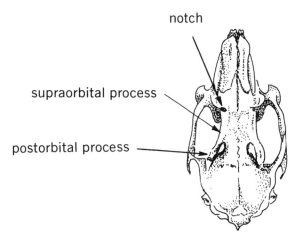

Fig. 12. Dorsal view of eastern cottontail *(Sylvilagus floridanus)* skull. (For lateral view, see fig. 5.) (Drawing by Elizabeth Dalvé.)

Edwards (1962) and Negus (1958, 1959), among others, have reported on the reproductive habits of Ohio cottontails. The breeding season normally extends from late February through September, with a peak of activity during the first two weeks of March. Ecke (1955) reports that a short but vigorous courtship may precede mating. A male and a female cottontail face each other on the ground; the male approaches the female and leaps straight up into the air, turning completely around (180 degrees) as he does so. While the male is "airborne," the female runs beneath him and immediately turns around so that when the male lands on the ground, the rabbits are again facing each other. These acrobatics are repeated over and over again, with the sexes alternating their roles; one time the male jumps into the air, the next time the female. When both members have been sufficiently stimulated, mating occurs. Coition stimulates ovulation in the female rabbit, and eggs are released from the ovaries shortly after copulation. The gestation period is 29 or 30 days. Negus (1959) examined the litters of 12 adult female cottontails from Ohio and determined the mean litter size to be 5.5 young. Mean litter size determined for the cottontail rabbit in Michigan was 5.1 (Allen, 1938) and 5.4 (Haugen, 1942a); in Illinois it was 5.6 (Ecke, 1955). Litter sizes range from 2 to 7, with each adult female in Ohio producing 3 or occasionally 4 or 5 litters each year (fig. 13).

Although there have been numerous reports of young female cottontails breeding during the season in which they were born (Cooley, 1946; Hendrickson, 1947; Ecke, 1955; Lord, 1961), this has been considered the exception rather than the rule. For example, Cooley (1946) states that in six years of work he found only 3 first-year females giving birth to young. However, Negus (1959) found 50 percent of a sample of 36 subadult females from the Unglaciated Allegheny Plateau of Ohio to be reproductively active during their first year. The same author found 10 of 30 (33 percent) subadult females from the Till Plain region of Ohio to be reproductively active. The following season, however, only 2 of 17 young females examined were reproductively active. More work is needed to clarify further this phase of the reproductive habits of the cottontail in Ohio. Unlike the females, young males, three to six months of age, are commonly found to be sexually mature.

The young are born in a fur-lined nest prepared by the female. Nests are hollow depressions in the ground dug by the female and lined first with leaves and/or grass and then with fur plucked from her belly and breast. These nests are covered and well concealed with grass and leaves; except for the short time each day when the young are being fed, the female usually stays completely away from them. Nestling young may be fed only once during each 24-hour period. The newborn are blind and covered with fine hair. They move actively about the nest. The eyes open between the fifth and eighth days, and the young rabbits leave the nest when they are about 26 days old. Although they do eat vegetation when they leave the nest, they may continue to nurse for as long as 25 days (Ecke, 1955). Newborn rabbits weigh about 30 g and gain 2 to 3 g daily while in the nest. Edwards (1963) discovered a nest containing 15 young cottontails of two different age groups. He observed two different females feeding the young at different times, thus substantiating earlier reports that female cottontails at times share the same nest.

The eastern cottontail lives in a wide variety of habitats, the most common of which are old fields containing tall grass, brush piles, and bramble thickets. Open woods with brushy areas bordering old fields are also good rabbit areas in Ohio. Only heavily forested and rather barren sections of land are usually ignored. However, Negus (1959) and Dusi (1952) have shown that rabbits are most plentiful in those areas of Ohio where the soil is most fertile. Although weed-grown and brushy, thousands of acres of agriculturally depleted land and strip mine spoil banks in the state have low rabbit populations. The necessary cover is present, but not

Fig. 13. Female eastern cottontail *(Sylvilagus floridanus)* nursing young. (Photograph by Ronald A. Austing.)

enough of the proper foods are growing on these depleted soils to support a good cottontail population (Bookhout et al., 1968; DeCapita and Bookhout, 1975).

Eastern cottontails are strictly herbivorous, feeding on all kinds of leaves, grass, roots, berries, twigs, and bark. Bluegrass is especially important in the year-round diet of Ohio cottontails (Dusi, 1952). Wild rye, timothy, winter wheat, soy beans, and garden crops are important seasonal foods. Nursery stock and the bark of trees may be eaten when snow covers more commonly used foods. It has long been known that domestic rabbits *(Oryctolagus* sp.) produce both hard and soft pellets and that the soft pellets are taken directly from the anus and reingested (coprophagy). Watson and Taylor (1955) suggested that reingestion might be a common habit among North American lagomorphs. Kirkpatrick (1956) reported the first observed case of coprophagy in *S. floridanus,* and Dexter (1959) reported a second case, the first to be based on stomach analysis, from a rabbit shot at Edinburg in Portage County, Ohio, 23 November 1957. The dietary significance of reingestion may involve vitamin nutrition enhancement and an increase in the efficiency of food utilization.

Seton (1929), Hendrickson (1936), Dalke and Sime (1938), Schwartz (1941), Haugen (1942b), and others have calculated the home range of the eastern cottontails, with little agreement in their findings. Haugen states that adult males may range over more than 41 hectares (100 acres) during the breeding season, whereas Hendrickson found individual summer ranges of less than 0.4 hectare (1 acre) per rabbit. Differences in the amount of cover and the availability of food are undoubtedly reflected in each of these studies. Sex, age, and season are also known to affect the movements of cottontails. Population densities in Ohio probably average about one rabbit per 1 to 1.5 hectares (2 to 3 acres) of land. Adult females defend their home territories during the breeding season, and there is little, if any, overlapping of home ranges (Haugen, 1942b). Adult male and juvenile home ranges broadly overlap in all seasons.

The eastern cottontail is the most popular game animal in Ohio. Estimates by the Ohio Division of Wildlife (David Urban, 1975, personal communication) indicate that the total rabbit population in the state ranges between 6 and 8 million and that approximately 2 million are killed by hunters in Ohio each year. It is difficult to transpose this number into dollars and cents, but if only one 20ᶜ shotgun shell was fired to kill each rabbit, the cost would be $400,000 for ammunition alone! (And how many rabbit hunters have perfect aim?) Add the additional costs for guns, gasoline, hunting licenses, additional ammunition, food, and special clothing, and it becomes apparent that rabbit hunting involves several million dollars of the Ohio economy each year. The Ohio Division of Wildlife constantly encourages good conservation practices to ensure a good rabbit harvest. Farmers are urged to establish fencerows, to leave part of their land unplowed, or to develop cover strips. Plant material for these conservation practices is usually available at reduced cost from certain governmental agencies. Wild rabbits are sometimes live-trapped in areas where they are overabundant and released elsewhere. Ohio has been notably successful in maintaining high cottontail populations, and ranks as one of the five leading states in total number of cottontails taken each year.

Rabbit fur because of the poor quality is not important to Ohio trappers. Also, it is illegal in Ohio to sell wild rabbit for meat, though the flesh is tender and good-tasting. Tularemia is a blood disease once common in wild rabbits (Foshay, 1950), and because of it some hunters will not handle or eat a rabbit after they have shot it. One should never handle a sick or slow-moving cottontail, and gloves should be worn when skinning a wild rabbit because the infective agent can be picked up through open cuts in one's hands. Cooking, however, destroys the tularemia organism, a bacteria living in the blood, so the cooked meat of even an infected animal is perfectly safe to eat. Development of several drugs has also made this disease less dangerous to man than it once was.

Man is not the only predator of rabbits. Surplus rabbits support most wild predator populations (such as large hawks, owls, foxes, and weasels) and constitute a "buffer" when rodents are not readily available as a food source.

SELECTED REFERENCES

Albrecht, C. W. 1962. Helminth parasites of the cottontail rabbit, *Sylvilagus floridanus mearnsii,* from selected areas in Ohio. M. S. Thesis, Ohio State Univ., Columbus. 69 pp.

Allen, D. L. 1938. Breeding of the cottontail rabbit in southern Michigan. Amer. Midl. Nat. 20:464–69.

Atzenhoefer, D. R. 1940. Population studies of the cottontail in northern Ohio. M. S. Thesis, Ohio State Univ., Columbus.

Bookhout, T. A., C. P. Stone, J. D. Bittner, R. A. Tubb, S. H. Taub, and R. E. Deis. 1968. Potential of a strip-mined area for fish and wildlife reclamation. Ohio State Univ. Research Found. Final Report, Project 2296, for the U.S. Dept. Interior, Fish & Wildl. Serv. 84 pp.

Conaway, C. H., and H. M. Wight. 1963. Age at sexual maturity of young male cottontails. J. Mammal. 44: 426–27.

Conaway, C. H., K. C. Sadler, and D. H. Hazelwood. 1974. Geographic variation in litter size and onset of breeding in cottontails. J. Wildl. Manage. 38:473–81.

Cooley, M. E. 1946. Cottontails breeding in their first summer. J. Mammal. 27:273–74.

Crites, J. L., and G. J. Phinney. 1958. *Dirofilaria scopiceps* from the rabbit (*Sylvilagus floridanus mearnsii*) in Ohio. Ohio J. Sci. 58:128–30.

Dalke, P. D., and P. R. Sime. 1938. Home and seasonal ranges of the eastern cottontail in Connecticut. Trans. Third North Amer. Wildl. Conf. pp. 659–69.

DeCapita, M. E. 1975. Evaluation of strip-mine reclamation for terrestrial wildlife restoration. M. S. Thesis, Ohio State Univ., Columbus. 134 pp.

DeCapita, M. E., and T. A. Bookhout. 1975. Small mammal populations, vegetational cover, and hunting use of an Ohio strip-mined area. Ohio J. Sci. 75: 305–13.

Dexter, R. W. 1959. Another record of coprophagy by the cottontail. J. Mammal. 40:250–51.

Dubos, R. J., and J. G. Hirsch, eds. 1965. Bacterial and mycotic infections of man. 4th ed. Lippincott Co., Philadelphia. pp. 681–89.

Dusi, J. L. 1952. The food habits of several populations of cottontail rabbits in Ohio. J. Wildl. Manage. 16:180–86.

Ecke, D. H. 1955. The reproductive cycle of the Mearns cottontail in Illinois. Amer. Midl. Nat. 53:294-311.

Edwards, W. R. 1962. Age structure of Ohio cottontail populations from weight of lenses. J. Wildl. Manage. 26:125-32.

_____. 1963. Fifteen cottontails in a nest. J. Mammal. 44:416-17.

Foshay, L. 1950. Tularemia. Annu. Rev. Microbiol. 4:313-30.

Hall, E. R. 1951. A synopsis of the North American Lagomorpha. Univ. of Kans. Publ., Mus. Nat. Hist. 5(10):119-202.

Hamilton, W. J., Jr. 1940. Breeding habits of the cottontail rabbit in New York State. J. Mammal. 21:8-11.

_____. 1955. Coprophagy in the swamp rabbit. J. Mammal. 36:303-4.

Haugen, A. E. 1942a. Life history studies of the cottontail rabbit in southwestern Michigan. Amer. Midl. Nat. 28:204-44.

_____. 1942b. Home range of the cottontail rabbit. Ecology 23:354-67.

Hendrickson, G. O. 1936. Summer studies on the cottontail rabbit (*Sylvilagus floridanus mearnsi* (Allen)). Iowa State Coll. J. Sci. 10:367-71.

_____. 1947. Cottontail breeding in its first summer. J. Mammal. 28:63.

Huber, J. J. 1959. Trappability of the cottontail rabbit and its effects on the censusing of cottontail rabbit populations. M. S. Thesis, Ohio State Univ., Columbus, Ohio. 126 pp.

_____. 1962. Trap response of confined cottontail populations. J. Wildl. Manage. 26:177-85.

Katz, D. T. 1941. Population studies and the ecology of the cottontail rabbit. M. S. Thesis, Ohio State Univ., Columbus.

Kirkpatrick, C. M. 1956. Coprophagy in the cottontail. J. Mammal. 37:300.

Lechleitner, R. R. 1957. Reingestion in the black-tailed jack rabbit. J. Mammal. 38:481-85.

Lee, R. C. 1940. The rectal temperature and the metabolism of the wild cottontail rabbit. J. Nutrition 19:173-77.

Lord, R. D., Jr. 1959. The lens as an indicator of age in cottontail rabbits. J. Wildl. Manage. 23:358-60.

_____. 1961. Magnitudes of reproduction in cottontail rabbits. J. Wildl. Manage. 25:28-33.

Martin, E. D., and D. R. Atzenhoefer. 1952. Rabbits have their ups and downs. Ohio Conserv. Bull. 11(3): 10-11, 31.

Negus, N. C. 1956. A regional comparison of cottontail rabbit reproduction in Ohio. Ph.D. Dissertation, Ohio State Univ., Columbus. 151 pp.

_____. 1958. Pelage stages in the cottontail rabbit. J. Mammal. 39:246-52.

_____. 1959. Breeding of subadult cottontail rabbits in Ohio. J. Wildl. Manage. 23:451-52.

Phinney, G. J. 1956. The epiphyseal gap as an aging criterion in cottontail rabbits. M. S. Thesis, Ohio State Univ., Columbus. 29 pp.

Rose, G. B. 1977. Mortality rates of tagged adult cottontail rabbits. J. Wildl. Manage. 41:511-14.

Schwartz, C. W. 1941. Home range of the cottontail in central Missouri. J. Mammal. 22:386-92.

_____. 1942. Breeding season of the cottontail in central Missouri. J. Mammal. 23:1-16.

Seton, E. T. 1929. Lives of game animals. Doubleday, Doran & Co., New York. Vol. 4, Part 2, pp. 779-815.

Stevens, V. C. 1962. A study of certain endocrine and physiological aspects of reproduction in the cottontail rabbit (*Sylvilagus floridanus mearnsii*). Ph. D. Dissertation, Ohio State Univ., Columbus. 202 pp.

Thacker, E. J., and C. S. Brandt. 1955. Coprophagy in the rabbit. J. Nutrition 55:375-85.

Thompson, H. V. 1955. The wild European rabbit and possible dangers of its introduction into the U. S. A. J. Wildl. Manage. 19:8-13.

Urban, D. 1976. Farm game. Pp. 1-4 to 1-7 *in* Ohio Division of Wildlife. Reasons for seasons, hunting and trapping. Ohio Dept. Nat. Resources Publ. 68 (R 1276). 58 pp.

Watson, J. S., and R. H. Taylor. 1955. Reingestion in the hare, *Lepus europaeus* Pal. Science 121(3139):314.

Wight, H. 1959. Eleven years of rabbit-population data in Missouri. J. Wildl. Manage. 23:34-39.

ORDER RODENTIA
Rodents

This group of generally small to medium-sized mammals shows great diversity, with more genera, species, and individuals than any other order. Mice, rats, squirrels, jumping mice, and a host of others belong to this order. Rodents are gnawing animals, and the single pair of prominent incisors in the upper and lower jaws are well suited for this function. The incisors are large and deeply rooted. They have open pulp cavities and continue to grow throughout life. The rotary action of the loosely hinged jaws during food chewing results in wear at the extremities of the incisors and keeps them sharp and chisellike. Rodents also have a diastema between the incisors and the cheek teeth (fig. 5), a maximum of 22 teeth, four clawed toes on each front foot, and five on each hind foot. Only Antarctica, New Zealand, and some arctic and oceanic islands do not have native species, and most areas now harbor some forms that have been introduced by man. Sometimes rodents are serious economic pests. Many harbor important disease vectors. There are five families of rodents in Ohio.

KEY TO THE FAMILIES OF RODENTIA IN OHIO

1. Size large, adults sometimes weighing 23 kg (50 lb); tail naked, flattened dorsoventrally.. .... CASTORIDAE (Beaver) p. 80
1'. Size of adults seldom exceeding 5½–6 ½ kg (12–14 lb); usually much smaller; tail not flattened dorsoventrally 2
2. Hind feet and legs elongated, modified for jumping; tail one and one-third the length of the head and body; upper incisors deeply grooved ..ZAPODIDAE (Jumping Mice) p. 112
2'. Hind feet and legs not elongated; tail not one and one-third the length of head and body .. 3
3. Tail well furred or bushy; skull with a prominent postorbital process ...... SCIURIDAE (Chipmunk, Squirrels, Woodchuck) p. 60
3'. Tail naked or nearly so (short hairs may be present but never bushy); skull without prominent postorbital processes 4
4. Upper cheek teeth with tubercles arranged in two longitudinal rows (fig. 14b-L) or with enamel patterns in loops, whorls, and triangles (fig. 14a); tail with hair except where laterally compressed .... CRICETIDAE (New World Mice and Rats, Muskrat) p. 83
4'. Upper cheek teeth with tubercles arranged in three longitudinal rows and enamel patterns not in loops, whorls, and triangles (fig. 14b-R); tail with sparse hairs, noticeably scaly................................. ... MURIDAE (Old World Mice and Rats) p. 107

Enamel pattern in loops, whorls, triangles

a

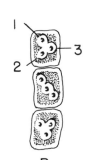

L   R

Tubercles arranged in 2 rows (L), 3 rows (R)

b

Fig. 14. Tooth enamel patterns in rodents. (Drawing by Elizabeth Dalvé.)

FAMILY SCIURIDAE
Woodchucks, Prairie Dogs, Ground Squirrels, Chipmunks, Tree Squirrels

With the exception of the woodchuck, all members of this family are squirrel-like mammals that are familiar to almost everyone. The squirrel-like forms are conveniently and arbitrarily divided into "ground squirrels" and "tree squirrels." The ground squirrels forage for food on the surface of the ground and build their homes and nests belowground. The tree squirrels spend as much time, or more, in the trees as on the ground, and they have their homes and nests in trees. Typically, the members of this family sleep through the night and are active during the day. Only flying squirrels are strictly nocturnal. Tree squirrels have tails that are large, showy, and bushy, whereas the woodchuck and other burrowers have less busy, brush-like tails.

In addition to the woodchuck, we have one species of chipmunk, one species of ground squirrel, and four species of tree squirrels in Ohio.

KEY TO THE SPECIES OF SCIURIDAE IN OHIO

1. Tail less than one-fourth the body length, not bushy; feet black; incisors white on front surfaces...... *Marmota monax* (Woodchuck p. 63
1'. Tail longer than one-fourth the body length, bushy or brush-like; feet usually not all black; incisors orange or yellow on front surfaces............................... 2
2. Fold of skin extending between fore and hind limbs; fur soft and silky in texture .... *Glaucomys volans* (Southern Flying Squirrel) p. 77
2'. Without fold of skin between fore and hind limbs; fur not soft and silky............... 3
3. Longitudinal stripes on back............. 4
3'. No longitudinal stripes on back 5
4. At least 10 (10–13) longitudinal white or yellowish stripes from rump to neck on back; alternate stripes broken into a series of spots; color of back uniform ....*Spermophilus tridecemlineatus* (Thirteen-lined Ground Squirrel p. 66
4'. Two white stripes bordered by black extending down center of back from rump to nape, fading as it reaches the head; rump red .... *Tamias striatus* (Eastern Chipmunk) p. 60
5. Back and dorsal surface of tail reddish; distinct dark line separating reddish-brown side fur from white belly fur; length less than 350 mm................................. .. *Tamiasciurus hudsonicus* (Red Squirrel) p. 74
5'. Back and dorsal surface of tail not red; no dark line separating side and belly fur; length more than 350 mm 6
6. Dorsal surface of feet, ventral surface of tail, belly, and ears rich burnt-orange or rufous; usually gray fur on back mixed with orange- or yellow-tipped hair presenting a grizzled appearance; usually top of head and nose with black markings; four cheek teeth in upper jaw on each side (fig. 15a) .... *Sciurus niger* (Fox Squirrel) p. 73
6'. Dorsal surface of feet, ventral surface of tail, and ears usually gray; belly not burnt-orange or rufous, usually whitish; dorsal pelage gray, not markedly grizzled; five cheek teeth in upper jaw on each side (fig. 15b); melanistic forms found in some areas .... *Sciurus carolinensis* (Gray Squirrel) p. 69

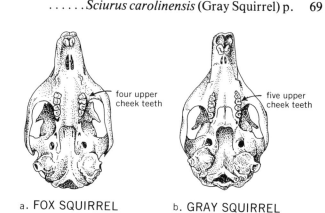

a. FOX SQUIRREL   b. GRAY SQUIRREL

Fig. 15. Ventral view of skulls of fox squirrel *(Sciurus niger)* and gray squirrel *(Sciurus carolinensis)*. (Drawing by Elizabeth Dalvé.)

Eastern Chipmunk
Tamias striatus (Linnaeus)

The eastern chipmunk, in one form or another, occupies almost the entire eastern portion of the United States. Its range extends roughly from the Atlantic Coast to eastern North Dakota, Oklahoma, and Texas and from southern Canada to northern Florida. Deciduous woods and forests are its favorite habitats, but it adapts well to civilization; almost every hedgerow, city park, and cemetery has its population of these colorful, frisky animals.

The pattern of black and cream-colored stripes on the reddish-brown body and head makes the eastern chipmunk an easy animal to recognize. Two black stripes on each side of the body extend from the shoulders to the rump and enclose a conspicuous cream-colored stripe. A fifth narrow, black line extends down the center of the back. This median black line becomes russet or red as it extends

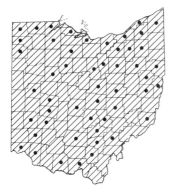

Tamias striatus (Linnaeus)

onto the neck and head. There is a light line bordered by an indistinct dark, brownish-red stripe above and below each eye. A dark line extends through the eye. The rump and thighs are bright fulvous or even dark red in color. The relatively short, flattened, brush-like tail is fulvous below and gray or black above. The sides, cheeks, and top of the feet are tan or straw-colored, and the belly is white. Broad bands of gray fur occupy the areas between the black stripes on the back; this gray fur may predominate on the head although during most of the year the head is fulvous. Chipmunks molt in the spring and again in the fall, but the basic color pattern remains the same thoughout the year. I have seen an albino specimen from Wooster, Ohio, and melanistic specimens are not uncommon in the northeastern counties (Goodpaster and Hoffmeister, 1968).

Each toe has a well-developed claw that makes it possible for the animal to dig and to climb with great skill. Two large internal cheek pouches are present: they are formed by an infolding of the lips and are not fur-lined. The skull is identified by its small size (less than 40 mm) and the presence of 20 teeth. The dental formula is as follows:

$$i\frac{1-1}{1-1}, c\frac{0-0}{0-0}, pm\frac{1-1}{1-1}, m\frac{3-3}{3-3} = 20$$

Following are measurements of 20 adult eastern chipmunks (*Tamias striatus*) from Ohio:

	Total Length (mm)	Tail (mm)	Hind Foot (mm)
Extremes	225–69	75–104	26–38
Mean	247	87	34

Chipmunks are more like ground squirrels than tree squirrels in their habits, and they live belowground in burrows that they dig. Most of these burrows are simple, unbranched tunnels that extend several feet below the surface and are used for retreats when danger threatens or as temporary resting sites. Other burrows are more elaborate, however, having a main tunnel with numerous side passages that lead to food caches, middens, or outside entrances. Somewhere within the system there is an enlarged chamber that has a snug nest made of shredded leaves or grass. Panuska and Wade (1956) laboriously traced one chipmunk burrow that was almost 9 m (29 ft) long. The nesting chamber was located 85 cm (34 in) below the surface of the ground. Many times the entrance or entrances to the burrow system are concealed beneath a stone or a log, but at other times they are located in the open. Except around newly excavated holes, there is never much loose dirt evident; it is either scattered about over the surface of the ground or stuffed into a previously excavated side chamber below ground. Digging is done primarily with the claws on the front feet. The transportation of soil from within the tunnel to the outside is accomplished by pushing, backward-kicking, or carrying it in the cheek pouches.

Chipmunks may be seen during any month of the year in Ohio, but they are least active during extended cold periods in the winter. There appears to be a great deal of individual variation, however, regarding their reaction to cold temperature. In North Carolina, where winter temperatures are relatively mild, Engels (1951) noted that certain chipmunks kept in outdoor pens remained belowground and presumably in an inactive state for as long as 25 weeks, while other chipmunks in the pens continued to be active every day. Condrin (1936) reported great individual differences in hibernation among the animals that he studied near Perrysburg, Wood County, Ohio. Davis and Beer (1959) reported seeing chipmunks as late as 17 December in Minnesota during subfreezing weather. I have observed active chipmunks around Cincinnati on some of the coldest days. Hibernating chipmunks exhibit the characteristics normally associated with a true hibernator: they roll into a tight ball, are cold to the touch, and have all metabolic activity slowed to a minimum. Captured chipmunks brought into the laboratory and placed in a cold chamber may or may not be induced to hibernate, and those that do exhibit different degrees of sleep: some remain torpid for weeks, some for only a day or two, and others wake up daily to move about and feed. Clearly, there is much that we do not understand about the winter sleep of chipmunks.

There is, nevertheless, a decided and obvious decline in chipmunk activity during the winter months; my observations around Cincinnati agree with those of Condrin (1936) in northwestern Ohio. Chipmunk activity decreases in November, and most of the animals do not become active again until late February or early March. Breeding occurs in late March and early April, and the first litters are born in April and May following a gestation period of 31 days. Many, but not all, mature females produce a second litter in August or early September. Both males and females mate for the first time in the spring following the year of their birth. It has been suggested that chipmunks may aestivate. This should not, however, be confused with the diminished activity evident during April, May, and June, when the mature females are busy caring for their spring litters and the young chipmunks have not yet left the nest.

Although some insects and insect larvae are included in the diet, chipmunks are primarily vegetarians; nuts, seeds, and fruits make up the bulk of the diet. October is an unusually busy month for Ohio chipmunks. Seeds and fruits of oak, buckeye, beech, persimmon, and other favorite food-producing trees are maturing, and chipmunks are busy collecting and storing them. Several bushels (approximately a hectoliter) of acorns and/or seeds may be stored belowground and used during winter. As fall progresses, it is interesting to watch the piles of husks and shells grow larger and larger at favorite feeding stations.

In favorable areas where there is abundant food and plenty of cover, many chipmunks may live together. Twelve to fourteen chipmunks make their homes within a 0.1-hectare (0.25-acre) plot that I have studied for three years. The species is not social, however. Except for the times when a female is raising a litter, each chipmunk occupies its own burrow, and no other chipmunk is allowed entrance. Favorite feeding stations are carefully guarded, and intruders are driven off. The home range of an individual may include as much as 1 hectare (2.5 acres), but in favorable situations 0.2 hectare (0.5 acre) is sufficient. Home ranges of various age groups and sexes broadly overlap.

Chipmunk populations in city parks and residential areas may become so high that the animals are pests, causing extensive property damage. Chipmunks occasionally invade home gardens and eat bulbs and fruit. Fortunately, these animals are easily attracted by walnut meats, peanut butter, peanuts, sunflower seeds, or rolled oats. A few rat traps or live traps baited with any of these will soon catch most of the population. Poison baits are also effective but are dangerous to use around the home or in public areas. Chipmunks become relatively tame in situations where they constantly encounter people, and their aesthetic value probably more than compensates for the damage they actually do. Because of their small size they are seldom used for food, and their fur has no value on the fur market.

Three subspecies of chipmunks are found in Ohio. *Tamias striatus ohionensis,* described by Bole and Moulthrop in 1942, occupies most of the state. This is not a well-differentiated subspecies but is recognized by its large size and dull coloration. *T. s. rufescens,* also described by Bole and Moulthrop (1942), is a short-tailed, brightly-colored chipmunk that occupies the northern two tiers of counties in Ohio. Most of the Unglaciated Allegheny Plateau in the eastern part of the state is inhabited by the light-colored *T. s. fisheri.*

SELECTED REFERENCES

Allen, E. G. 1938. The habits and life history of the eastern chipmunk *Tamias striatus lysteri.* Bull. N.Y. State Mus. 314:1–122.

Blair, W. F. 1942. Size of home range and notes on the life history of the woodland deer-mouse and eastern chipmunk in northern Michigan. J. Mammal. 23:27–36.

Bole, B. P., Jr., and P. N. Moulthrop. 1942. The Ohio recent mammal collection in the Cleveland Museum of Natural History. Sci. Publ. Cleveland Mus. Nat. Hist. 5:83–181, esp. 130–37.

Condrin, J. M. 1936. Observations on the seasonal and reproductive activities of the eastern chipmunk. J. Mammal. 17:231–34.

Davis, W. H., and J. R. Beer. 1959. Winter activity of the eastern chipmunk in Minnesota. J. Mammal. 40:444–45.

Engels, W. L. 1951. Winter inactivity of some captive chipmunks (*Tamias s. striatus*) at Chapel Hill, North Carolina. Ecology 32:549–55.

Goodpaster, W. W., and D. F. Hoffmeister. 1968. Notes on Ohioan mammals. Ohio J. Sci. 68:116–17.

Hough, F., and D. Smiley. 1963. Albinism in the chipmunk, *Tamias striatus.* J. Mammal. 44:577.

Howell, A. H. 1929. Revision of the American chipmunks. North Amer. Fauna 52:1–145.

———. 1932. Notes on range of the eastern chipmunk in Ohio, Indiana, and Quebec. J. Mammal. 13:166-67.

Layne, J. N. 1954. The os clitoridis of some North American Sciuridae. J. Mammal. 35:357-66.

———. 1957. Homing behavior of chipmunks in central New York. J. Mammal. 38:519-520.

Mumford, R. E., and C. O. Handley, Jr. 1956. Notes on the mammals of Jackson County, Indiana. J. Mammal. 37:407-12.

Panuska, J. A. 1959. Weight patterns and hibernation in *Tamias striatus*. J. Mammal. 40:554-66.

Panuska, J. A., and N. J. Wade. 1956. The burrow of *Tamias striatus*. J. Mammal. 37:23-31.

———. 1957. Field observations on *Tamias striatus* in Wisconsin. J. Mammal. 38:192-96.

———. 1960. Captive colonies of *Tamias striatus*. J. Mammal. 41:122-24.

Preble, N. A. 1936. Notes on New Hampshire chipmunks. J. Mammal. 17:288-89.

Schooley, J. P. 1934. A summer breeding season in the eastern chipmunk, *Tamias striatus*. J. Mammal. 15:194-96.

Seidel, D. R. 1961. Homing in the eastern chipmunk. J. Mammal. 42:256-57.

Smith, L. C., and D. A. Smith. 1972. Reproductive biology, bredding seasons, and growth of eastern chipmunks, *Tamias striatus* (Rodentia: Sciuridae), in Canada. Can. J. Zool. 50:1069-85.

Smith, W. P. 1931. Calendar of disappearance and emergence of some hibernating mammals at Wells River, Vermont. J. Mammal. 12:78-79.

Stebbins, L. L., and R. Orich. 1977. Some aspects of overwintering in the chipmunk, *Eutamias amoenus*. Can. J. Zool. 55:1139-46.

Stevenson, H. M. 1962. Occurrence and habits of the eastern chipmunk in Florida. J. Mammal. 43:110-11.

Verme, L. J. 1957. Acorn consumption by chipmunks and white-footed mice. J. Mammal. 38:129-32.

Woodward, A. E., and J. M. Condrin. 1945. Physiological studies on hibernation in the chipmunk. Physiol. Zool. 18:162-67.

Yahner, R. H. 1975. The adaptive significance of scatter hoarding in the eastern chipmunk. Ohio J. Sci. 75:176-77.

———. 1977. Activity lull of *Tamias striatus* during the summer in southeast Ohio. Ohio J. Sci. 77:143-45.

Yahner, R. H., and G. E. Svendsen. 1978. Effects of climate on the circannual rhythm of the eastern chipmunk, *Tamias striatus*. J. Mammal. 59:109-17.

Yerger, R. W. 1953. Home range, territoriality, and populations of the chipmunk in central New York. J. Mammal. 34:448-58.

———. 1955. Life history notes on the eastern chipmunk, *Tamias striatus lysteri* (Richardson) in central New York. Amer. Midl. Nat. 53:312-23.

Woodchuck or Groundhog
Marmota monax Linnaeus

The woodchuck, or groundhog, occupies the central eastern portion of the United States, and its range is extending south and west. Indications are that before the advent of the white man, woodchucks were scarce in Ohio, and they lived primarily in forests. As the land was cleared and crops planted, the woodchuck changed its habits somewhat. Today its favorite habitat is woodland edges that border meadows, pastures, and weedy or cultivated fields. The clearing of land and planting of crops have obviously benefited the woodchuck, as is evidenced by its spread into new areas and increasing numbers.

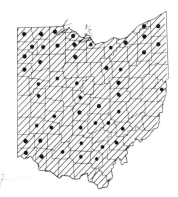

Marmota monax Linnaeus

The woodchuck is the largest member of the squirrel family. Adults in Ohio commonly weigh 4 to 4.5 kg (approximately 9 to 10 lb); occasionally an old male weighs as much as 6.4 kg (14 lb). Within groups males are usually larger than females. Individual woodchucks exhibit an annual weight cycle related to hibernation, gaining weight from about the middle of March until October and losing weight from November until March.

The underfur on the back and sides is black at the base and either white, yellow, or orange on the upper one-third. Each of the long guard hairs overlying the underfur is annulated, black-cream-black-white, in that order from the base outward. The overall effect is a salt-and-pepper coloration with a predominantly yellowish, reddish, or gray appearance, depending upon which of these colors predominates in the underfur. The top of the head, nose, and tail are dark brown; the fur on the legs is often rich burnt-orange, and the feet are black. There is a sparse covering of black or orange fur on the belly. Both albino and melanistic woodchucks have been reported, but not from Ohio.

The tail of a woodchuck does not exceed one-fourth the total body length and is proportionally shorter than that of any other sciurid. The legs are short and powerful, and each front foot bears four clawed toes. A thumb bearing a flattened nail is located some distance up on the inside of the leg. The ears project only slightly above the fur and can be folded forward, closing the ear openings when the animal is burrowing.

The woodchuck skull is distinctive and easily recognized. It may be as much as 102 mm (4 in) long, although 75 to 90 mm (3.0 to 3.5 in) is more common. Among Ohio rodents only the beaver sometimes has a skull larger than this. Unlike the skulls of other members of the family, the woodchuck skull is flattened and presents a straight rather than rounded profile. Also, the front surfaces of the incisors are white or only slightly yellow rather than orange or brown. Some older individuals may have dark stains on the incisors. Large, sharp-pointed, postorbital processes are distinctive features of the skull. The dental formula is as follows:

$$i\frac{1-1}{1-1}, c\frac{0-0}{0-0}, pm\frac{2-2}{1-1}, m\frac{3-3}{3-3} = 22$$

Following are measurements of 15 adult woodchucks (*Marmota monax*) from Ohio:

	Total Length (mm)	Tail (mm)	Hind Foot (mm)
Extremes	522–655	108–59	68–91
Mean	566	130	85

Throughout the summer and fall, woodchucks are constantly eating large amounts of food and storing it as fat, and an individual may increase its weight as much as 20 to 25 percent. Although the animals may be seen sitting beside their burrows or moving about in open fields and pastures at almost any hour of the day, Bronson (1962) has shown that there are peaks of activity from 8 to 9 A.M. and 4 to 5 P.M. The remaining hours are spent in the burrow or aboveground sunning. During October there is a gradual decrease in daily activity, and ordinarily by early November in Ohio woodchucks have started their winter sleep. Hibernation dates vary from year to year depending primarily upon weather conditions. One specimen in our collection was caught in northern Ohio on 27 November 1943.

The burrow containing the hibernation chamber is usually different from the one used during the summer. The summer burrow often has several entrances that lead to an extensive system of tunnels with numerous side passageways and chambers. It is generally located in an open field or pasture. In contrast, the winter burrow is most often located in the forest and is rather simple in construction. There is a single entrance to the main passageway that may descend several feet before leveling off. The hibernation chamber is located either at the end of the main burrow or at the end of a side passageway. Leaves and/or dry grass are used to line the chamber, which may be several feet in diameter.

Woodchucks are truly deep sleepers, and once they have entered the hibernation state, it is difficult to arouse them. However, studies by Davis (1967) show that they do occasionally wake and eat. A hibernating woodchuck resembles a round ball of fur: the head is bent down between the hind legs, the tail rests on the head, and the eyes are tightly closed. All metabolic processes are drastically reduced: the heart beat slows to several beats per minute, breathing is sporadic and infrequent, blood flow is correspondingly reduced, and the animal is cold to the touch. Whereas the temperature of an active woodchuck is between 37° and 40° C, the temperature drops to 8° to 14° C during hibernation. Even in a warm room, it takes a hibernating woodchuck several hours to return to a normal state.

Although the interaction of factors determining when a woodchuck will emerge from hibernation in the spring is not clearly understood, increasing average daily temperatures appear to be the most critical factor. At any rate, woodchucks in Ohio begin to emerge from their dormant state in February, though not necessarily on 2 February (Groundhog Day). Some late sleepers do not appear until March, and mature males precede the females and immature males by three to four weeks. Woodchucks usually are in good physical condition following the hibernation period because little of their stored fat has been used. During the first month after emerging from hibernation, however, the woodchuck eats little and places a heavy drain on the reserve fat supply. According to Snyder et al. (1961), individuals will have lost 20 to 37 percent of their prehibernal weight by the end of this fasting period.

Adult males and females are in breeding condition when they emerge from hibernation, and as soon as the females appear, mating takes place. Although females mate only once, males are promiscuous, and fights between two males over the same female are not uncommon during this period (Bronson, 1963). The gestation period is 31 or 32 days, so most woodchucks in Ohio are born during the first two weeks of April. The young are blind and naked at birth and weigh from 20 to 30 g. Hamilton (1934) gives the average weight of newborn young as 26.5 g. A single litter may contain two to nine young, but three, four, and five seem to be the most frequent numbers. The young first venture forth from the burrow when about one month old and start eating vegetation. Eventually they dig burrows of their own some distance from the parents' burrow and become independent. In Ohio new burrows appear during July and August. Groundhogs usually live alone except in the breeding season. They do not defend territories, but there is a dominance hierarchy among individuals whose ranges overlap.

Woodchucks are primarily vegetarians. A single animal may consume over 0.5 kg (1.5 lb) or more of green plants per day, including alfalfa, clover, most grasses, many herbs, peas, beans, corn, apples, other fruits, and the bark of trees. An occasional insect and small amounts of carrion, collected usually along the highways, make up the less-than-one percent of items other than vegetable material in their diet.

The two pairs of large incisors are used in digging as well as in food-getting and are especially subject to injury and malformation. Stones sometimes get lodged between the incisors, causing them to spread apart and they will not properly occlude. If the teeth are not kept worn down, they can grow so long that they are useless for handling food; eventually, if the animal does not first die of starvation, the teeth form a loop and grow back through the jaw or skull, killing it (Williams, 1922).

The swimming ability of woodchucks is well documented (Phillips, 1923). They have been known to swim regularly across a stream to reach favored food, and they will enter water to escape danger. They can and do climb trees. I once discovered a full-grown woodchuck watching me from a tree limb 6 m (20 ft) above the ground.

There is some question as to the economic status of the woodchuck. Its fur is long and coarse and is little used in the fur industry. (The fur that is used does not come from the North American woodchuck.) However, thousands of woodchucks are shot for sport in Ohio each year, and the revenue from guns, shells, hunting licenses, and gasoline amounts to thousands of dollars. Grain and stock farmers wage a constant war against the woodchuck because it eats their feed crops and digs holes that are responsible for broken legs among farm animals. It is well known, however, that woodchuck burrows are utilized by many species of game animals and birds during inclement weather. Cottontails especially make use of these burrows during winter. Groundhog flesh is good-tasting, but few are used as food in Ohio. They are known to harbor a variety of internal and external parasites, and at times they have been carriers of Rocky Mountain spotted fever as well as several other diseases. Control methods consist of trapping, shooting, and poisoning. Automobiles, hunters, hawks, and mammalian predators take their toll.

Ohio has two subspecies of woodchucks. *Marmota monax monax* is found throughout most of the state, and *M. m. rufescens* occupies the northeastern corner. However, large series of specimens from all parts of the state are needed to determine the exact status of these two subspecies in Ohio.

SELECTED REFERENCES

Bailey, E. D. 1965. The influence of social interaction and season on weight change in woodchucks. J. Mammal. 46:438–45.

Benedict, F. G., and R. C. Lee. 1938. Hibernation and marmot physiology. Carnegie Inst., Washington, D. C. Publ. 497. 239 pp.

Bronson, F. H. 1962. Daily and seasonal activity patterns in woodchucks. J. Mammal. 43:425–27.

———. 1963. Some correlates of interaction rate in natural populations of woodchucks. Ecology 44:637–43.

———. 1964. Antagonistic behavior in woodchucks. Anim. Behav. 12:470–78.

Butterfield, R. T. 1954. Some raccoon and groundhog relationships. J. Wildl. Manage. 18:433–37.

Cram, W. E. 1923. Another swimming woodchuck. J. Mammal. 4:256.

Currier, W. W. 1949. The effect of den flooding on woodchucks. J. Mammal. 30:429–30.

Davis, D. E. 1964. Evaluation of characters for determining age of woodchucks. J. Wildl. Manage. 28:9–15.

———. 1967. The role of environmental factors in hibernation of woodchucks (*Marmota monax*). Ecology 48:683–89.

———. 1977. Role of ambient temperature in emergence of woodchucks (*Marmota monax*). Amer. Midl. Nat. 97:244–29.

Davis, D. E., J. J. Christian, and F. Bronson. 1964. Effect of exploitation on birth, mortality, and movement rates in a woodchuck population. J. Wildl. Manage. 28:1-9.

DeCapita, M. E., and T. A. Bookhout. 1975. Small mammal populations, vegetational cover, and hunting use of an Ohio strip-mined area. Ohio J. Sci. 75: 305-13.

Fisher, W. H. 1893. Investigations of the burrows of the American marmot. J. Cincinnati Soc. Nat. Hist. 16: 105-23.

Grizzell, R. A., Jr. 1955. A study of the southern woodchuck, *Marmota monax monax.* Amer. Midl. Nat. 53:257-93.

Hamilton, W. J., Jr. 1934. The life history of the rufescent woodchuck, *Marmota monax rufescens* Howell. Ann. Carnegie Mus. (Pittsburgh) 23:85-178.

Hamlin, H. E. 1929. The blood pressure of the common wood-chuck. Ohio J. Sci. 29:176.

Howell, A. H. 1915. Revision of the American marmots. North Amer. Fauna 37:1-80.

Hoyt, S. F. 1952. Additional notes on the gestation period of the woodchuck. J. Mammal. 33:388-89.

Hoyt, S. Y., and S. F. Hoyt. 1950. Gestation period of the woodchuck, *Marmota monax.* J. Mammal. 31:454.

Johnson, C. E. 1923. Aquatic habits of the woodchuck. J. Mammal. 4:105-7.

Lee, J. O. 1951. A survey of host parasite relationships of woodchucks inhabiting a limited range. M. A. Thesis, Kent State Univ., Kent, Ohio.

Merriam, H. G. 1963. An unusual fox:woodchuck relationship. J. Mammal. 44:115-16.

―――. 1971. Woodchuck burrow distribution and related movement patterns. J. Mammal. 52:732-46.

Moss, A. E. 1940. The woodchuck as a soil expert. J. Wildl. Manage. 4:441-43.

Phillips, J. C. 1923. A swimming woodchuck. J. Mammal. 4:256.

Rausch, R., and J. Tiner. 1946. *Obeliscoides cuniculi* from the woodchuck in Ohio and Michigan. J. Mammal. 27:177-78.

Snyder, R. L. 1962. Reproductive performance of a population of woodchucks after a change in sex ratio. Ecology 43:506-15.

Snyder, R. L., and J. J. Christian. 1960. Reproductive cycle and litter size of the woodchuck. Ecology 41: 647-56.

Snyder, R. L., D. E. Davis, and J. J. Christian. 1961. Seasonal changes in the weights of woodchucks. J. Mammal. 42:297-312.

Thorpe, M. R. 1930. A remarkable woodchuck skull. J. Mammal. 11:69-70.

Wallace, F. G. 1942. The stomach worm, *Obeliscoides cuniculi,* in the woodchuck. J. Wildl. Manage. 6:92.

Wilber, C. G., and G. H. Weidenbacher. 1961. Swimming capacity of some wild mammals. J. Mammal. 42:428-29.

Williams, S. R. 1922. A case of unhindered growth of the incisor teeth of the woodchuck. Ohio J. Sci. 22: 170-72.

Thirteen-lined Ground Squirrel or Striped Gopher
Spermophilus tridecemlineatus (Mitchill)

The genus *Spermophilus,* to which the ground squirrels belong, occupies a large portion of the North American continent from Alaska to Mexico. Being animals of mountain slopes and prairies, however, they are absent from the forested areas, especially in the eastern United States. The one representative in Ohio of this large genus, *Spermophilus* (formerly *Citellus*) *tridecemlineatus,* reaches its easternmost natural limits in our state. Several colonies are reported in western Pennsylvania, but they represent introduced stock (Doutt et al., 1966).

Spermophilus tridecemlineatus (Mitchill)

The first specimens in Ohio were apparently reported by Enders (1930), who commented: "The ground squirrel is extending its range to the east and south with the advent of prairie conditions. Specimens were taken at Columbus and Circleville and seen in Sandusky and Morrow counties. Thus, it appears to be confined to the Till Plains region of the western part of the state. Further extension of the range is to be expected." In 1933 Goslin reported collecting 13 specimens between 1929 and 1933 near Lancaster, Fairfield County. He believed that the thirteen-lined ground squirrel could be expected to extend its range east and northeast. Preble (1942) reported three specimens from near the Morrow-Knox county line and expressed the belief that this is near the eastern limits of this species in the United States. Karl Maslowski has observed a population of ground squirrels living near the outskirts of Cincinnati, and Dr. Robert Strecker has told me about a colony living at the Oxford airport in Butler

County, Ohio. The ground squirrel now inhabits the entire western half of the state, although not in great numbers. The Unglaciated Allegheny Plateau with its forested areas will undoubtedly serve as an ecological barrier to further movement east.

The name "thirteen-lined ground squirrel" is somewhat misleading, for, as Burt (1957) has pointed out, there are actually twenty-three stripes, twelve dark ones and eleven light ones, running the length of the body. Five of the light lines break up into a series of spots as they progress down the back and over the rump. Five of the light lines and four of the dark ones continue to the top of the head and end between the eyes. The cheeks, sides of the body, and legs are tan, yellowish, or tan with an orange cast. The entire ventral surface has a sparse covering of light tan fur. Melanistic ground squirrels have been reported from Ross County (Goslin, 1959; Goodpaster and Hoffmeister, 1968). The tail is equal to, or slightly less than, one-half the body length and is brush-like rather than bushy. The ears are short, rounded, and inconspicuous. Each of the four toes on the front foot has a long, slender claw that is useful for digging. There are five clawed toes on each hind foot. These rat-sized squirrels seldom weigh more than 227 g (0.5 lb). The skull is not likely to be confused with other members of the family, except perhaps *Glaucomys*, the flying squirrels. Both genera have 22 teeth, but the front of the incisors of *Spermophilus* are light yellow rather than orange. Both genera also have a conspicuous notch immediately in front of the postorbital processes. The incisors of *Glaucomys* are laterally compressed, whereas they are not compressed in *Spermophilus*. The dental formula is as follows:

$$i\frac{1-1}{1-1}, c\frac{0-0}{0-0}, pm\frac{2-2}{1-1}, m\frac{3-3}{3-3} = 22$$

Following are measurements of 19 adult thirteen-lined ground squirrels (*Spermophilus tridecemlineatus*) from Ohio:

	Total Length (mm)	Tail (mm)	Hind Foot (mm)
Extremes	205-77	70-98	33-44
Mean	248	82	36

One of the most characteristic features of the thirteen-lined ground squirrel is its ability to hibernate for as long as six months each year. In Ohio these squirrels make their last appearance aboveground in October and do not reappear until April. Prior to entering hibernation, the thirteen-lined ground squirrel stores great quantities of excess fat; indeed, this is a prerequisite for hibernation. Unlike the woodchuck the ground squirrel draws heavily on its reserve fuel supply during the winter; individuals will lose between 30 and 40 percent of their prehibernal weight before emerging in the spring. This suggests that even though the ground squirrel appears to be a deep sleeper, it may periodically arouse and move around in the hibernation chamber. Lyman (1954), following a series of experiments with different hibernators, states: "In their periodic wakenings the ground squirrels ate and drank almost nothing and in consequence lost much of their stored fat. . . . Exposure to cold brings on hibernation quickly and the animal lives almost exclusively on his stored fat during the hibernation period." Johnson (1928) made the following comparisons between hibernating and nonhibernating ground squirrels:

	Hibernating	Normal
Respiratory (breathing) rate	0.5-4/min	100-200/min
Heart beat	17.4/min (mean); 5/min (lowest)	200-350/min
Temperature	Usually within 3°C of air temperature	32°-41°C

Beer (1962) found that males emerge from hibernation a week to ten days before the females, and the younger females appear before the older, heavier females. Mating occurs soon after the females emerge. Females remain receptive for only about two weeks, and individual males are sexually active for an even shorter period. The females deliver a litter of 5 to 14 young 27 or 28 days following copulation. Each female has but one litter each year. The newborn are blind and naked. At the end of five weeks, they are taking short trips away from the home burrow and soon dig their own tunnels and become completely independent.

The burrows are seldom more than 2 m (6 ft) long and may extend from 15 cm to 1 m below the surface of the ground. The opening to the burrow is hard to detect, since it seldom exceeds 5 cm in diameter, and the squirrels scatter the soil about rather than piling it in a mound around the entrance. Each squirrel usually builds several burrows, some of which are used only as temporary retreats or storage areas. Although it is almost a certainty that this ground squirrel does not eat during the hibernation period, it is a curious fact that it nevertheless stores great quantities of food in these underground burrows. If and when this stored food is eaten is not known.

The thirteen-lined ground squirrels have rather peculiar eating habits. Perhaps as much as half of their diet consists of animal matter. They consume great quantities of insect larvae, adult insects, worms, spiders, and small crustaceans. They even have been seen feeding on small chickens as well as on carrion along the highways, including the dead of their own species. Succulent green plants of all kinds, grasses, seeds, berries, nuts, and grains are included in the diet.

"Striped gophers" are not sufficiently abundant in Ohio to be of significant economic importance. However, in parts of the upper Mississippi Valley they have been serious pests in grainfields, where they either dig up the seeds before germination or destroy the heads before harvest. Occasionally they invade golf courses and cause damage to greens. Since they eat tremendous numbers of injurious insects and insect larvae, however, they are probably more beneficial than harmful when not interfering directly with man's endeavors.

Ground squirrels can be caught in ordinary rat traps, live traps, or noose snares. Poisoned grain is readily eaten. Hawks, cats, foxes, weasels, and owls prey on them, although their strictly diurnal activity and their habit of sitting upright to survey the surroundings make it difficult to get within striking distance.

SELECTED REFERENCES

Bailey, V. 1893. The prairie ground squirrels or spermophiles of the Mississippi Valley. U. S. Dept. Agric., Div. Ornith. & Mammal., Bull. 4:1-69.

Baldwin, F. M., and K. L. Johnson. 1941. Effects of hibernation on the rate of oxygen consumption in the thirteen-lined ground squirrel. J. Mammal. 22:180-82.

Beer, J. R. 1962. Emergence of thirteen-lined ground squirrels from hibernation. J. Mammal. 43:109.

Blair, W. F. 1942. Rate of development of young spotted ground squirrels. J. Mammal. 23:342-43.

Bridgwater, D. D., and D. F. Penny. 1966. Predation by *Citellus tridecemlineatus* on other vertebrates. J. Mammal. 47:345-46.

Burt, W. H. 1957. Mammals of the Great Lakes Region. Univ. of Mich. Press, Ann Arbor. 246 pp.

Chiasson, R. B. 1953. Distribution of *Citellus tridecemlineatus* in Illinois. J. Mammal. 34:510.

Doutt, J. K., C. A. Heppenstall, and J. E. Guilday. 1966. Mammals of Pennsylvania. Pa. Game Commission, Harrisburg, Pa. 288 pp.

Enders, R. K. 1930. Some factors influencing the distribution of mammals in Ohio. Occ. Papers, Univ. of Mich. Mus. Zool. 212:1-27.

Evans, F. C. 1951. Notes on a population of the striped ground squirrel (*Citellus tridecemlineatus*) in an abandoned field in southeastern Michigan. J. Mammal. 32:437-49.

Fitzpatrick, F. L. 1925. The ecology and economic status of *Citellus tridecemlineatus*. Iowa State Univ. Studies Nat. Hist. 11:1-40.

Foster, M. A. 1934. The reproductive cycle in the female ground squirrel, *Citellus tridecemlineatus* (Mitchill). Amer. J. Anat. 54:487-511.

Goodpaster, W. W., and D. F. Hoffmeister. 1968. Notes on Ohioan mammals. Ohio J. Sci. 68:116-17.

Goslin, R. M. 1933. The striped spermophile in Fairfield County, Ohio. J. Mammal. 14:369.

———. 1959. Melanistic ground squirrels from Ohio. J. Mammal. 40:145

Howell, A. H. 1938. Revision of the North American ground squirrels. North Amer. Fauna 56:1-256.

Johnson, G. E. 1917. The habits of the thirteen-lined ground squirrel (*Citellus tridecemlineatus*) with special reference to the burrows. Quart. J., Univ. Of N. Dakota 7:261-71.

———. 1928. Hibernation of the thirteen-lined ground-squirrel, *Citellus tridecemlineatus* (Mitchell) [sic]. J. Exp. Zool. 50:15-30.

———. 1931. Early life of the thirteen-lined ground squirrel. Trans. Kans. Acad. Sci. 34:282-90.

Johnson, G. E., M. A. Foster, and R. M. Coco. 1933. The sexual cycle of the thirteen-lined ground squirrel in the laboratory. Trans. Kans. Acad. Sci. 36:250-69.

Leedy, D. L. 1947. Spermophiles and badgers move eastward in Ohio. J. Mammal. 28:290-92.

Lishak, R. S. 1977. Censusing 13-lined ground squirrels with adult and young alarm calls. J. Wildl. Manage. 41:755-59.

Lyman, C. P. 1954. Activity, food consumption and hoarding in hibernators. J. Mammal. 35:545-52.

McCarley, H. 1966. Annual cycle, population dynamics and adaptive behavior of *Citellus tridecemlineatus*. J. Mammal. 47:294-316.

Phillips, P. 1936. The distribution of rodents in overgrazed and normal grasslands of central Oklahoma. Ecology 17:673-79.

Preble, N. A. 1942. Notes on the mammals of Morrow County, Ohio. J. Mammal. 23:82-86.

Rongstad, O. J. 1965. A life history study of thirteen-lined ground squirrels in southern Wisconsin. J. Mammal. 46:76-87.

Streubel, D. P., and J. P. Fitzgerald. 1978. *Spermophilus tridecemlineatus*. Amer. Soc. Mammal., Mammalian Species No. 103:1-5.

Wade, O. 1927. Breeding habits and early life of the thirteen-striped ground squirrel. *Citellus tridecemlineatus* (Mitchill). J. Mammal. 8:269-76.

———. 1930. The behavior of certain spermophiles with special reference to aestivation and hibernation. J. Mammal. 11:160-88.

———. 1950. Soil temperatures, weather conditions, and emergence of ground squirrels from hibernation. J. Mammal. 31:158-61.

Wells, L. J. 1935. Seasonal sexual rhythm and its experimental modification in the male of the thirteen-lined ground squirrel (*Citellus tridecemlineatus*). Anat. Rec. 62:409-47.

Zimny, M. L. 1965. Thirteen-lined ground squirrels born in captivity. J. Mammal. 46:521-22.

Gray Squirrel
Sciurus carolinensis Gmelin

The range of six subspecies of the gray squirrel, *Sciurus carolinensis,* extends from the Atlantic Coast to central Montana and central Texas. This animal is naturally an inhabitant of our eastern forests. Within Ohio, squirrel habitat "may be divided into two general categories: (1) the woodlot country of western and northeastern Ohio, inhabited almost exclusively by the fox squirrel; (2) the forestland of southeastern and south-central Ohio, inhabited by the gray squirrel and to a lesser degree by the fox squirrel" (Donohoe, 1965).

Sciurus carolinensis Gmelin

Although not as large as the fox squirrel, but larger than the red squirrel, the gray squirrel can properly be called a big tree squirrel. The average weights of 4,050 mature squirrels from West Virginia were 523 g for males and 518 g for females (Uhlig, 1955b). The largest individuals in a population may exceed 700 g (approx. 1.5 lb), but usually weigh around 470 g (about 1 lb) in Ohio.

The color in both sexes is predominantly gray, although there is much variation in the species. I have seen single litters that contained albino, melanistic, and "typically" gray animals. Hardly a year passes that I do not get several phone calls from citizens asking for identification of the odd brown and white, black and brown, or white squirrel-like animal in their yards. The town of Olney, Illinois, is rather famous for its population of entirely white gray squirrels (Hoffmeister and Mohr, 1972). Creed and Sharp (1958) present an interesting historic account of melanism in the gray squirrel and suggest several causative factors including the obvious presence of several genes affecting the distribution of black pigment. Partial or complete melanism is common in the northern part of the range. In Ohio melanism is most abundant in northeastern Ohio and South Bass Island of western Lake Erie. Milton Trautman (personal communication) noted that the melanistic squirrels of South Bass Island were introduced there from southern Michigan between 1945 and 1950.

Usually, however, the gray squirrel is basically gray above with a grayish-white, orange-white, or rusty belly. The fur on the belly is rather sparse and in nursing females may be completely absent around each of the eight teats. Guard hairs on the back and sides, though gray at the base, may have orange tips, resulting in an overall orange-gray color. An orange-brown color may extend down the center of the back and along either side; it is present on the legs, feet, and top of the head as well. This orange-brown tinge is especially noticeable in older squirrels. The tail is flat, long, and bushy with some individual guard hairs 7.5 to 10 cm (3 to 4 in) long. Each of the long guard hairs is multicolored, so that the tail, when flattened and viewed from above or below, appears to be predominantly orange-gray with a distinct black and then an outer white border. The well-furred ears are erect and prominent, about 2.5 cm (1 in) from base to tip. Many, but by no means all, gray squirrels have white fur on the backs of their ears. There is apparently one annual molt in the fall or late summer, when the short summer coat is replaced by longer, more woolly, brighter, and thicker fur. Animals in winter are grayer and more frosted in appearance. There is a conspicuous, light, straw-colored ring around the eye at all seasons.

The skull is not likely to be confused with that of the smaller ground squirrels. Also, it can be separated from the fox squirrel by the presence of five rather than four molar-like teeth; the first (actually the third premolar) is small, inconspicuous, and peg-like (fig. 15). Although Burt (1957) declares this tooth to be missing in about 1 percent of the skulls that he examined from Michigan, my impression is that practically all the gray squirrels in Ohio have five upper molariform teeth. If only four molariform teeth should be present, the skull can be further identified by looking at the infraorbital foramen. In the gray squirrel the foramen is somewhat oval in shape, whereas in the fox squirrel it is only a narrow slit. The dental formula is as follows:

$$i\frac{1-1}{1-1}, c\frac{0-0}{0-0}, pm\frac{2-2 \text{ (or } 1-1)}{1-1 \text{ (or } 1-1)}, m\frac{3-3}{3-3} = 22 \text{ (or } 20)$$

Following are measurements of 39 adult gray squirrels (*Sciurus carolinensis*) from Ohio:

	Total Length (mm)	Tail (mm)	Hind Foot (mm)
Extremes	396–560	153–300	53–70
Mean	470	215	64

With the increased clearing of forest land soon after the Civil War, the gray squirrel rapidly decreased in numbers. Some were left in the hickory, oak, and beech stands that remained, however, and their numbers gradually recovered. Today every county in Ohio probably has a population of gray squirrels, although in northwestern Ohio, except in the cities, populations are still very sparse. In city parks gray squirrels became rather tame and will take food from the hand. In country woodlots they remain secretive and shy, and only the careful observer is aware that they are even there. A typical population of gray squirrels in Ohio today is one animal per 1.2 hectares (3 acres) of suitable habitat (Donohoe, 1971, personal communication).

Perhaps because of the clearing of forest land and a formerly open hunting season, gray squirrels do not occur in the numbers that they once did. The early literature contains a number of interesting accounts of the emigration of thousands or even millions of gray squirrels from one area to another (Kennicott, 1857; Audubon and Bachman, 1846). In the early days in Ohio, gray squirrels were so abundant that citizens were encouraged to pay their property taxes in squirrel skins at the rate of three cents per skin. According to Seton (1920), Dr. P. R. Hoy, of Racine, Wisconsin, in 1857 knew of an Ohio hunter who killed 160 in one day in an off-season; and in 1819 Bachman "saw 130 miles [208 km] of the Ohio [River] 'strewed' with them." Hoy commented that it took a month for an army of squirrels to pass. Such movements, although less common today and in much smaller numbers, still do occur. Larson (1962) states: "Some authors contend that failure of food supplies such as acorns and other mast initiates these movements, while others hold that over-population is the catalyst. A combination of these two factors may be responsible, since varying conditions have been reported."

Gray squirrels are agile tree dwellers but come to the ground when burying nuts, seeds, and acorns or when searching for these same morsels at a later date. Occasionally an individual will be seen running along the ground from tree to tree, but they much prefer to travel high among the branches and often make spectacular leaps in getting from one tree to another. Many times I have watched one of these acrobats hang by the claws of the two hind feet from the outermost branches of a tree while manipulating seeds or tender buds into its mouth with the two front feet. It is not unusual in cities to see a gray squirrel nimbly walking along a telephone wire high above a busy thoroughfare. Occasionally they err; they have been seen to fall 12 to 15 meters (40 to 50 feet) to the ground. However, as the animal falls, the legs are spread wide, the tail is extended and flattened, and the fall is largely cushioned. After falling, most squirrels immediately pick themselves up and continue as though nothing unusual had occurred.

Gray squirrels do not hibernate, although they are less active during periods of inclement weather than at other times. They live in nests usually made of leaves, grass, and/or fur. I have also seen pieces of paper and cloth incorporated into the nesting materials. Probably most nests are made in tree cavities or enlarged woodpecker holes, but large outside nests are also commonly constructed high up in the branches of trees. These outside nests are constructed primarily of leaves so tightly woven together that the nest is virtually waterproof. Gray squirrels are more social than red squirrels or chipmunks; several of them may share the same nest. Birdhouses are convenient, dry, and safe, and gray squirrels often use them as nesting sites.

In February and March it is not uncommon to see three or four gray squirrels rapidly chasing each other up, down, and around the trunk and branches of a tree. Although I have never seen two squirrels mate during these chases, I presume that this is part of the courtship. Most females are in estrus during this period, and the testes of the male are scrotal and contain active sperm. Most mature female gray squirrels (2 years old or older) produce two litters each year, one in March or April and the other in August or September. Most females do not produce litters during the year in which they are born, although Donohoe (1965) says that his Ohio squirrel harvest data indicate that juveniles are capable of having young. Smith and Barkalow (1967) report a female breeding when she was only 124 days old.

Two or three young are usually produced by a female following a gestation period of 43 to 45 days. Barkalow (1967) reports a record litter size of eight, and Uhlig (1955a) states that in almost every instance where there were four young in the litters he observed, one of the young squirrels was a runt. Males are promiscuous and, once they have mated with a female, take no part in rearing the young.

The young are born toothless and naked, with the eyes and ears tightly closed. The eyes open from three and one-half to five weeks after birth, and the ears unfold during the third or fourth week. Young squirrels seldom venture onto the ground before they are ten weeks old (about the time they are weaned). By the time they are fourteen or fifteen weeks old, they are completely independent of the female, although the fall litter may remain with her throughout their first winter. According to Uhlig (1955a), "The inspection of den trees, boxes, and leaf nests in order to ascertain the peak parturition period of the gray squirrel, *Sciurus carolinensis,* has been undertaken by almost every investigator studying the species."

Titles such as "Gray Squirrel feeding on *Crataegus*" (Dambach, 1942); "Gray squirrels feeding on insects in car radiators" (Layne and Woolfenden, 1958); and "Anting by gray squirrels" (Hauser, 1964) suggest that the gray squirrel makes use of a wide variety of foods. As with many highly successful species, gray squirrels generally eat what is available (Nixon et al., 1968). Beechnuts, acorns, walnuts, pignuts, and hickory nuts are favored foods in the fall and winter, and young shoots of maple, oak, elm, and various other trees play a prominent role in the spring. Insects and insect larvae (Nixon, 1970), an occasional bird egg, and even their own dead, plus all kinds of grasses, seeds, fungi, and fruits probably complete the menu. The cutting of leaves and twigs accompanies their nut-gathering activities in the fall; anyone passing beneath a tree where a busy squirrel is at work may be showered with debris. Many acorns and nuts are buried singly by the squirrels in the fall and early winter months. Although some of these are hunted and reclaimed the following spring, many are never recovered. Gray squirrels may play an important part in maintaining our forests through their inadvertent replanting habits. A limited amount of bulk storage of acorns and nuts occurs in hollow trees and tree cavities. Habeck (1960) reports "aerial caching" of butternuts in the uppermost whorls of pine saplings in Wisconsin, and he calculates a 99 percent recovery rate.

Approximately 1.5 million squirrels are taken in Ohio each year by hunters. Donohoe (1965) states that 822,261 fox squirrels and 606,615 gray squirrels were killed in 1962. At least 300,000 people hunt squirrels in Ohio each year, and, as in the case of the eastern cottontail, the squirrel contributes significantly to the state's economy. Gray squirrels occasionally damage corn crops and eat some of the apples in an orchard, but on the whole they are more beneficial than harmful to man.

Presumably there are two subspecies of gray squirrels in Ohio: *Sciurus carolinensis pennsylvanicus,* occupying the northern and eastern half of Ohio, and *S. c. carolinensis* in southern Ohio. Obviously, a large series of study skins from all areas of the state is needed for comparison before the exact distribution of each of the subspecies can be accurately determined.

SELECTED REFERENCES

Allen, J. M. 1952. Gray and fox squirrel management in Indiana. Ind. Dept. Conserv. (Indianapolis), Pittman Robertson Bull. 1:1–112.

Audubon, J. J., and J. Bachman. 1846. The viviparous quadrupeds of North America. 3 vols. J. J. and V. G. Audubon, New York.

Barkalow, F. S., Jr. 1967. A record gray squirrel litter. J. Mammal. 48:141.

Bertram, E. C. 1952. Kentucky squirrel investigations: a three year report. Ky. Dept. Fish & Wildlife Resources, Frankfort, Ky.

Bole, B. P., Jr., and P. N. Moulthrop. 1942. The Ohio recent mammal collection in the Cleveland Museum of Natural History. Sci. Publ. Cleveland Mus. Nat. Hist. 5:83–181, esp. 138–39.

Bouffard, S. H., and D. Hein. 1978. Census methods for eastern gray squirrels. J. Wildl. Manage. 42:550-57.

Brown, L. G., and L. E. Yeager. 1945. Fox squirrels and gray squirrels in Illinois. Bull. Ill. Nat. Hist. Surv. 23:449-536.

Burt, W. H. 1957. Mammals of the Great Lakes Region. Univ. of Mich. Press, Ann Arbor. 246 pp.

Carson, J. D. 1961. Epiphyseal cartilage as an age indicator in fox and gray squirrels. J. Wildl. Manage. 25:90-93.

Chapman, F. B. 1938. The development and utilization of the wildlife resources of unglaciated Ohio. 2 vols. Ph.D. Dissertation, Ohio State Univ., Columbus. 791 pp.

Creed, W. A., and W. M. Sharp. 1958. Melanistic gray squirrels in Cameron County, Pennsylvania. J. Mammal. 39:532-37.

Dambach, C. A. 1942. Gray squirrel feeding on *Crataegus*. J. Mammal. 23:337.

Doebel, J. H., and B. S. McGinnes. 1974. Home range and activity of a gray squirrel population. J. Wildl. Manage. 38:860-67.

Donohoe, R. W. 1965. Squirrel harvest and population studies in Ohio. Game Research in Ohio 3:65-93.

Donohoe, R. W., and R. Stoll, Jr. 1976. Forest game. Pp. 2-1 to 2-17 *in* Ohio Div. Wildlife. Reasons for seasons, hunting and trapping. Ohio Dept. Nat. Resources Publ. No. 68 (R 1276).

Habeck, J. R. 1960. Tree-caching behavior in the gray squirrel. J. Mammal. 41:125-26.

Hall, E. R., and K. R. Kelson. 1959. The mammals of North America. Ronald Press Co., New York. 2 vols. 1162 pp.

Hauser, D. C. 1964. Anting by gray squirrels. J. Mammal. 45:136-38.

Hefner, J. 1971. Age determination of the gray squirrel. M.S. Thesis, Ohio State Univ., Columbus. 51 pp.

Hoffmeister, D. F., and C. O. Mohr. 1957. Fieldbook of Illinois mammals. Manual 4, Illinois Natural History Survey. Reprinted 1972 by Dover Publications, New York. 233 pp.

Katz, J. S. 1938. A survey of the parasites found in and on the fox squirrel (*Sciurus niger rufiventer* Geoffroy) and the southern gray squirrel (*Sciurus carolinensis carolinensis* Gmelin) in Ohio. M.S. Thesis, Ohio State Univ., Columbus. 38 pp.

Kennicott, R. 1857. Excerpt, Document 32, 35th Congress, House of Representatives, 64. U.S. Patent Office Report for 1856 (1857).

Kirkpatrick, C. M., and E. M. Barnett. 1957. Age criteria in male gray squirrels. J. Wildl. Manage. 21:341-47.

Larson, J. S. 1962. Notes on a recent squirrel emigration in New England. J. Mammal. 43:272-73.

Layne, J. N., and G. E. Woolfenden. 1958. Gray squirrels feeding on insects in car radiators. J. Mammal. 39:595-96.

Middleton, A. D. 1931. The gray squirrel. Sidgwick and Jackson, London. 107 pp.

Mosby, H. S., R. L. Kirkpatrick, and J. O. Newell. 1977. Seasonal vulnerability of gray squirrels to hunting. J. Wildl. Manage. 41:284-89.

Mossman, H. W., R. A. Hoffman, and C. M. Kirkpatrick. 1955. The accessory genital glands of male gray and fox squirrels correlated with age and reproductive cycles. Amer. J. Anat. 97:257-301.

Nixon, C. M. 1970. Insects as food for juvenile gray squirrels. Amer. Midl. Nat. 84:283.

Nixon, C. M., R. O. Beal, and R. W. Donohoe. 1968. Gray squirrel litter movement. J. Mammal. 49:560.

Nixon, C. M., and M. W. McClain. 1969. Squirrel population decline following a late spring frost. J. Wildl. Manage. 33:353-57.

_____. 1975. Breeding seasons and fecundity of female gray squirrels in Ohio. J. Wildl. Manage. 39:426-38.

Nixon, C. M., M. W. McClain, and R. W. Donohoe. 1975. Effects of hunting and mast crops on a squirrel population. J. Wildl. Manage. 39:1-25.

Nixon, C. M., D. M. Worley, and M. W. McClain. 1968. Food habits of squirrels in southeast Ohio. J. Wildl. Manage. 32:294-305.

Redman, R. H. 1953. Analysis of gray squirrel breeding studies and their relation to hunting seasons, gunning pressure and habitat conditions. Trans. 18th North Amer. Wildl. Conf., pp. 378-89.

Seton, E. T. 1920. Migrations of the gray squirrel (*Sciurus carolinensis*). J. Mammal. 1:53-58.

Sharp, W. M. 1958. Aging gray squirrels by use of tail-pelage characteristics. J. Wildl. Manage. 22:29-34.

Smith, N. B., and F. S. Barkalow, Jr. 1967. Precocious breeding in the gray squirrel. J. Mammal. 48:328-30.

Thompson, D. C. 1977a. Diurnal and seasonal activity of the grey squirrel (*Sciurus carolinensis*). Can. J. Zool. 55:1185-89.

_____. 1977b. Reproductive behavior of the grey squirrel. Can. J. Zool. 55:1176-84.

Uhlig, H. G. 1955a. The determination of age of nestling and sub-adult gray squirrels in West Virginia. J. Wildl. Manage. 19:479-83.

_____. 1955b. Weights of adult gray squirrels. J. Mammal. 36:293-96.

_____. 1957. Gray squirrel populations in extensive forested areas of West Virginia. J. Wildl. Manage. 21:335-41.

Whittenberger, R. 1968. A gastrointestinal helminth survey of the gray and fox squirrels in Athens and Champaign Counties, Ohio. M.S. Thesis, Ohio Univ., Athens.

Fox Squirrel
Sciurus niger Linnaeus

Fox squirrels are often confused with gray squirrels, but they can be easily distinguished by differences in size, pelage, habitat, and behavior. *Sciurus niger* is a bigger animal; adults may weigh as much as 1,362 g (3 lb), but usually about 680 g (1.5 lb). It is in fact the largest species of tree squirrel in the United States. It has a dark head and nose, and the overall color is decidedly orange or straw-yellow. The belly and undersurface of the tail are bright orange, burnt orange, or orange suffused with pink. The fur is coarser and the tail less fluffy and full than those of the gray squirrel. The fox squirrel is clumsy in trees, and when surprised on the ground, it usually escapes by bounding away for a considerable distance rather than immediately taking to a tree. When the animal is running, the tail is held straight out behind rather than curled over the back. Fox squirrels are indeed striking and beautiful animals, particularly in the fall just after they have acquired their new coats. Like the gray squirrel they display a great variety of individual color differences, and again albino and melanistic individuals have been reported in Ohio.

Sciurus niger Linnaeus

The general characteristics of the skull are the same as those described for the gray squirrel, and the differences between the two have been delineated in the discussion of that species (p. xxx). The dental formula is as follows:

$$i\frac{1-1}{1-1}, c\frac{0-0}{0-0}, pm\frac{1-1}{1-1}, m\frac{3-3}{3-3} = 20$$

Following are measurements of 41 adult fox squirrels (*Sciurus niger*) from Ohio:

	Total Length (mm)	Tail (mm)	Hind Foot (mm)
Extremes	438–575	214–73	65–79
Mean	528	245	73

Fox squirrels may produce two litters per year, but this is normally true only for those females two years of age or older. Females are capable of fertile mating when six to eight months of age, but only about 60 percent or fewer reproduce at this early age in Ohio (Nixon, 1965). Nixon states: "More young are produced in the western areas than in the east, paralleling the situation seen in the cottontail rabbit." Nixon further says: "The causative mechanism may be related to reduced soil fertility in the southeastern area." This is an interesting observation that bears further investigation. Development of the young is similar to that of the gray squirrel.

Although the fox squirrel and the gray squirrel frequently occur together, the fox squirrel is more commonly found in the more open deciduous woods and is absent from, or rare in, the more heavily forested areas of Ohio. Occasionally, fox squirrels are found denning in isolated nut trees considerable distances from woodlots. Woodlots of western Ohio are inhabited almost exclusively by fox squirrels (Donohoe, 1965), except in the extreme southwestern portions, where the fox squirrel has all but disappeared and gray squirrels are numerous. It has only been since about 1970 that the fox squirrel has started to reestablish itself in any numbers in the southwestern corner of Ohio.

Trautman (personal communication) claims that the fox squirrel was far less numerous in the western half of Ohio before 1900, when it was restricted largely to prairie edges and oak-hickory "islands." In Union County an uncle of Trautman's had a 90-acre woods that before 1914 contained only gray squirrels. About 1914 the woods was lumbered, the canopy was broken, and the gray squirrels began to decrease in numbers. Prior to 1914 this woods supplied a barrel of pickled gray squirrels annually for winter use. The fox squirrel then moved in, becoming increasingly numerous. As the trees were removed, the gray squirrel population decreased, and the last were seen in 1924. The red squirrel population, formerly abundant, also decreased steadily, and none were seen after 1945. The fox squirrel continues to be present but not in the abundance of the gray squirrel before 1914.

Baumgartner (1940a, p. 23) states that the present range of the fox squirrel "is considerably larger today than it was before the advent of man. The extension east and north of the Mississippi River was largely due to the production of more small open woodlots with the accompanying production of domestic grain crops." Fox squirrels, however, do not roam from their home territory as much as gray squirrels, and there is little evidence of the large migrations described in earlier accounts of gray squirrels.

SELECTED REFERENCES

Allen, D. L. 1943. Michigan fox squirrel management. Mich. Dept. Conserv. (Lansing), Game Div. Publ. 100:1-404, 212 figs.

Baumgartner, L. L. 1939. Fox squirrel dens. J. Mammal. 20:456-65.

―――. 1940a. The fox squirrel: its life history, habits, and management. Ph. D. Dissertation, Ohio State Univ., Columbus. 257 pp.

―――. 1940b. Trapping, handling, and marking fox squirrels. J. Wildl. Manage. 4:444-50.

―――. 1943a. Fox squirrels in Ohio. J. Wildl. Manage. 7:193-202.

―――. 1943b. Pelage studies of fox squirrels (*Sciurus niger rufiventer*). Amer. Midl. Nat. 29: 588-90.

Donohoe, R. W. 1965. Squirrel harvest and population studies in Ohio. Game Research in Ohio 3:65-93.

Donohoe, R. W., and R. Stoll, Jr. 1976. Forest game. Pp. 2-1 to 2-17 *in* Ohio Div. Wildlife. Reasons for seasons, hunting and trapping. Ohio Dept. Nat. Resources Publ. No. 68 (R 1276).

Greene, H. C. 1950. A record of fox squirrel longevity. J. Mammal. 31:454-55.

Husband, T. P. 1976. Energy metabolism and body composition of the fox squirrel. J. Wildl. Manage. 40:255-63.

Katz, J. S. 1938. A survey of the parasites found in and on the fox squirrel *Sciurus niger rufiventer* Geoffroy) and the southern gray squirrel (*Sciurus carolinensis carolinensis* Gmelin) in Ohio. M. S. Thesis, Ohio State Univ., Columbus. 38 pp.

Moore, J. C. 1956. Variation in the fox squirrel in Florida. Amer. Midl. Nat. 55:41-65.

―――. 1957. The natural history of the fox squirrel, *Sciurus niger* Shermani. Bull. Amer. Mus. Nat. Hist. 113, Art. 1. 71 pp.

Mossman, H. W., R. A. Hoffman, and C. M. Kirkpatrick. 1955. The accessory genital glands of male gray and fox squirrels correlated with age and reproductive cycles. Amer. J. Anat. 97:257-301.

Nixon, C. M. 1965. Productivity rates of gray and fox squirrels in Ohio. Game Research in Ohio 3:93-106.

Nixon, C. M., R. W. Donohoe, and T. Nash. 1974. Overharvest of fox squirrels from two woodlots in western Ohio. J. Wildl. Manage. 38:67-80.

Thoma, B. L., and W. H. Marshall. 1960. Squirrel weights and populations in a Minnesota woodlot. J. Mammal. 41:272-73.

Whittenberger, R. 1968. A gastrointestinal helminth survey of the gray and fox squirrels in Athens and Champaign Counties, Ohio. M. S. Thesis, Ohio Univ., Athens.

(See also Selected References for *Sciurus carolinensis*.)

Red Squirrel or Chickaree
Tamiasciurus hudsonicus (Erxleben)

Few mammals have been so carefully studied as the red squirrel (fig. 16). Klugh (1927), Hatt (1929), Hamilton (1939), Layne (1954), Smith (1968, 1970), and Kemp and Keith (1970) have published detailed life history studies, and a host of other biologists have meticulously investigated almost every phase of the animal's life. Although no detailed study has been published in Ohio, I believe the habits of the species are basically the same here as throughout the rest of its range. Because the fur of the red squirrel is of little commercial value and its flesh is seldom eaten by humans, Hatt (1929) was prompted to comment: "It is by reason of an undeniable aesthetic attraction and its biotic interest rather than because of its economic status, perhaps, that the red squirrel has long held the attention of the naturalist, the nature essayist, and the layman."

Tamisciurus hudsonicus (Erxleben)

The red squirrel's primary habitat is evergreen forests, and it is most numerous in the northern United States and Canada. It does well in hardwood forests, however, and at one time was probably common throughout much of Ohio. As land was cleared, the squirrel was possibly extirpated from the state. It recovered, however, and again reestablished itself. The planting of berry-, seed-, and nut-producing trees and bushes for ornamental and shade purposes, especially in urban and suburban areas, has been responsible in part for the squirrel's recovery. Indeed, the red squirrel is now extending its range southward, and in many instances is more common in wooded city areas than in rural farming communities.

The red squirrel is the second smallest of our tree squirrels and might be easily overlooked except for its habit of loudly scolding intruders in its domain. Its olive brown color blends well with the background and makes the animal difficult to see when it is sitting still. The undersurface of the throat and belly is grayish-white. A black line runs between the fore and hind legs along either side. This black lateral line is most conspicuous in the summer. A broad band of red fur extends down the center of the back from between the ears to the rump and continues along the dorsal surface of the tail to the distal end. The stripe is brightest and most distinct during the fall and winter but also is quite evident in summer specimens in Ohio. There is a black tip on the end of the tail. Lighter, burnt-orange hairs cover the hips, thighs, and dorsal surfaces of the legs and feet. The underside of the tail is olive brown. There is a white ring around the eye.

Layne (1954) found that red squirrels molt the body fur twice each year, once in the fall (August to December) and again in the spring (March to July). The fur on the tail is molted only in the fall. Red squirrels have a brush-like rather than a bushy tail that is equal to, or less than, the length of the body. The presence of 20 teeth serves to identify the skull. A small pair of premolars is sometimes present, bringing the tooth count to 22. The dental formula is as follows:

$$i\frac{1-1}{1-1}, c\frac{0-0}{0-0}, pm\frac{1-1 \text{ (or 2-2)}}{1-1 \text{ (or 1-1)}}, m\frac{3-3}{3-3} = 20 \text{ (or 22)}$$

Following are measurements of 15 adult red squirrels *(Tamiasciurus hudsonicus)* from Ohio:

Fig. 16. Red Squirrel *(Tamiasciurus hudsonicus)*. (Photograph by Karl H. Maslowski.)

	Total Length (mm)	Tail (mm)	Hind Foot (mm)
Extremes	273–360	119–42	47–51
Mean	320	126	49

Unlike ground squirrels, tree squirrels do not hibernate. Even in the subarctic spruce forests of Alaska, the red squirrel remains active throughout the winter. Pruitt and Lucier (1958) state: "In the subarctic taiga red squirrels become subterranean and subnivean [under the snow] animals in winter. Extensive burrow systems are constructed under and through the kitchen midden. It is evident . . . that squirrels are virtually never active above the snow surface when the ambient air temperature is below −25°F [−32°C] and may or may not be active when the ambient air temperature is above that." In the more southern parts of its range, including Ohio, red squirrels are active aboveground throughout the year, although they may disappear for several days at a time during severe rain, snow, or wind storms.

Like most tree squirrels the red squirrel is most active for one to two hours just after dawn and again for about one hour preceding dusk. The remainder of the day is spent in the nest or quietly sitting in a tree. The red squirrel builds a nest of leaves and shredded bark in a natural tree cavity or an abandoned flicker hole and uses this for a home. Rock piles and holes in the ground will occasionally be selected for home sites, especially along old hedgerows. Globular external nests made of leaves and usually located next to the tree trunk may be made high up in trees, although these are not used as often as inside nests. A family may also take up residence in the rafters of a barn or an abandoned building.

The breeding season in Ohio extends roughly from January through September. Most adult females probably produce two litters each year, one in April or May and another in August or early September. It is not known whether females born in the spring produce their first litter later that summer or wait until the following spring. Both situations probably exist, depending upon the individual squirrel. During the breeding periods several males may be seen actively pursuing a single female. Actual mating takes place either in the trees or on the ground. In both sexes there is a pair of scent glands that produce a yellow musk-smelling liquid. These open on nearly naked ridges on either side of the anus and are so located that the scent is left on almost every surface where the squirrels rest. Perhaps these glands are more active during the breeding season and play some part in bringing the sexes together.

The gestation period is estimated to be 40 days. A single young may be born, or there may be as many as eight; four or five is the most common litter size. Newborn squirrels are blind, earless, pink, and completely naked. Hamilton (1939) and Svihla (1930) both give 7.5 g as the average weight for newborn individuals. Layne (1954), who has made the most detailed study on the development of the red squirrel, reports: "The external auditory meatus becomes patent [evident] at about 18 days. The eyes open between 26 and 35 days. The pelage is almost completely developed at 40 days. Lower incisors appear above the gums at approximately 21 days, and the full permanent dentition is attained by 18 or 19 weeks. At the latter age external and skeletal measurements fall within the range of the adult structures. Average adult weight is generally not reached until some time later." The young are weaned at about two months, and they sever all family ties when two and one-half to three months old. Males take no part in raising or providing for the family.

Red squirrels are primarily vegetarians, utilizing the most available foods during each season of the year. Nuts, berries, acorns, fruits, seeds, and buds of all kinds are eaten in quantity. Not only do they eat fresh mushrooms but they impale individual fungi on twigs and branches and let them dry in the sun, after which they store them. They are especially fond of tree sap of any kind and will often strip the bark from the trunk of a tree, discard it, and avidly lap the sap as it oozes from the wound. Young birds and birds' eggs are occasionally included in their diet, and they probably consume quantities of adult insects and insect larvae. In the fall of the year, they cut green cones from evergreen trees and store them in large piles or heaps.

Food availability determines the size of the home range of the red squirrel, but there is also much individual variation. Some individuals use only 0.2 hectare (0.5 acre) of forest, whereas others wander over 2 hectares (5 acres). Few squirrels, however, include more than 0.4 hectare (1 acre) in their range. Bole (1939) reported peak populations of 4.3 squirrels per hectare (1.7 per acre) in beech-maple forests of Ohio. Hamilton (1939) has suggested a red squirrel cycle of eight years, meaning that within any population a maximum number would appear every eight years.

Red squirrels will defend a small area around a favorite feeding station or nesting site against outside intrusion. Kilham (1954) believes that the birdlike calls made by red squirrels that he observed burying white pine cones were used to announce occupancy of territory, warning other squirrels to keep away. Certainly, the characteristic rapid-fire "churr" of this perky squirrel given as soon as one enters its domain is a voice of disapproval of those who invade tamiasciuran property.

Because of its small size, the red squirrel is not as economically important in Ohio as the other squirrels. However, it destroys many harmful insects, and many of the seeds and nuts that it buries, but never reclaims, eventually grow into new trees. Ardent bird watchers have always looked upon the red squirrel as a dedicated killer of birds, but this has been overemphasized. In Canada, Oregon, Washington, and some New England states, red squirrels have caused great damage to conifer forests by cutting and eating the terminal and lateral buds during severe winters when other foods were not available, but this has not happened in Ohio. Hawks, owls, carnivorous mammals, automobiles, and guns take their toll of red squirrels each year.

SELECTED REFERENCES

Allen, J. A. 1898. Revision of the chickarees, or North American red squirrels (subgenus Tamiasciurus). Bull. Amer. Mus. Nat. Hist. 10:249-98.

Ballou, W. H. 1927. Squirrels as mushroom eaters. J. Mammal. 8:57-58.

Benton, A. H. 1958. Melanistic red squirrels from Cayuga County, New York. J. Mammal. 39:445.

Bole, B. P., Jr. 1939. The quadrat method of studying small mammal populations. Sci. Publ. Cleveland Mus. Nat. Hist. 5:15-77.

Chapman, F. B. 1938. The development and utilization of the wildlife resources of unglaciated Ohio. 2 vols. Ph. D. Dissertation, Ohio State Univ., Columbus. 791 pp.

Clarke, C. H. D. 1939. Some notes of hoarding and territorial behavior of the red squirrel, *Sciurus hudsonicus* (Erxleben). Can. Field Nat. 53:42-43.

Cole, L. J. 1922. Red squirrels swimming a lake. J. Mammal. 3:53-54.

Dapson, R. W. 1963. Color aberration in the red squirrel. J. Mammal. 44:123.

Hamilton, W. J., Jr. 1934. Red squirrel killing young cottontail and young gray squirrel. J. Mammal. 15:322.

_____. 1939. Observations on the life history of the red squirrel in New York. Amer. Midl. Nat. 22:732-45.

Hatfield, D. M. 1937. Notes on Minnesota squirrels. J. Mammal. 18:242-43.

Hatt, R. T. 1929. The red squirrel: its life history and habits with special reference to the Adirondacks of New York and the Harvard Forest. Roosevelt Wildlife Annals (N. Y. State Coll. Forestry) 2:3-146.

Hosley, N. W. 1928. Red squirrel damage to coniferous plantations and its relation to changing food habits. Ecology 9:43-48.

Kemp, G. A., and L. B. Keith. 1970. Dynamics and regulation of red squirrel (*Tamiasciurus hudsonicus*) populations. Ecology 51:763-79.

Kilham, L. 1954. Territorial behaviour of red squirrel. J. Mammal. 35:252-53.

_____. 1958. Red squirrels feeding at sapsucker holes. J. Mammal. 39:596-97.

Kirtland, J. P. 1838. Report on the zoology of Ohio. Second Annu. Rep., Geol. Surv. Ohio. 2:160-61, 175-77.

Klugh, A. B. 1927. Ecology of the red squirrel. J. Mammal. 8:1-32.

Layne, J. N. 1952. The os genitale of the red squirrel, *Tamaisciurus*. J. Mammal. 33:457-59.

_____. 1954. The biology of the red squirrel, *Tamiasciurus hudsonicus loquax* (Bangs) in central New York. Ecol. Monogr. 24:227-67.

Mayfield, H. 1948. Red squirrel nesting on the ground. J. Mammal. 29:186.

Nelson, B. A. 1945. The spring molt of the northern red squirrel in Minnesota. J. Mammal. 26:397-400.

Nice, M. M., C. Nice, and D. Ewers. 1956. Comparison of behavior development in snowshoe hares and red squirrels. J. Mammal. 37:64-74.

Pruitt, W. O., Jr. and C. V. Lucier. 1958. Winter activity of red squirrels in interior Alaska. J. Mammal. 39:443-44.

Smith, C. C. 1968. The adaptive nature of social organization in the genus of tree squirrels, *Tamiasciurus*. Ecol. Monogr. 38:31-63.

_____. 1970. The coevolution of pine squirrels (*Tamiasciurus*) and conifers. Ecol. Monogr. 40:349-71.

Svihla, R. D. 1930. Development of young red squirrels. J. Mammal. 11:79-80.

Tiner, J. D., and R. Rausch. 1949. *Syphacia thompsoni* (Nematoda: Oxyuridae) from the red squirrel. J. Mammal. 30:202-3.

Yeager, L. E. 1937. Cone-piling by Michigan red squirrels. J. Mammal. 18:191-94.

Southern Flying Squirrel
Glaucomys volans (Linnaeus)

Although not difficult to identify, *Glaucomys* is so completely nocturnal in habit that a special effort must be made to see one of these attractive animals in its natural habitat. Its daylight hours are spent asleep in a nest at the bottom of an abandoned woodpecker hole or in a hollow tree. Soon after dark, however, it emerges from the nest and spends the remainder of the night gliding from tree to tree, scampering among the branches, and occasionally coming down to the ground to hunt for, and feed on, various nuts and seeds on the forest floor. The flying squirrel is neither skillful nor fast on the ground and spends most of its time in trees, where it is completely at ease, making even the agile red squirrel seem clumsy by comparison.

Flying squirrels, of course, cannot actually fly, but they are excellent gliders. Launching themselves into space from a height of 10 to 12 meters (30 to 40 feet), they may glide as far as 50 meters (approx. 160 feet) before landing on the trunk of a tree or on the ground. Most of their flights, however, are in the range of 20 to 30 meters (approx. 65 to 100 feet). A thin, loose fold of skin extends between the fore and hind limbs and provides a broad gliding surface when the limbs are fully extended. The tail is flattened dorsoventrally, and in addition to providing additional gliding surface, it is used as a rudder during "flight." Flying squirrels maneuver quite well while gliding, turning and twisting to avoid obstacles in their paths. Just before landing, the tail is raised, forcing the body down and head up, thus bringing the feet and legs into position for absorbing the shock of landing and for grasping the landing surface.

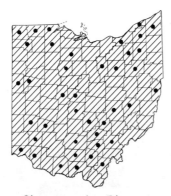

Glaucomys volans (Linnaeus)

The fur of the flying squirrel is dense, long, soft, and silky in texture. It is gray, fawn, or light brown on the head, back, and tail; the ventral surface of the head and body are immaculately white or cream-colored. There is a narrow black line extending along either side between the fore and hind limbs, separating the fur on the back from that on the gliding membrane and belly. Males and females have the same coloring. Mature males have a thin covering of hair on the conspicuous, dark-colored scrotum, and mature females have eight conspicuous, elongated, roughened nipples on the venter. The long vibrissae extending from either side of the nose, plus the unusually large, round, jet-black eyes that flash wine red when illuminated at night further attest to the nocturnal habits of this animal. The highly arched brain case, slightly upturned nasal bones, and five upper cheek teeth (the first, small and inconspicuous) separate the skull of this species from those of the chipmunk and thirteen-lined ground squirrel, which are about the same size. The dental formula is as follows:

$$i\frac{1-1}{1-1}, c\frac{0-0}{0-0}, pm\frac{2-2}{1-1}, m\frac{3-3}{3-3} = 22$$

Following are measurements of 35 adult flying squirrels (*Glaucomys volans*) from Ohio:

	Total Length (mm)	Tail (mm)	Hind Foot (mm)
Extremes	205–50	85–124	20–36
Mean	232	102	31

Perhaps because of its nocturnal and secretive habits, this squirrel has escaped close investigation by all but a few mammalogists; no one has studied it in detail in Ohio. Trapping records indicate that the animal is found throughout the state. Sollberger (1940), among others, states that the southern flying squirrel is most common in mature beech-maple forests. Bole and Moulthrop (1942) state that the species is highly cyclical, with peak populations occurring at three- or four-year intervals. In one study Jordan (1948) recorded a density of a little greater than 2.5 squirrels per hectare (1.0 per acre); a few years later he (Jordan, 1956) found an average of 4 squirrels per hectare (1.6 per acre) in a black oak-hickory woodland in Illinois. Burt (1940) estimated 3.4 to 4.0 squirrels per hectare (1.3 to 1.6 per acre) in Michigan. The presence of flying squirrels in a woods can be determined by observing discarded nut shells: flying squirrels usually get the meat from the nut by cutting through the bottom of the shell, whereas other squirrels cut through the sides. It should be pointed out that population density estimates for this species must take into account the time of year. In the north especially, flying squirrels are known to aggregate during cold weather; as many as 50 individuals have been found occupying a single nest. A female and her litter often form the nucleus of a group, although Muul (1969) states that the young must be at least 60 days old before the female will allow other individuals into the group. As spring approaches and temperatures rise, the groups generally disperse, and each individual lives alone throughout the spring and summer months. It is believed that the winter aggregations help individual squirrels conserve heat.

Nuts, seeds, insects, and insect larvae probably are the most important items in the diet; but, according to Sollberger (1940), corn, salt, sugar, birds, bird eggs, beef, berries, acorns, and birch catkins are also eaten. Enders (1930) says that specimens were not captured until it was learned that meat was the best bait. I once caught a flying squirrel in a live trap baited with peanut butter and oats. Donohoe (personal communication) reports that he and his coworkers have live-trapped several hundred flying squirrels over a 12-year period using English walnuts in the shell. Storing of food is apparently a well-developed behavior in the flying squirrel. In the fall many individual nuts are buried just below the surface of the ground, and modest caches are made in hollow stumps and branches. In either case part of the stored food is later recovered and eaten when other food is scarce or unavailable. Muul (1968) discovered that storing activity decreases as the photoperiod increases, a relationship which lessens the probability that the squirrels will waste energy hunting for nuts when there are none available.

Although litters may be born any time from March through September, most females produce young during April and August (Muul, 1968). Since the gestation period is 39 to 40 days, most matings must occur in January-February and June-July. Muul (1968) determined in a series of controlled experiments that the reproductive cycle of the flying squirrel is apparently keyed to coincide with increasing hours of daylight in the spring and decreasing hours of light in the fall. Artificially regulating and changing the temperature had little or no significant effect on the normal reproductive cycle, but changing the photoperiod had a direct effect on the animals. Occasionally a single young will be born, but three or four in a litter is more common and as many as six have been recorded. The young at birth are hairless, their eyes and ears are closed, and they weigh from three to five grams. At about two weeks of age, the body is completely covered with hair; and by the end of four weeks, the eyes and ears are open. Sollberger (1943) says that by the time they are six weeks old they seem able to take care of themselves.

SELECTED REFERENCES

Barkalow, F. S., Jr. 1956. A handicapped flying squirrel, *Glaucomys volans*. J. Mammal. 37:122-123.

Bole, B. P., Jr., and P. N. Moulthrop. 1942. The Ohio recent mammal collection in the Cleveland Museum of Natural History. Sci. Publ. Cleveland Mus. Nat. Hist. 5:83-181, esp. 140-141.

Burt, W. H. 1940. Territorial behavior and populations off some small mammals in southern Michigan. Misc. Publ., Univ. of Mich. Mus. Zool. 45:1-58.

Chapman, F. B. 1938. The development and utilization of the wildlife resources of unglaciated Ohio. 2 vols. Ph. D. Dissertation, Ohio State Univ., Columbus. 791 pp.

Dolan, P. G., and D. C. Carter. 1977. *Glaucomys volans*. Amer. Soc. Mammal., Mammalian Species No. 78:1-6.

Enders, R. K. 1930. Some factors influencing the distribution of mammals in Ohio. Occ. Papers, Univ. of Mich. Mus. Zool. 212:1-27.

Gupta, B. B. 1966. Notes on the gliding mechanism in the flying squirrel. Occ. Papers, Univ. of Mich. Mus. Zool. 645:1-7.

Howell, A. H. 1918. Revision of the American flying squirrels. North Amer. Fauna 44. 64 pp.

Jordan, J. S. 1948. A midsummer study of the southern flying squirrel. J. Mammal. 29:44-48.

———. 1956. Notes on a population of eastern flying squirrels. J. Mammal. 37:294-95.

Kelker, G. 1931. The breeding time of the flying squirrel (*Glaucomys volans volans*). J. Mammal. 12:166-67.

Moore, J. C. 1946. Mammals from Welaka, Putnam County, Florida. J. Mammal. 27:49-59.

Muul, I. 1968. Behavioral and physiological influences on the distribution of the flying squirrel, *Glaucomys volans*. Misc. Publ., Univ. of Mich. Mus. Zool. 134 1-66.

———. 1969. Photoperiod and reproduction in flying squirrels, *Glaucomys volans*. J. Mammal. 50:542-49.

Pearson, O. P. 1947. The rate of metabolism of some small mammals. Ecology 28:127-45.

Sollberger, D. E. 1940. Notes on the life history of the small eastern flying squirrel. J. Mammal. 21:282-93.

———. 1943. Notes on the breeding habits of the eastern flying squirrel (*Glaucomys volans volans*). J. Mammal. 24:163-73.

Stack, J. W. 1925. Courage shown by flying squirrel, *Glaucomys volans*. J. Mammal. 6:128–29.

Svihla, R. D. 1930. A family of flying squirrels. J. Mammal. 11:211–13.

Uhlig, H. G. 1956. Reproduction in the eastern flying squirrel in West Virginia. J. Mammal. 37:295.

FAMILY CASTORIDAE
Beaver

The members of this family, the beavers, are presently represented in Europe, Asia, and North America. The characteristics of the family are essentially those listed below for the only species occurring throughout North America.

Beaver
Castor canadensis Kuhl

The beaver is the largest rodent in North America. Individuals weighing 32 kg (approx. 70 lb) have been reported, but an average adult weight of 18 kg (approx. 40 lb) is more representative of the species. In addition to their large size, beaver are easily recognized by their large, webbed hind feet and broad, paddle-like, dorsoventrally flattened tail. The following adaptive characteristics are beneficial to this aquatic species: valves in the nose and ears that close when the beaver dives below the

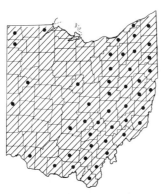

Castor canadensis Kuhl

water surface; underfur that is unusually thick and fine and is overlaid with coarse guard hairs; small eyes that are adapted to see under water; and lips that meet behind the incisor teeth, allowing the animal to gnaw and chew under water. Internal modifications include a large liver and oversized lungs that provide for the storage of large quantities of oxygen, allowing the beaver to remain below the surface for as long as 10 to 15 minutes. The front feet are relatively small and play little part in swimming; the larger webbed hind feet supply most of the power for movement through the water. Also, each of the two inner toes of the hind feet has a specially modified split claw (one half immediately above the other) that is used to comb the hair, oil the fur, and presumably remove dirt and parasites from the coat. The conspicuous and characteristic flat tail is furred at its base only, and the remainder is covered with large flat scales. The tail is used as a rudder and to a lesser extent as a sculling organ in swimming. As it submerges, an alarmed beaver may raise its tail and bring the broad flat surface down with a resounding whack on the water as a warning signal to other beaver. Both sexes are rich dark brown in color with a somewhat lighter colored belly and head. Beaver molt once each year, and the adult coat is in "prime" condition during the winter and early spring.

The skull can be recognized by its large size, total of 20 teeth, and complete absence of postorbital processes. The front surfaces of the large incisors are orange. The dental formula is as follows:

$$i\frac{1\text{-}1}{1\text{-}1},\ c\frac{0\text{-}0}{0\text{-}0},\ pm\frac{1\text{-}1}{1\text{-}1},\ m\frac{3\text{-}3}{3\text{-}3} = 20$$

Barbour and Davis (1974) give the following measurements for the beaver in Kentucky: total length 900–1,170 mm; tail 300–400 mm; hind foot 165–85 mm.

Although abundant throughout Ohio at one time, the American beaver was extirpated by 1830 (Bole and Moulthrop, 1942; Chapman, 1949). Over the years a few have wandered back into the state from Pennsylvania in the northeast and Michigan in the northwest. Others have been deliberately introduced, and a few have accidentally escaped from confinement. According to Chapman (1949), "What is believed to have been the first evidence of beavers in Ohio during the twentieth century was a number of peeled aspen trees found on the Ohio side of the Pymatuning Reservoir in Ashtabula County on or about May 15, 1936." A report in 1946 that beaver were causing damage in Columbiana County by causing the flooding of crop and pasture lands led the Ohio Division of Wildlife to conduct a survey of beaver activity in the state. The survey disclosed that in 1947, in 11 counties, there were 25 active beaver colonies containing approximately 100 beaver (Chapman, 1949). By 1972, twenty-five years later, approximately 4,420 beaver occupied 884 colonies in 37 counties, according to Bednarik (1973). By 1976 Bednarik and Warhurst (1977) reported that an estimated 7,518 beaver occupied 1,564 colonies in 40 counties. Beaver continue to become increasingly abundant, especially in old strip-mined areas of eastern and southeastern Ohio.

Beaver are herbivores, probably best known for eating the bark and cambium layer of such trees as aspen, alder, willow, and maple. Their dietary preferences, however, are quite broad. Nixon and Ely (1969), studying beaver in Athens and Vinton counties, and Henry and Bookhout (1970), working in Ashtabula and Columbiana counties, found about twenty different woody species were being utilized. Included in their lists are alder, aspen, hop-hornbeam, ash, willow, dogwood, hickory, oak, tupelo, beech, apple, red maple, black cherry, hawthorn, witch-hazel, ironwood, basswood, elm, ninebark, black locust, buttonbush, Virginia pine, sumac, and sassafras. In each of the studies, the use of a particular species was usually determined by its availability, although in some instances it appeared that a beaver traveled farther to reach one particular kind of food. The beaver seldom traveled more than 100 m from the lodge to make their cuttings. In Ohio, as elsewhere, the plant stems most frequently cut are those that are less than 7.5 cm (3 in) in diameter. Bradt (1938) estimated that one beaver cuts approximately 0.6 trees per day, or between 215 and 225 trees each year. Nixon and Ely (1969) calculated that one Ohio colony averaged between 245 and 295 cut or barked stems per beaver over a ten-month period. During the late summer and early fall, small tree trunks and branches that are not immediately stripped of bark are cut into convenient lengths (1 to 2 m) and carried to an underwater winter food cache near the lodge entrance. This cache serves as the principal source of food during the winter months when the beaver pond is frozen over. In Ohio winter food storage is not critical because the ponds usually do not remain frozen for periods long enough to interfere seriously with the beaver's normal food-getting activities. Nevertheless, winter caches are made and used.

In addition to woody plants, herbaceous plants are also included in the diet. During summer, for example, when water plants are abundant, they are used as food, sometimes even in preference to woody plants (Bradt, 1938). Also, young beaver prefer leafy vegetation (Tevis, 1950).

The engineering feats of the beaver are well documented; their elaborately constructed dams and lodges attest to their ability. The dams maintain the proper level of water for the lodge entrance (or entrances). In Ohio, as elsewhere, a dam is usually constructed on one of the side channels of a stream rather than in the main stream. The dam causes flooding and the development of a pond where the beaver then build their lodge or lodges. The dam, made by one or several beaver, is constructed of sticks, stones, aquatic vegetation, and bottom debris held together by mud. Some dams are truly impressive structures 16 m or more in length and up to 1.5 m in height. For as long as a beaver colony lives in a pond, work will continue on the dam. In some instances several dams may be maintained by a colony. If one part of a dam is damaged or removed, it will immediately be repaired or replaced. If the water level in the pond drops, the dam will be tightened; if the level gets too high, a temporary opening will be created. One of the basic problems faced by Ohio beaver is that they are continually forced to move because of rapid fluctuation of water levels created by periodic summer droughts and spring flood conditions (Henry and Bookhout, 1969; Nixon and Ely, 1969).

Beaver in Ohio live either in dens along the banks of streams and lakes or in lodges that are usually located in quiet ponds. Because of the rapidly changing water levels, most beaver live in bank dens. These are chambers that are 1 m or more in diameter and possibly 1 m high. They are reached by a tunnel that may be 3 to 10 m long, opening into a stream or lake bank 1 m or so below the water surface. The tunnel slopes upward toward the den, and the floor of the den itself is above water level.

Lodges are constructed out of the same kinds of materials that are used to build dams. They too are held together by mud that is plastered to the outside. A beaver lodge can be as large as 12 m in diameter and over 2 m high. The dome and walls are so well constructed that it is almost impossible to collapse them. Would-be predators soon discover that they cannot rip or tear their way through them. There is a single chamber within the lodge with a raised platform that serves as a floor. The lodge is entered and exited by one or more passageways that lead down to openings a meter or more below the surface of the water. The openings allow the beaver to come and go without being observed, and when the ponds freeze over, the animals can still move in and out of their quarters.

A beaver colony consists of a single family and theoretically could include as many as 18 individuals. The individuals in a typical colony, however, are the adult male and female parents, 4 to 5 one-year-olds, and 4 to 5 two- or almost two-year-olds. The kits remain with the adults for almost two years, leaving just before the birth of a new litter in their second spring. At that time the two-year-olds either leave home voluntarily or are driven out by the parents. Once they have left the lodge, they cannot return. They move overland to new streams and ponds or to remote areas of their home pond. Most soon find a mate and start a colony of their own. Limiting the size of the colony has some definite advantages for the beaver: (1) it prevents overuse of limited but essential food sources; (2) it prevents close inbreeding, which is genetically undesirable; and (3) it ensures dispersal of the species into new and unoccupied territory.

Once a mate has been chosen, the pair probably remains together for life, although it is known that certain males mate with more than one female during the breeding season. Beaver mate during February and March in Ohio. Henry and Bookhout (1969) have shown that beaver in Ohio do not breed before they are one and a half years old; here, as elsewhere, most probably breed for the first time in the spring of their third year. The gestation period is between three and four months. The exact number of young born to any one female is difficult to determine. Henry and Bookhout (1969) report a mean ovulation rate of 4.3 and mean placental scar count of 3.8 for 105 female beaver trapped in Ohio. A reasonable conclusion from these data is that the average litter size in Ohio is about 4. Litters of from 2 to 8 have been recorded for the beaver throughout its range. Newborn beaver are fully furred, their eyes are open, they weigh approximately .45 kg (about 1 lb), and they can swim. They usually remain in the nest for at least a month before venturing out with their mother. For a short time after the young are born, the adult males take up residence outside the lodge. Upon returning, they help care for, and raise, the family. In their 1969 study Henry and Bookhout found a sex ratio of 100 females to 84 males. Individuals may live 20 or more years (Larson, 1967).

The reestablishment of the beaver in Ohio, brought about by the sound conservation practices established and enforced by the Ohio Division of Wildlife, is an excellent example of how meaningful such a program can be. Bednarik (1973) writes: "Ohio's beaver status is dynamic. In some areas, particularly northeast Ohio, nuisance beaver cause flooding of farm crops, roads, and rural home sites. Controlling these beaver is a problem. Beaver have been trapped and removed from nuisance colonies to alleviate damage problems. Despite such efforts, damage has continued in some areas. In these areas, a beaver trapping season is clearly justified. Other reasons for a harvest season are (1) to provide recreation and (2) to provide a harvest of animals where, in a high population, the annual increment might be lost to natural mortality."

The first beaver trapping season in Ohio in 130 years occurred in January 1961 with three counties open. In 1979 a total of 3,318 beaver were trapped in 32 counties. Beaver pelts reported by Ohio fur buyers (Appendix Table 2) rose in total number from 140 in 1961 to 1,747 in 1973. Average price per pelt fluctuated from $10.18 in 1971 to $17.96 in 1979.

Another potential source of income is castoreum, the material found within the paired castor glands that are located just within and on either side of the anus. When properly removed and dried, the castor glands have been sold for as much as $10 to $20 a pound. Castoreum is used as a base in some perfumes, and trappers have long used it for attracting animals to their traps.

Because of their dam-building and pond-forming habits, beaver have always had an important influence on the plants and animals that also inhabit these areas. The influence of beaver on the environment may be particularly important in Ohio in the ravaged strip-mined counties. Nixon and Ely (1969) have made a very interesting and important observation: "Based on the occupancy by beavers in the Hewitt's Fork and Raccoon Creek watersheds, highly acid waters such as often occur in old strip-mined areas are acceptable for beavers if suitable foods are present."

Except for man, the beaver has no important predators in Ohio.

Ohio is surrounded by three subspecies of *Castor canadensis: C. c. michiganensis* to the north, *C. c. canadensis* to the east, and *C. c. carolinensis* to the south and west. With the reinvasion of Ohio by the beaver, it is not clear as to how extensive the movements of each have been. Much work is needed to clarify subspecies distribution within the state.

SELECTED REFERENCES

Barbour, R. W., and W. H. Davis. 1974. Mammals of Kentucky. Univ. Press of Ky., Lexington. 322 pp.

Bednarik, K. E. 1965. The beaver: an empire builder. Ohio Conserv. Bull. 29(1):7-9, 26.

_____. 1973. Ohio beaver harvest, 1961-1973. Ohio Dept. Nat. Resources, Div. Wildlife, Note 210.

_____. 1976. Furbearers. Pp. 3-1 to 3-6 in Ohio Div. Wildlife. Reasons for seasons, hunting and trapping. Ohio Dept. Nat. Resources Publ. No. 68 (R 1276).

Bednarik, K. E., and R. A. Warhurst. 1977. Ohio beaver harvest, 1976 and 1977. Ohio Dept. Nat. Resources, Div. Wildlife, In-Service Note 358. 12 pp.

Benson, S. B. 1936. Notes on the sex ratio and breeding of the beaver in Michigan. Occ. Papers Univ. of Mich. Mus. Zool. 355:1-6.

Bole, B. P., Jr., and P. N. Moulthrop. 1942. The Ohio recent mammal collection in the Cleveland Museum of Natural History. Sci. Publ. Cleveland Mus. Nat. Hist. 5:83-181, esp. 141-42.

Bradt, G. W. 1938. A study of beaver colonies in Michigan. J. Mammal. 19:139-62.

_____. 1939. Breeding habits of beaver. J. Mammal. 20:486-88.

_____. 1947. Michigan beaver management. Mich. Dept. Conserv., Lansing. 56 pp.

Brenner, F. J. 1964. Reproduction of the beaver in Crawford County, Pennsylvania. J. Wildl. Manage. 28:743-47.

Chapman, F. B. 1949. The beaver in Ohio. J. Mammal. 30:174-79.

Dalke, P. D. 1947. The beaver in Missouri. Missouri Conserv. 8:1-3.

Friley, C. E., Jr. 1949. Use of the baculum in age determination of Michigan beaver. J. Mammal. 30:261-67.

Henry, D. B. 1967. Age structure, productivity, and habitat characteristics of the beaver in northeastern Ohio. M. S. Thesis, Ohio State Univ., Columbus. 68 pp.

Henry, D. B., and T. A. Bookhout. 1969. Productivity of beavers in northeastern Ohio. J. Wildl. Manage. 33:927-32.

_____. 1970. Utilization of woody plants by beavers in northeastern Ohio. Ohio J. Sci. 70:123-27.

Irving, L. 1937. The respiration of beaver. J. Cell. & Compar. Physiol. 9:437-51.

Larson, J. S. 1967. Age structure and sexual maturity within a western Maryland beaver (*Castor canadensis*) population. J. Mammal. 48:408-13.

Nixon, C. M., and J. Ely. 1969. Foods eaten by a beaver colony in southeast Ohio. Ohio J. Sci. 69:313-19.

Novak, M. 1977. Determining the average size and composition of beaver families. J. Wildl. Manage. 41:751-54.

Osborn, D. J. 1953. Age classes, reproduction, and sex ratios of Wyoming beaver. J. Mammal. 34:27-44.

Rutherford, W. H. 1964. The beaver in Colorado: its biology, ecology, management, and economics. Colo. Game, Fish & Parks Dept., Tech. Bull. 17. 49 pp.

Shadle, A. R. 1930. An unusual case of parturition in a beaver. J. Mammal. 11:483-85.

Shadle, A. R., A. M. Nauth, E. C. Gese, and T. S. Austin. 1943. Comparison of tree cuttings of six beaver colonies in Allegany State Park, New York. J. Mammal. 24:32-39.

Stegeman, L. C. 1954. The production of aspen and its utilization by beaver on the Huntington Forest. J. Wildl. Manage. 18:348-58.

Tevis, L., Jr. 1950. Summer behavior of a family of beavers in New York State. J. Mammal. 31:40-65.

Van Nostrand, F. C., and A. B. Stephenson. 1964. Age determination for beavers by tooth development. J. Wildl. Manage. 28:430-34.

Warren, E. R. 1927. The beaver: its works and its ways. Williams and Wilkins, Baltimore, Md. 177 pp.

FAMILY CRICETIDAE
Native Rats and Mice, Voles, and Lemmings

Except for the jumping mice, the common house mouse, and the Norway rat, all mouse-like mammals in Ohio belong to this large family of rodents. The family is conveniently divided into two rather well-defined groups, the cricetines and the microtines. The cricetines have long tails, large eyes, pointed noses, large ears, conspicuous feet and legs, and (except for the woodrat) molar teeth with longitudinal rows of tubercles or cusps. Cricetine rodents are almost completely nocturnal in habit. Microtine rodents in Ohio (excluding the muskrat) have short tails, small beady eyes, short rounded muzzles, small well-hidden ears, short legs that are hidden in the body integument, and flat-crowned molars with enamel patterns arranged in loops and triangles. Microtines are active both day and night.

The family Cricetidae is represented throughout the world except in Australia and Antarctica. There are 17 genera and 58 species in the continental United States. In Ohio we have 4 species of cricetine and 6 species of microtine rodents.

KEY TO THE SPECIES OF CRICETIDAE IN OHIO

1. Tail length equal to or greater than one-half the body length, not flattened laterally . . . . .

.... CRICETINAE 2
1'. Tail either shorter than one-half the body length or flattened laterally . MICROTINAE 5
2. Length less than 130 mm; belly with some orange or buff hairs, not pure white; upper incisors with a deep groove down the front .. .. *Reithrodontomys humulis* (Eastern Harvest Mouse) p. 84
2'. Length greater than 130 mm; belly white; no grooves in upper incisors 3
3. Length more than 300 mm ("rat size"); molars flat-crowned with loops and triangles; interorbital region depressed; infraorbital canal triangular in outline when viewed from front .... *Neotoma floridana* (Eastern Woodrat) p. 91
3'. Length less than 300 mm ("mouse size")' molars cuspidate; interorbital region not depressed; infraorbital canal small, not decidedly triangular in outline ...... *Peromyscus** 4
4. Tail usually less than 65 mm and distinctly bicolored; ear 19 mm or more from notch, and margin edged with white; upper parts slate gray or grayish-brown with a broad dark stripe down the center of the back; little or no rich fulvous coloration; hind foot less than 21 mm; greatest length of skull less than 25 mm; anterior end of palatine foramen rounded and blunt (fig. 17) .... *Peromyscus maniculatus* (Deer Mouse) p. 86
4'. Tail usually more than 70 mm and often not distinctly bicolored; ear 15–18 mm from notch and without white border; upper parts rich fulvous; hind foot 21 mm or more; greatest length of skull 25 mm or more; anterior end of palatine foramen narrow and pointed (fig. 17) .... *Peromyscus leucopus* (White-footed Mouse) p. 88
5. Tail longer than one-half the body length and compressed from side to side; hind feet with long stiff hairs between the toes; skull more than 50 mm in length .... *Ondatra zibethicus* (Muskrat) p. 102
5'. Tail less than one-half the body length, not compressed; no long hairs between the toes of the hind feet; skull less than 50 mm in length .. 6
6. Tail equal to, shorter than, or only exceeding by a few millimeters the length of the hind foot; foot usually less than 25 mm in length .. 7
6'. Tail definitely longer than hind foot; foot usually exceeding 25 mm in length 8
7. Fur on back uniform auburn or red and smooth in appearance; no grooves on upper incisors .... *Microtus pinetorum* (Woodland Vole) p. 99
7'. Fur on back gray or grayish-brown often with a green tinge and a decidedly grizzled appearance; a faint groove on the side of each upper incisor .... *Synaptomys cooperi* (Southern Bog Lemming), p. 105
8. A distinct reddish band down the center of the back, set off from the grey sides; limited to northeastern Ohio .... *Clethrionomys gapperi* (Southern Red-backed Vole) p. 94
8'. No reddish band down the center of the back; upper parts and sides gray or grayish-brown; not limited to northeastern Ohio .... *Microtus*† 9
9. Third molar of upper jaw with five or more irregular triangles (see fig. 18a); first lower molar with six or more enclosed triangles; second lower molar with five enclosed triangles. Tubercles on soles of hind feet usually six, sometimes five; mammae ten (six inguinal and four pectoral); belly gray or white; dorsum smooth, not grizzled, no salt-and-pepper fur . . *Microtus pennsylvanicus* (Meadow Vole) p. 95
9'. Third molar of upper jaw with four triangles (see fig. 18b); first lower molar with five enclosed triangles; second lower molar with four enclosed triangles. Tubercles on soles of hind feet five; mammae six (four inguinal, two pectoral); belly with some buff hairs; dorsum grizzled with salt-and-pepper fur .... *Microtus ochrogaster* (Prairie Vole) p. 98

*The two species of *Peromyscus* are difficult to separate; a combination of characteristics must be considered when making identification.
†Positive identification of the two species of *Microtus* can be made only by studying the tooth patterns (fig. 18).

Eastern Harvest Mouse
Reithrodontomys humulis (Audubon and Bachman)

The eastern harvest mouse inhabits fields and open areas from Maryland and central Ohio south to Florida and Texas. Harvest mice have rich dark brown coats with a broad, dark, mid-dorsal stripe. The belly is gray or buff gray, and most animals have a conspicuous orange or bright buff line extending along the side, separating the belly and dorsal fur. This orange stripe is not so conspicuous in museum skins, but it is the most noticeable characteristic of freshly trapped animals. The tail is darker above than below but not strikingly bicolored. The feet are white or yellowish above. In my collection the one or two specimens caught in November are considerably lighter than those collected in Decem-

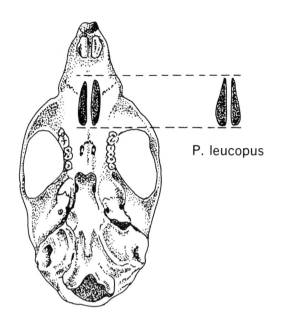

Fig. 17. Ventral view of the skull of the deer mouse *(Peromyscus maniculatus)*, comparing its anterior palatine foramina with that of the white-footed mouse *(P. leucopus)*. (Drawing by Elizabeth Dalvé.)

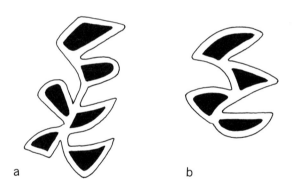

Fig. 18. Enamel patterns on the third molar of the upper jaw of Microtus. *a.* Meadow vole *(M. pennsylvanicus)*, showing 6 triangles; *b.* prairie vole *(M. ochrogaster)*, showing 4 triangles. (Drawing by Elizabeth Dalvé.)

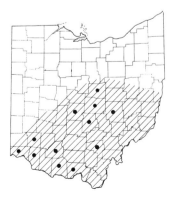

Reithrodontomys humulis (Aydybon and Bachman)

ber, January, and February, indicating a molt in late November and/or early December in Ohio. There may be another molt in the spring, but I have no evidence for this.

Occasionally the eastern harvest mouse is mistaken for a small deer mouse (*Peromyscus* spp.) or a buff-colored house mouse (*Mus musculus*), but close examination of the dentition will readily identify the harvest mouse. It is the only long-tailed mouse with grooved upper incisors; the groove down the center of each incisor is so deep that it appears to divide the tooth into two separate parts. The skull is smaller and more delicate than that of any other cricetine rodent. The dental formula is as follows:

$$i\frac{1-1}{1-1}, c\frac{0-0}{0-0}, pm\frac{0-0}{0-0}, m\frac{3-3}{3-3} = 16$$

Following are measurements of 24 adult eastern harvest mice (*Reithrodontomys humulis*) from Ohio:

	Total Length (mm)	Tail (mm)	Hind Foot (mm)
Extremes	104–29	42–62	14–17
Mean	112	53	16

Not included in the above measurements is a female taken by me in Highland County that is larger than any specimen ever recorded for this subspecies: total length, 144 mm; tail, 65 mm; foot, 16 mm; ear, 12 mm; weight, 13.5 g. When taken in early November, she was obviously nursing young, and there were six fresh placental scars in the uterus.

The eastern harvest mouse is not abundant and rarely enters traps. In one study in southwestern Ohio, only 20 were caught in 17,975 trap-nights, making up 2.5 percent of the total catch (Gottschang, 1965). They are most often found in abandoned fields with abundant grass cover, but their distribution is spotty. Most specimens have been caught in winter.

Brimley (1923) found embryos in female harvest mice in May and November and trapped young in July. There are no records of breeding animals in the spring or summer from Ohio. A female taken 6 November 1960 contained five 6-mm embryos with a combined weight of 1.7 g. Another female taken 8 November 1962 had just given birth to a litter of six. One juvenile in the University of Cincinnati collection was caught between 2 December and 10 December, indicating a birth date during the middle of November. None of the animals that I have taken after November were in breeding condition. A male taken 15 March 1952 had large testes (3 mm × 5 mm), indicating that he was approaching breeding condition. The gestation period is 21 to 22 days. Litter sizes range from 1 to a record of 8 (Dunaway, 1962), but 3 or 4 is average. According to Kaye (1961a), the mean weight of 27 young at birth was 1.2 g. He also states: "Harvest mice at birth appear unpigmented, are pink, have a translucent skin, and have tiny, white natal hairs present on most parts of the body. By the end of the first week, the brownish hair on the head and nape almost completely hides the underlying skin. All young are weaned by the end of the third week and are cared for only by the female parent." The eyes open from days 9 to 13, and the external ear opening is evident on days 7 to 11 (Layne, 1959; Kaye, 1961a). A single female probably produces two or three litters each year. One female in Layne's laboratory (1959) had eight litters in two years. Cannibalism seems to be a common practice: whole litters may be consumed by the mother, or brothers and sisters may eat each other.

In the summer harvest mice construct small, teacup-like nests of grass in which they sleep and rear their families. These nests are usually well hidden in tall grass with one or two inconspicuous paths leading to them. Since I have not been able to locate such nests during the winter in Ohio, I suspect that the mice live belowground during the colder months of the year. Brimley (1923) unearthed a nest containing two harvest mice in late fall. Kaye (1961a) described winter nests of harvest mice in North Carolina, where the winters are relatively mild.

To my knowledge no one has made a study of the food habits of eastern havest mice. In the laboratory they have accepted fresh fruits and vegetables, whole rolled oats, sunflower seeds, and wild-bird seed. In the wild, grass seeds and insects are probably main items in the diet.

Little is known about the home range or movements of this mouse. Kaye (1916b) tagged four harvest mice with small radioactive gold-198 wires and followed their movements with a portable Geiger counter. It was impossible for him to plot a home range for any one mouse because each one utilized several nests located around the periphery of a given area and constantly moved back and forth between the nests. Occasionally, however, a mouse traveled as far as 75 m from one nest to another. Harvest mice may use voles' runways; they do not construct runways of their own.

Because of their habitat preference, food habits, and low population density, harvest mice seldom, if ever, interfere with man's activities. Owls, hawks, snakes, and carnivorous mammals all take their toll of harvest mice.

SELECTED REFERENCES

Bole, B. P., Jr. 1932. An addition to the known list of Ohio mammals. Ohio J. Sci. 32:402.

Brimley, C. S. 1923. Breeding dates of small mammals at Raleigh, North Carolina. J. Mammal. 4:263-64.

Dunaway, P. B. 1962. Litter-size record for eastern harvest mouse. J. Mammal. 43:428-29.

Goslin, R. 1939. The harvest mouse in central Ohio. J. Mammal. 20:257.

Gottschang, J. L. 1965. Winter populations of small mammals in old fields of southwestern Ohio. J. Mammal. 46:44-52.

Holding, B. F., and O. L. Royal. 1952. The development of a young harvest mouse, *Reithrodontomys*. J. Mammal. 33:388.

Kaye, S. V. 1961a. Laboratory life history of the eastern harvest mouse. Amer. Midl. Nat. 66:439-51.

———. 1961b. Movements of harvest mice tagged with gold-198. J. Mammal. 42:323-37.

Layne, J. N. 1959. Growth and development of the eastern harvest mouse, *Reithrodontomys humulis*. Bull. Fla. State Mus. Biol. Sci. 4:61-82.

Deer Mouse
Peromyscus maniculatus (Wagner)

This species includes a large number of intergrading subspecies that occur throughout the United States and Canada and extend southward into Mexico and Central America. The one subspecies found in Ohio, *Peromyscus maniculatus bairdii,* is common in the open grasslands of the central United States. Its range extends east to western New York, west to central Nebraska, Wyoming, and Montana, north to Lake Winnipeg, Manitoba, and south to northern Tennessee and Arkansas. The deer mouse is strictly an animal of the grasslands in Ohio, even avoiding the thick tangle of shrubs and tall weeds so often found along fencerows. As would be expected, *P. maniculatus* is more common in the Glaciated Till Plains of western Ohio than on the Allegheny Plateau. The deer mouse could possibly have lived in Ohio in small natural prairie areas, remnants of the once larger prairie peninsula of several thousand years ago (Transeau, 1935), or it could be a fairly recent arrival that moved into and across the state from the West only after the forests were cleared in the nineteenth century. Brayton (1882) did not include the deer mouse in his list of Ohio mammals. Osgood (1909) lists two specimens from London, Madison County, probably the first record of the species for Ohio. It is interesting to note that Madison County had numerous natural prairie areas (Gordon, 1969), but most of them now have been eliminated by agriculture. Bole and Moulthrop (1942) examined many specimens from 14 counties in Ohio.

Peromyscus maniculatus (Wagner)

Our two species of *Peromyscus* are often difficult to identify properly in the field, but the deer mouse is smaller and has a shorter tail (not more than 65 mm and usually less than 60 mm). Coloration is variable, but in most populations the upper parts are decidedly gray or grayish-brown. Most individuals have a dark gray, mid-dorsal stripe, although this is not always evident. A few individuals in all populations are dark brown or reddish-brown above, with or without the dark mid-dorsal stripe, and in some populations brown is actually the predominant color. An occasional adult is uniform gray. Deer mice seldom show the bright rufous or cinnamon coloration so typical of *P. leucopus*. Adults in Ohio molt once a year in November and December, and winter specimens are darker than summer ones. Jackson (1961, p. 213) states that in Wisconsin the annual molt occurs during the summer. The tail is decidedly bicolored, gray above and white below. The belly is white. Juveniles and subadults (less than 117 mm in total length) are uniform light gray above and white below. The deer mouse has a smaller ear (less than 16 mm) and a shorter foot (less than 20 mm) than the white-footed mouse. Skull characteristics of the two species are contrasted as follows:

	P. maniculatus	*P. leucopus*
Total length	Less than 25 mm	25 mm or more
Interorbital distance	Less than 4 mm	4 mm or more
Anterior palatine foramen	Rounded anteriorly	Pointed anteriorly

The dental formula is as follows:

$$i\frac{1-1}{1-1}, c\frac{0-0}{0-0}, pm\frac{0-0}{0-0}, m\frac{3-3}{3-3} = 16$$

Following are measurements of 30 adult deer mice (*Peromyscus maniculatus*) from Ohio:

	Total Length (mm)	Tail (mm)	Hind Foot (mm)
Extremes	128–48	44–61	17–21
Mean	136	53	18

Clark (1938), Dice and Bradley (1942), and Svihla (1935) have studied the reproductive habits of the deer mouse in the laboratory, and Blair (1940) studied them in the field. Females become sexually mature at the average age of 48 days, but some precocious individuals are ready to breed when 28 days old. Some 40-day-old males examined by Clark (1938) had motile sperm in the epididymis, but as a group males reached sexual maturity about 10 days later than females. Some females breed throughout the year, but the peaks of reproductive activity occur from February to June and from August to November. I have taken reproductively active males in late February and pregnant females as early as 23 March. In our latitude the first litters are probably born usually during the last of March or in early April. Adult females exhibit a postpartum estrus and mate within 24 to 48 hours after giving birth. The gestation period is 21 to 23 days. A female probably rears three litters in the spring and two or three more in the fall. Litter sizes range from one to eight: four is average. The pregnant females that I have examined from Ohio have contained from four to six embryos. The young at birth are 40 to 50 mm long and weigh 1 to 2 g. They are hairless, pink, and blind, and have the external pinnae folded against the sides of the head. Within 24 hours the mice are darkly pigmented. Hair is evident and the pinnae have unfolded within 2 to 4 days. The eyes open on about day 14 and the young are weaned about day 25. Growth is rapid for the first 4 to 5 weeks, after which it gradually decreases. Dice and Bradley (1942) showed that some growth may continue for as long as 2 years. As soon as a new litter is born, members of the previous litter are forced out of the nest and must fend for themselves.

Deer mouse populations fluctuate with the seasons, reflecting the reproductive activities of the species. Thus, numbers are highest in the late spring and fall and lowest at the end of the winter and in midsummer. The fluctuations in numbers that occur from year to year are not cyclic (see *Microtus*, p. xx), but are unpredictable and appear to be governed by a number of ecological and environmental factors. Population density seldom exceeds 22.5 individuals per hectare (9.0 per acre) and at times is as low as 2.5 per hectare (1.0 per acre) according to Blair (1940). He also found that in southern Michigan the average home range of both immature and adult males and females comprises about 0.2 hectare (0.5 acre). The ranges of both sexes and all age groups broadly overlap; in winter 12 to 15 animals may occupy the same nest.

Surprisingly little information is available concerning the food habits of deer mice. In the laboratory they will accept a wide variety of food, including seeds of all types, chicken scratch, fruits, nuts, berries, pieces of raw liver and other meat, bulbs, oats, insects, and spiders. Starting in October, these mice store large quantities of seeds and nuts. Howard and Evans (1961) took advantage of this to study their winter food habits. From 136 nest boxes distributed over 120 hectares (300 acres) of grassland in Michigan, they collected and analyzed 155 food stores. Although many different species of seeds were collected, only ragweed (*Ambrosia*), oak (*Quercus*), clover (*Lespedeza*), and panic grass (*Panicum*) were present in significantly large amounts. Other items found by Howard and Evans in the food caches were remains of grasshoppers, beetles, other insects, deer pellets, a fox scat, a few leaves in most boxes, flower parts, and lichens. The largest cache, accumulated in less than one month, contained 150 cc of weed seeds, 565 acorns, and six deer pellets. Each cache represented the combined efforts of from four to ten mice. The variation from one food store to another shows that the mice are not highly selective in their food habits. They appear to store and use foods that are readily available to them.

Deer mice destroy some grain but not enough to make them an economically important pest to man. Mortality rate is high. Blair (1948) found the mean life expectancy to be 4.88 ± 0.2 months, with most mortality attributed to predation. Certainly the deer mouse is an important item in the normal food chain of its environment.

SELECTED REFERENCES

Blair, W. F. 1940. A study of prairie deer-mouse populations in southern Michigan. Amer. Midl. Nat. 24:273–305.

———. 1948. Population density, life span, and mortality rates of small mammals in the blue-grass meadow and blue-grass field associations of southern Michigan. Amer. Midl. Nat. 40:395–419.

Bole, B. P., Jr., and P. N. Moulthrop. 1942. The Ohio recent mammal collection in the Cleveland Museum of Natural History. Sci. Publ. Cleveland Mus. Nat. Hist. 5:83–181, esp. 144–145.

Brayton, A. W. 1882. Report on the Mammalia of Ohio. Rep. Geol. Surv. Ohio 4:1–185.

Browne, R. A. 1976. Relative population densities and adrenal gland weights as related to island populations of *Peromyscus maniculatus*. Ohio J. Sci. 76:114–15.

Clark, F. H. 1938. Age of sexual maturity in mice of the genus *Peromyscus*. J. Mammal. 19:230–34.

Dice, L. R., and R. M. Bradley. 1942. Growth in the deer-mouse, *Peromyscus maniculatus*. J. Mammal. 23:416–27.

Gordon, R. B. 1969. The natural vegetation of Ohio in pioneer days. Ohio Biol. Surv. Bull. N.S. 3(2):54–60.

Hine, J. S. 1929. Distribution of Ohio mammals. Annu. Rep., Proc. Ohio Acad. Sci. 8(6):267.

Howard, W. E., and F. C. Evans. 1961. Seeds stored by prairie deer mice. J. Mammal. 42:260–63.

Jackson, H. H. T. 1961. Mammals of Wisconsin. Univ. of Wisc. Press, Madison, Wisc. 504 pp.

King, J. A., ed. 1968. Biology of *Peromyscus* (Rodentia). Spec. Publ. No. 2. Amer. Soc. Mammal. 593 p.

Osgood, W. H. 1909. Revision of the mice of the American genus *Peromyscus*. North Amer. Fauna 28:1–285, esp. 79–83.

Svihla, A. 1935. Development and growth of the prairie deermouse, *Peromyscus maniculatus bairdii*. J. Mammal. 16:109–15.

Transeau, E. N. 1935. The prairie peninsula. Ecology 16:423–37.

White-footed Mouse
Peromyscus leucopus (Rafinesque)

Only 1 of the 17 recognized subspecies of *Peromyscus leucopus* occurs in Ohio: *P. l. noveboracen-*

sis. This subspecies is found throughout the entire northeastern United States, a range that broadly overlaps that of the deer mouse (*P. maniculatis bairdii*) west of the Atlantic seaboard states. This attractive mouse is found throughout Ohio, and readily adapts to many diverse ecological environments. It is most commonly found in deciduous woods, but throughout its range it also lives in abandoned and cultivated fields, roadside ditches, barns and warehouses, the busiest city parks, and even university buildings. At least 25 percent of the "house mice" caught in private homes in Cincinnati and sent to the University of Cincinnati for identification are white-footed mice.

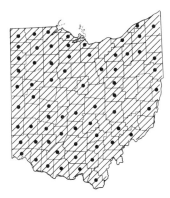

Peromyscus leucopus (Rafinesque)

Typically, adult white-footed mice are rich fulvous, reddish-brown, or cinnamon-colored on the upper parts with a broad, dark-brown band extending down the center of the back. Dorsal coloration, however, is subject to much variation. The belly and feet are immaculately white. The upper surface of the tail is brown and the undersurface white or more often grayish-white, so that the tail is not as distinctly bicolored as it is in *P. m. bairdii*. However, this is a slight difference at best, and I have not been able to separate consistently the two species using this characteristic. Juvenile mice are dull mouse-gray above and grayish-white below. About 95 percent of the animals start to replace the juvenile coat with the adult pelage when they are between 40 and 50 days of age. In approximately 3½ weeks, the molt is completed (Gottschang, 1956). Adults in Ohio molt in November and December in what appears to be a single annual molt. The large eyes and well-developed vibrissae on the muzzle are important sensory organs for this strictly nocturnal animal. Skull characteristics and dental formula are discussed under *P. m. bairdii*. Following are measurements of 40 adult white-footed mice (*Peromyscus leucopus*) from Ohio:

	Total Length (mm)	Tail (mm)	Hind Foot (mm)
Extremes	152–86	66–85	19–22
Mean	166	76	20

The breeding habits of the white-footed mouse are similar to those of the deer mouse. Females are polyestrous with a five- to six-day estrus cycle. They mate within 48 hours after giving birth. The gestation period for a nonlactating female averages 22 days, but for a female nursing a litter the time may be extended to as long as 37 days (Svihla, 1932). White-footed mice in Ohio seldom have more than five in a litter, but one to eight have been reported; four is average. Growth rate of the young is comparable to that for the deer mouse. The young are weaned when 22 to 37 days old, and they leave the nest permanently in 3 weeks or less after weaning. My field data indicate that females in Ohio produce their first litters in late March or early April (the earliest being 22 March) and their last litter at the end of October or during the first week in November. A "rest or recovery period" for breeding females of four to six weeks during midsummer is indicated. Each female produces four to six litters each year. In the laboratory or in favorable environments such as barns and houses, the mice may breed throughout the year. One female that I caught in a house on 5 February 1965 contained five half-grown embryos.

The young are born and raised in globular nests made of finely shredded leaves, grass, plant fibers, fur, feathers, paper, or any soft material that is available. These materials are used alone or in any combination. The nest may be belowground, but more often it is located in or under a hollow log, in a hollow stump or fence post, in a reconstructed bird or squirrel nest, or in an old woodpecker hole. White-footed mice are excellent climbers, and their nests are often located in trees or bushes. Burt (1940) discovered a nest on the battery of his "often used" automobile, and I have found them in empty cans and bottles in woodlots, sometimes containing as many as 15 to 20 mice snuggled together.

The white-footed mouse is one of the most common animals in the wooded areas of Ohio. Ruffer (1961) found 19 resident mice in a 0.29-hectare (0.71-acre) plot in Wood County. Metzger (1955) states that the white-footed mouse is the most common mammal in Perry County. Burt (1940) found the population varying from 7.7 to 27.2 mice per hectare (3.08 to 10.87 per acre) in southern Michigan woodlots. Bole (1939) reported great fluctuations in the white-footed mouse population in the Cleveland area with three years of low population density followed by a year of peak population. Blair (1948) reported noncyclic fluctuations in Michigan populations, and I have made similar observations in southwestern Ohio.

Numerous home range studies have been conducted with the white-footed mouse. The following are of interest here. In Ruffer's study (1961) male home ranges averaged 0.08 hectare (0.19 acre) and females 0.08 hectare (0.20 acre). In a 9.3-hectare (23-acre) woodlot that I studied from 1947 to 1950, adult male home ranges averaged 0.68 hectare (1.44 acres) and those of adult females 0.22 hectare (0.54 acre). Burt (1940) estimated the home range of adult males to be 0.11 hectare (0.27 acre) and adult females 0.09 hectare (0.21 acre).

White-footed mice may remain within a given home range only for several months and then move. I have found a single individual using three different home ranges within a five-month period. Adult males and females establish territories that they defend against other mice during the breeding season. White-footed mice do not make runways of their own, but those living in fields frequently use vole runs. There is some evidence too that they use certain logs and branches more frequently than others for moving about their home territories in the woods, but further investigation is needed to supply more complete information on this subject.

White-footed mice eat large quantities of seeds, nuts, berries, fruits, insects, and insect larvae. Snails, spiders, centipedes, and carrion also appear in their diets, and, like most other kinds of mice, they occasionally become cannibalistic. In the fall they store quantities of nuts and seeds that are used throughout the winter. Hamilton (1941) and Whitaker (1963) have made detailed studies and reviewed the literature on the food of white-footed mice.

At times white-footed mice can be troublesome for man. In certain areas where seeds are planted for reforestation purposes, these mice may dig up and eat the seeds almost as fast as they are planted. White-footed mice often enter summer cottages left unattended for the winter, eating stored foods and using mattress stuffing for building nests in drawers and cupboards. Families of white-footed mice that have taken up residence in a barn may eat the stored grain; if they move into a house, they will raid the pantry. These same mice, however, eat or destroy myriads of undesirable insects and seeds, and they provide the chief food supply for many predatory birds, mammals, and reptiles. Since the species readily enters traps, a harmful population can usually be controlled by setting simple snap traps baited with oats and peanut butter.

SELECTED REFERENCES

Blair, W. F. 1948. Population density, life span and mortality rates of small mammals in the blue-grass meadow and blue-grass field associations of southern Michigan. Amer. Midl. Nat. 49:395–419.

Bole, B. P., Jr. 1939. The quadrat method of studying small mammal populations. Sci. Publ. Cleveland Mus. Nat. Hist. 5:15–77.

Burt, W. H. 1940. Territorial behavior and populations of some small mammals in southern Michigan. Misc. Publ., Univ. of Mich. Mus. Zool. 45:1–58.

DeCapita, M. E., and T. A. Bookhout. 1975. Small mammal populations, vegetational cover, and hunting use of an Ohio strip-mined area. Ohio J. Sci. 75:305–13.

Eberhard, K. J. 1950. An ecological study of mammals with special reference to the woodland white-footed mouse. M.S. Thesis, Ohio Univ., Athens. 53 pp.

Fall, M. W., W. B. Jackson, and M. L. Carpenter. 1968. The occurrence and origin of small mammals on the islands and peninsulas of western Lake Erie. Ohio J. Sci. 68:109–16.

Getz, L. L. 1961. Notes on the local distribution of *Peromyscus leucopus* and *Zapus hudsonius*. Amer. Midl. Nat. 65:486–500.

Gottschang, J. L. 1950. A life history study of the deer mouse, *Peromyscus leucopus noveboracensis* (Fischer), in the Ithaca, New York region. Ph. D. Dissertation, Cornell Univ., Ithaca, N.Y. 70 pp.

―――. 1956. Juvenile molt in *Peromyscus leucopus noveboracensis*. J. Mammal. 37:516–20.

Hamilton, W. J., Jr. 1941. The food of small forest mammals in eastern United States. J. Mammal. 22:250–63.

King, J. A., ed. 1968. Biology of *Peromyscus* (Rodentia). Spec. Publ. No. 2, Amer. Soc. Mammal. 593 pp.

Metzger, B. 1955. Notes on mammals of Perry County, Ohio. J. Mammal. 36:101–5.

Nicholson, A. J. 1941. The homes and social habits of the wood-mouse (*Peromyscus leucopus noveboracensis*) in southern Michigan. Amer. Midl. Nat. 25: 196–223.

Pagniano, R. P. 1958. An ecological study of the helminths and arthropod parasites of the woodland white-footed mouse, *Peromyscus leucopus* of South Bass Island. M.S. Thesis, Ohio State Univ., Columbus.

Ruffer, D. G. 1961. Effect of flooding on a population of mice. J. Mammal. 42:494–502.

Stickel, L. F. 1968. Home range and travels. Pp. 373–411 in John A. King, ed. Biology of *Peromyscus* (Rodentia). Spec. Publ. No. 2. Amer. Soc. Mammal.

Svihla, A. 1932. A comparative life history study of the mice of the genus *Peromyscus*. Misc. Publ., Univ. of Mich. Mus. Zool. 24:1–39.

Whitaker, J. O., Jr. 1963. Food of 120 *Peromyscus leucopus* from Ithaca, New York. J. Mammal. 44:418–19.

Eastern Woodrat or
Allegheny Woodrat or Pack Rat
Neotoma floridana (Ord)

The eastern woodrat has a limited and restricted distribution in Ohio. Few recent records exist, and all these are from Adams County. In part this is probably a reflection of the animal's secretive and nocturnal habits in rugged terrain, but also this species is simply not abundant in the state. The Ohio Division of Wildlife (1976) lists it as an endangered species in the state. Suitable habitat consists of limestone and sandstone cliffs and rocky outcrops, which do occur in portions of unglaciated Ohio. Schwartz and Odum (1957) state: "It seems likely that *Neotoma* invaded the eastern United States from the southwest, . . . with one segment moving north along the Appalachian Plateau." The Ohio River could have been a factor in preventing the more rapid movement of this rat into our state. Two early unsubstantiated records, one from Ashland County on the Mohican bluffs (Kirtland, 1838, p. 177) and the other from Tuscarawas County (Brayton, 1882, p. 136), indicate that the range may have originally extended northward in the unglaciated plateau, but no known specimens have been collected from these areas since the time of these reports. Hine (1929) reported collecting the species in the Hocking Hills in Hocking and Fairfield counties and in Adams County. Neotoma Valley in Hocking County, in which considerable ecological research has been conducted, derived its name from Hine's collection (Wolfe et al., 1949, p. 37). Chapman (1938) observed the species in Scioto County. Bole and Moulthrop (1942, p. 145) describe the range in Ohio as being "from Adams County northeastward to Hocking, Fairfield, and Washington counties."

Neotoma floridana (Ord)

Though superficially the eastern woodrat resembles the more familiar and unpopular Norway rat in body shape and form, it in fact has little in common with Old World rats and mice. The woodrat has a coat of long, soft, smooth fur (fig. 19). The back and sides are brown to gray in color, and the belly and feet are immaculately white. The tail, which is fat and somewhat shorter than the body, is covered with short, bristle-like hairs, white below and gray or brown on the upper surface. There is a definite pencil of hairs on the end of the tail. Five plantar tubercles are present on each front foot and six on each hind foot. The ears are large, rounded, thin, and covered over the entire outer surface with inconspicuous hair. The long whiskers surrounding the nose and the large, round, black, bright eyes are adaptations that enable the animal to move around freely and rapidly in the semi- or complete darkness of the caves and crevices where it is most often found.

The dental formula is the same as for all cricetines, but the cheek teeth are flattened and have the enamel in patterns of loops and whorls as in the microtine rodents. The skull is similar to *Peromyscus* except that it is longer and more angular, and the bones of the cranium are thicker (not transparent). The infraorbital foramen is triangular, large, and conspicuous. Following are measurements of 7 adult eastern woodrats (*Neotoma floridana*) from Ohio:

Fig. 19. Eastern woodrat *(Neotoma floridana).* (Photograph by Karl H. Maslowski.)

	Total Length (mm)	Tail (mm)	Hind Foot (mm)
Extremes	355–419	118–203	39–44
Mean	395	167	42

It appears that the woodrat in Ohio produces a single litter each year between the months of February and May. However, other subspecies of *N. floridana* produce several litters per year. Litters of one to four are reported, with two or three being the usual number. The young at first have a sparse covering of fine silver-gray hair that is replaced within three weeks by the addition of a full coat of light gray fur. Eyes and ears are closed at birth, the pinnae unfold by the ninth day, and eyes are fully opened by about the twentieth day (Poole, 1936). Weaning occurs on or about the twenty-first day.

Young woodrats are known for the tenacity with which they cling to the teats of the female. If the mother is forced from the nest, the young, even when only three days old, stay attached to the nipples and are banged and jostled about as she awkwardly runs for safety. According to Svihla and Svihla (1933), the tips of the incisors in the milk teeth are spread apart, forming a diamond shape between them, so that the harder the jaws are pushed together, the firmer is the hold on the teat. Young woodrats seldom release their hold on the nipple during the pre-weaning period.

Although woodrats are gregarious in the sense that a number of them may occupy the same cave or have their nests close together in a thicket, they are not truly social animals. Frequently when individuals are placed together in a cage, regardless of sex, they immediately start to fight. No matter how long they remain together, fighting and bickering continue to be a regular part of their relationship. Each of the several individuals inhabiting a single cave has its own territory and probably has little to do with the affairs of its neighbors. Even adult males and females are found together only during the breeding season. At this time there is a well-developed skin gland, 100 mm or more in length, on the belly of both sexes. This gland secretes a greasy, strong, and characteristic-smelling substance that may serve to attract the sexes. The gland is less developed and less conspicuous during the rest of the year.

In the southern and western United States, woodrats characteristically live in large nests occupying a space in the center of a large cactus patch or perhaps a meter or two up in a tree or in a tangle of bushes. In Texas these great heaps of sticks are loosely constructed 1.0 to 1.5 m (4 to 5 ft) high and up to 3 m (8 or more ft) in diameter. Usually there are several well-worn pathways (runs) leading to round openings in the pile of sticks. These openings lead to the several rooms in the nest that serve as sleeping quarters, storage chambers, fecal depository areas, and feeding sites. All the rooms are globular in shape and very clean. In contrast, Poole (1940a) describes the eastern woodrat nest as "shaped much like that of birds of the size of a blue jay or a grackle. It is bulky, and well made of the shred bast fibers of chestnut, hemlock, red cedar, basswood, wild grape, or some similar material." Various grasses, moss, feathers, sticks, and leaves are other items used to construct these open nests. Newcombe (1930), working in West Virginia, found the nests quite uniform in size and composition and having the general shape of a robin's nest with a diameter of about 250 mm and a depth of 75 mm. Nests are most often found on ledges or among boulders in the deep recesses of caves and crevices. It is not unusual to find a number of them within a relatively small area. Kinsey (1971, written communication) found captive eastern woodrats constructing houses of sticks and straw similar to western species. He believes these larger nests may occur in nature but deeper in caves and crevices than where the smaller nests are found.

Woodrats are better known in the West and Southwest as pack rats because of their familiar habit of collecting a variety of loose items and carrying them back to the nest. Poole's list (1940a) of items found in nests includes nails, coins, bits of tin, china, scraps of leather, bones, a woodchuck skull, watch, spectacles, crow, turkey, and chicken feathers, rubber from an old tire, paper, snail shells, shotgun shells, shells of duck eggs, rubber bands, and fox and horse dung. I can add to this list corncobs, eating utensils, flashlight batteries, museum labels, and elbow macaroni, or almost anything not under lock and key.

The eastern woodrat is primarily a vegetarian. Nuts and fungi are stored in quantity. Woodrats are suspected of eating birds, bird eggs, and carrion, but there is little direct evidence to support this. Neither have I been able to find any evidence that these animals eat insects, a rather surprising conclusion in that insects occur in the diet of most other wild rodents. In captivity the animals frequently chew on bones and pices of hardwood, possibly indicating their need for certain minerals not commonly found in a vegetable diet. There are conflicting reports concerning their utilization of drinking water; some captive animals drink often, but others never have been observed to drink.

Poole (1940a) lists skunks, snakes, hawks, owls, wildcats, and man as the primary enemies of the woodrat. I would have to add the fox to this list in Ohio. Since woodrats are so uncommon, and live in such out-of-the-way places, they create no economic problems to man but, rather, become a very interesting and unusual member of Ohio's fauna.

SELECTED REFERENCES

Bole, B. P., Jr., and P. N. Moulthrop. 1942. The Ohio recent mammal collection in the Cleveland Museum of Natural History. Sci. Publ. Cleveland Mus. Nat. Hist. 5:83–181.

Brayton, A. W. 1882. Report on the mammals of Ohio. Rep. Geol. Surv. Ohio 4:1–185.

Chapman, A. O. 1951. The estrous cycle in the woodrat, Neotoma floridana. Univ. of Kans. Sci. Bull. 34, Pt. 1(7):267–99.

Chapman, F. B. 1938. The development and utilization of the wildlife resources of unglaciated Ohio. 2 vols. Ph. D. Dissertation, Ohio State Univ., Columbus. 791 pp., esp. p. 561.

Feldman, H. W. 1935. Notes on two species of wood rats in captivity. J. Mammal. 16:300–303.

Goldman, E. A. 1910. Revision of the wood rats of the genus Neotoma. North Amer. Fauna 31:82–86.

Hamilton, W. J., Jr. 1930. Notes on the mammals of Breathitt County, Kentucky. J. Mammal. 11:306–11.

———. 1953. Reproduction and young of the Florida

wood rat, *Neotoma f. floridana* (Ord). J. Mammal. 34:180–89.

Hickie, P. F., and T. Harrison. 1930. The Allegheny wood rat in Indiana. Amer. Midl. Nat. 12:169–74.

Hine, J. S. 1929. Distribution of Ohio mammals. Annu. Rep., Proc. Ohio Acad. Sci. 8(6):260–68.

Howell, A. B. 1926. Anatomy of the wood rat; comparative anatomy of the subgenera of the American wood rat (genus *Neotoma*). Williams and Wilkins Co., Baltimore, Md. 225 pp.

Howell, A. H. 1910. Notes on the mammals of the middle Mississippi Valley, with description of a new wood rat. Proc. Biol. Soc. Washington 23:23–34.

Kinsey, K. P. 1972. Social organization in confined populations of the Allegheny wood rat, *Neotoma floridana magister*. Ph. D. Dissertation, Bowling Green State Univ., Bowling Green, Ohio. 149 pp.

Kirtland, J. P. 1838. Report on the zoology of Ohio. Second Annu. Rep., Geol. Surv. Ohio, 2:157–200.

Nawrot, J. R., and Klimstra, W. D. 1976. Present and past distribution of the endangered southern Illinois woodrat (*Neotoma floridana illinoensis*). Chicago Acad. Sci. Nat. Hist. Miscellanea 96:1–12.

Newcombe, C. L. 1930. An ecological study of the Allegheny cliff rat *(Neotoma pennsylvanica* Stone). J. Mammal. 11:204–11.

Ohio Division of Wildlife. 1976. Endangered wild animals in Ohio. Ohio Dept. Nat. Resources Publ. 316(R576). 3 pp.

Parks, H. E. 1922. The genus *Neotoma* in the Santa Cruz mountains. J. Mammal. 3:241–53.

Patterson, R. C. 1933. Notes on *Neotoma pennsylvanica*, with special reference to the genital organization. Proc. W. Va. Acad. Sci., Ser. 33: 38–42.

———. 1934. Habits of *Neotoma pennsylvanica*. Proc. W. V. Acad. Sci., Ser. 34:32–35.

Pearson, P. G. 1952. Observations concerning the life history and ecology of the wood rat, *Neotoma floridana floridana* (Ord). J. Mammal. 33:459–63.

Poole, E. L. 1936. Notes on the young of the Allegheny wood rat. J. Mammal. 17:22–26.

———. 1940a. A life history sketch of the Allegheny woodrat. J. Mammal. 21:249–70.

———. 1940b. The technical name of the Allegheny woodrat. J. Mammal. 21:316–18.

Rainey, D. G. 1956. Eastern wood rat, *Neotoma floridana*: life history and ecology. Univ. of Kans. Publ., Mus. Nat. Hist. 8(10):535–646.

Rhoads, S. N. 1894. A contribution to the life history of the Allegheny cave rat, *Neotoma magister* Baird. Proc. Acad. Nat. Sci. Philadelphia. Pp. 213–21.

Schwartz, A., and E. P. Odum. 1957. The woodrats of the eastern United States. J. Mammal. 38:197–206.

Svihla, A., and R. D. Svihla. 1933. Notes on the life history of the wood rat, *Neotoma floridana rubida* Bangs. J. Mammal. 14:73–75.

Warren, E. R. 1926. Notes on the breeding of wood rats of the genus *Neotoma*. J. Mammal. 7:97–101.

Wetmore, A. 1923. The wood rat in Maryland. J. Mammal. 4:187–88.

Wolfe, J. N., R. T. Wareham, and H. T. Scofield. 1949. Microclimates and macroclimates of Neotoma, a small valley in central Ohio. Ohio Biol. Surv. Bull. 41. 267 pp.

Southern Red-backed Vole
Clethrionomys gapperi (Vigors)

The southern red-backed vole is a boreal mammal that ranges southward into the United States along mountain chains. In the east it has advanced as far south as northern Georgia along the Appalachian Mountains, but generally it is not found as far south as Ohio. There are some questions as to the presence of this species in Ohio. The first report was by Enders (1928) of a specimen taken at Farnham, Ashtabula County, in 1927. Bole and Moulthrop (1942) report five specimens from Padanaram, Ashtabula County, but later state: "This subspecies formerly occupied the great swampy hemlock forest of Pymatuning in Ashtabula County, Ohio, . . . [but] devastating changes have almost certainly eliminated the species on the Ohio side of the reservoir, since the actual site of the capture of the specimens listed below is now under water." The only "recent" Ohio specimen is in the Ohio State University Museum of Zoology, collected 10 October 1960 in Springfield Township, Jefferson County. Smith (1940) reported red-backed voles (at that time called *Evotomys*) from Knox and Licking counties in central Ohio; but in the absence of extant specimens, I strongly suspect these represent misidentified woodland voles (*Microtus pinetorum*).

Clethrionomys gapperi (Vigors)

This vole might be confused with a small *Microtus* except for the broad brick-red or mahogany mid-dorsal stripe that covers most of the back. The red hair gradually disappears along the sides; the belly, feet, and sides of the head and nose are gray. The belly may be frosted or silver. The tail is decidedly longer than the hind foot, but proportionally shorter than it is in *Microtus*. The skull is similar to *Microtus,* but the palatine border posteriorly ends abruptly and is straight. The dental formula is the same as for the other cricetines. Burt (1957, p. 225) gives the following measurements for this mouse: total length, 116–58 mm; tail length, 31–50 mm; hind foot, 17–21 mm.

Indications are that southern red-backed voles in our latitude breed from April to October and that a single female may produce two or three litters each year. Litter sizes of one to seven have been recorded by Patric (1962) and others, with four to five being most common. The gestation period is 17 to 19 days and the young are weaned when about 20 days old (Svihla, 1930). Apparently, it is not unusual for the female to mate immediately following parturition. The social habits, population dynamics, and other life history information have not been extensively studied for this species.

Coniferous woods or hardwood forests with plenty of damp sphagnum moss or wet swampy areas are the preferred habitats of this vole. It certainly must be considered a very rare species and may in fact be extirpated in Ohio.

SELECTED REFERENCES

Aldrich, J. W. 1943. Biological survey of the bogs and swamps in northeastern Ohio. Amer. Midl. Nat. 30:346–402.

Blair, W. F. 1941. Some data on the home ranges and general life history of the short-tailed shrew, red-backed vole, and woodland jumping mouse in northern Michigan. Amer. Midl. Nat. 25:681–85.

Bole, B. P., Jr., and P. N. Moulthrop. 1942. The Ohio recent mammal collection in the Cleveland Museum of Natural History. Sci. Publ. Cleveland Mus. Nat. Hist. 5:83–181, esp. 151–53.

Burt, W. H. 1957. Mammals of the Great Lakes region. Univ. of Mich. Press, Ann Arbor. 246 pp.

Enders, R. K. 1928. Two new records for Ohio. J. Mammal. 8:155.

Hine, J. S. 1929. Distribution of Ohio mammals. Annu. Rep., Proc. Ohio Acad. Sci. 8(6):267–68.

Muul, I., and F. W. Carlson. 1963. Red-back vole in trees. J. Mammal. 44:415–16.

Patric, E. F. 1962. Reproductive characteristics of red-backed mouse during years of differing population densities. J. Mammal. 43:200–205.

Smith, C. E. 1940. A study of small mammals of central Ohio. M. S. Thesis, Ohio State Univ., Columbus. 66 pp.

Svihla, A. 1930. Breeding habits and young of the red-backed mouse, *Evotomys*. Mich. Acad. Sci., Arts & Letters 11:485–90.

Meadow Vole or Field Mouse
Microtus pennsylvanicus (Ord)

This ubiquitous vole occurs in tremendous numbers over a vast territory including all of Canada, most of Alaska, and the northeastern part of the United States. Within its range 23 different subspecies are recognized, but only the 1 common to the entire eastern United States, *Microtus pennsylvanicus pennsylvanicus,* occurs in Ohio. The meadow vole is present in every county in Ohio.

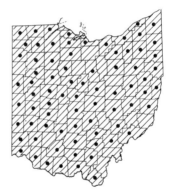

Microtus pennsylvanicus (Ord)

Meadow voles are predominantly brown-colored animals with short, smooth fur. The belly is uniform gray or silver. The tail is dark above and lighter below, but not distinctly bicolored, and always much longer than the hind foot. The head is short and rounded; the eyes are small and beady. The ears, although well developed, are mostly hidden by fur. Until they are about four weeks old, juvenile meadow voles are uniform gray. The skull is recognized as microtine by its small size and general shape but can only be identified to species by studying the enamel pattern of the molar teeth (fig. 18). The dental formula is as follows:

$$i\frac{1-1}{1-1}, c\frac{0-0}{0-0}, pm\frac{0-0}{0-0}, m\frac{3-3}{3-3} = 16$$

Following are measurements of 50 adult meadow voles (*Microtus pennsylvanicus*) from Ohio:

	Total Length (mm)	Tail (mm)	Hind Foot (mm)
Extremes	129–74	28–45	16–22
Mean	147	36	19

One large male in the University of Cincinnati collection, not included in the above measurements, has the following measurements: total length, 190 mm; tail, 48 mm; hind foot, 20 mm; weight, 60 g. Certainly this is a near record size for the species.

Meadow voles are the most common and abundant mammals inhabiting the open grassy fields in Ohio. In years when they are at the peak of abundance, 125 or more may inhabit a single hectare (50 per acre) of suitable habitat (Bole, 1939). Hamilton (1941) says that 500 or more may live in a single hectare (200 per acre). Blair (1948) found peak population densities of 74.5 mice per hectare (29.8 per acre) in Michigan. In various old fields that I have trapped in Ohio, I have estimated populations of 5, 20, 27.5, 41.3, and 82.5 voles per hectare (2, 8, 11, 16.5, and 33 per acre). Meadow voles have cyclic population levels and may be extremely abundant one year and scarce the next. Hamilton (1937a) has made the most complete studies of meadow voles in this country, and he has estimated a four- to five-year cycle for the species in New York State. Bole (1939) postulated a three-year cycle for Ohio meadow voles. My data for southwestern Ohio fit rather well into Bole's three-year cycle, although more trapping data are needed before a definite three-year cycle can be assigned to this vole in Ohio. Local ecological conditions strongly influence small-mammal populations from year to year; the population of a species may be high in one area and low in another.

Meadow voles are primarily inhabitants of grassy fields, but they occasionally live in deciduous forests and are often caught in areas that are overgrown with low shrubs, berry bushes, tall weeds, and scattered patches of short grass. Grassy strips along the edges of streams and ditches and the fringe of tangled grass along fences often have high populations. Usually they are not found in deep woods or in especially rocky situations. Most of the runways found in fields are made by meadow voles; they are made either on the surface, where they appear as miniature highways crisscrossing the fields, or just below the duff, dead leaves, and sticks that cover the ground. A labyrinth of runs is also made some distance belowground and used during inclement winter months, even though these hardy mice are active aboveground both day and night throughout the year. Active runways usually have some dark green or brown droppings deposited in them; at certain points large mounds of droppings may be found just to the side of the runway, indicating a common depository area. Cut pieces of grass are also normally found in an active run. Meadow voles leave their runs, and many have been caught in baited traps some distance from runways. In fields where the ground is soft and many mice are present, the runways may be worn into little troughs that fill with water during rains.

Meadow voles are prolific animals. One captive female gave birth to 17 litters in one year (Bailey, 1924). Hamilton (1937a) records a captured mouse producing 9 litters in eight months. Under ordinary field conditions females probably produce 6 to 8 litters each year. Litter sizes range from 1 to 11, and larger average litter sizes are produced in the spring than in the fall (Beer et al., 1957; Kott and Robinson, 1963). DeCoursey (1957) found the average litter size to be 4.48 (1 to 8) for females in Greene County, Ohio. Sixteen pregnant females chosen at random from my collection had an average litter size of 4.

Females mate immediately after producing a litter or as soon as they can find a mature male. The gestation period is 21 days. Since the young are weaned before the next litter is born, gestation time is never prolonged by the female nursing young, as it is in other species of mice (see *Peromyscus*, p. 88). The young are weaned when they are 2 weeks old, and leave the nest permanently within a week after weaning. Most males become reproductively active when they are 30 to 35 days old; females may reproduce at 25 days, but probably most are at least 4 weeks old before their first successful mating (Hamilton, 1941). The young are blind, hairless, and pink; the ear pinnae are closed; the average weight is 2.07 g. Hair develops by the fourth or fifth day; the eyes are open and the pinnae unfold on about the eighth day. At the end of the first week the young weigh about 8 g; at 14 days, 16 g; and by the age of 1 month, between 30 and 35 g.

Trapping results in Ohio show that some meadow voles in any one population are reproductively active during every month of the year, but that most individuals probably stop breeding in December, January, and February. There are peaks of reproductive activity in the spring and again in the fall. It is apparent that under favorable conditions a population of meadow voles can increase tremendously in a short period of time.

Anyone who has worked with small mammals, however, knows that life expectancy is short. For example, few animals that are caught, marked for identification, and released in the fall of one year are recovered the following spring. Both Getz (1960) and Hamilton (1941) have concluded that the most hazardous period for these mice is the first month of life, when 88 percent of the mortality occurs. Hamilton (1941) estimates the normal life span of the meadow vole at 10 to 16 months, but Blair (1948) found these animals remaining on his study plots an average of just over 4 months.

Meadow voles construct large oval or globular nests of loosely woven grass. In summer these nests are commonly found in depressions on the surface of the ground; in winter they have been unearthed from several inches belowground. Occasionally the baseball-sized chamber in the center of the nest will be lined with feathers or leaves or mouse fur. During the winter several adult mice of either or both sexes may huddle together in one nest for warmth, but during the spring and summer only a female and her young occupy the same nest. Adult field mice have home ranges of about 0.1 to 0.2 hectares (0.25 to slightly more than 0.5 acre). Individual home ranges may broadly overlap, but Getz (1961b) suggests that the meadow vole, especially the female, displays some territorial behavior.

Meadow voles are predominantly vegetarians and live almost exclusively on the green shoots, leaves, stems, and seeds of many kinds of plants. Various fruits, the bark of trees, roots, and tubers are included in their diet. Insects and other animal foods are usually excluded from the diet, although caged animals will occasionally eat their peers. Oat flakes, seeds, and vegetable greens of all kinds are accepted in captivity.

Because of their great numbers, meadow voles can be destructive. In Ohio the greatest damage probably occurs in orchards, where these voles damage trees by eating the roots or the cambium in the bark. Orchard growers and nurserymen have controlled meadow voles by keeping the weeds and grass cut, by removing grass from around each tree, by placing a protective metal shield around the trunk of each tree, by spraying the turf with certain poisons, or by distributing poison baits throughout the orchard. All of these are costly processes that require both time and energy of the orchardist. Eadie (1954, p. 138) has estimated that a modest population of 25 mice per hectare (10 per acre) living on 40 hectares (100 acres) of grassland would eat 10,000 kg (11 tons) of grass in one year, or the equivalent of 5,000 kg (5.5 tons) of cured hay! Grain crops of all kinds are attacked. Formerly, when corn was harvested and left standing in the shock, it often was severely damaged by meadow voles. Truck crops, vegetable gardens, and some field crops may suffer noticeable damage at high population levels of these voles. The exact amount of damage and the resulting monetary loss are difficult to evaluate, but figures run to thousands of dollars each year in Ohio alone. Poison baits are most effective in controlling large populations, but in the home garden or on small plots, mouse traps baited with a mixture of rolled oats and peanut butter and spaced eight to ten steps apart will catch most of the mice in a few days.

Most predators rely heavily on the meadow vole as a source of food; in this respect, at least, voles are of considerable value. Plagues of this animal have occurred regularly during the past several hundred years in various regions of Europe and Asia; several have also been reported in the United States. (For an interesting and authoritative historical account of small mammal plagues, read Charles Elton's book, *Voles, Mice and Lemmings,* 1942).

SELECTED REFERENCES

Bailey, V. O. 1900. Revision of American voles of the genus *Microtus*. North Amer. Fauna 17:1–88.

———. 1924. Breeding, feeding, and other life habits of meadow mice (*Microtus*). J. Agric. Res. 27:523–36.

Barrett, G. W. 1974. Occurrence of an albino *Microtus pennsylvanicus* in Ohio. Ohio J. Sci. 74:102.

Beer, J. R., and C. F. MacLeod. 1961. Seasonal reproduction in the meadow vole. J. Mammal. 42:483–89.

Beer, J. R., C. F. MacLeod, and L. D. Frenzel. 1957. Prenatal survival and loss in some cricetid rodents. J. Mammal. 38:392–402.

Blair, W. F. 1948. Population density, life span, and mortality rates of small mammals in the blue-grass meadow and blue-grass field associations of southern Michigan. Amer. Midl. Nat. 40:395–419.

Bole, B. P., Jr. 1939. The quadrat method of studying small mammal populations. Sci. Publ. Cleveland Mus. Nat. Hist. 5:15–77.

Brown, E. B., III. 1973. Changes in patterns of seasonal growth of *Microtus pennsylvanicus*. Ecology 54:1103–10.

Christian, J. J., and D. E. Davis. 1966. Adrenal glands in female voles (*Microtus pennsylvanicus*) as related to reproduction and population size. J. Mammal. 47:1-18.

Clough, G. C. 1965. Viability of wild meadow vole under various conditions of population density, season, and reproductive activity. Ecology 46: 119-34.

DeCoursey, G. E., Jr. 1957. Identification, ecology, and reproduction of *Microtus* in Ohio. J. Mammal. 38:44-52.

Eadie, W. R. 1954. Animal control in field, farm, and forest. Macmillan Co., New York. 257 pp.

Elton, C. S. 1942. Voles, mice, and lemmings: problems in population dynamics. Clarendon Press, Oxford. 496 pp.

Getz, L. L. 1960. A population study of the vole, *Microtus pennsylvanicus*. Amer. Midl. Nat. 64:392-405.

_____. 1961a. Factors influencing the local distribution of *Microtus* and *Synaptomys* in southern Michigan. Ecology 42:110-19.

_____. 1961b. Home ranges, territoriality, and movement of the meadow vole. J. Mammal. 42:24-36.

Goodpaster, W., and K. Maslowski. 1949. Meadow vole uses same nest for two litters. J. Mammal. 30:73-74.

Hall, E. R., and E. L. Cockrum. 1953. A synopsis of the North American microtine rodents. Univ. of Kans. Publ., Mus. Nat. Hist. 5:373-498.

Hamilton, W. J., Jr. 1937a. The biology of microtine cycles. J. Agric. Res. 54:779-90.

_____. 1937b. Growth and life span of the field mouse. Amer. Nat. 71:500-507.

_____. 1941. Reproduction of the field mouse, *Microtus pennsylvanicus* (Ord). Cornell Univ., Agric. Exp. Sta. Mem. 237:1-23.

_____. 1943. The mammals of eastern United States: an account of recent land mammals occurring east of the Mississippi. Comstock Publ. Co., Ithaca, N. Y. 438 pp.

Kott, E., and W. L. Robinson. 1963. Seasonal variation in litter size of the meadow vole in southern Ontario. J. Mammal. 44:467-70.

Krebs, C. J. 1970. *Microtus* population biology: behavioral changes associated with the population cycle in *M. ochrogaster* and *M. pennsylvanicus*. Ecology 51:34-52.

Krebs, C. J., B. L. Keller, and R. H. Tamarin. 1969. *Microtus* population biology: demographic changes in fluctuating populations of *M. ochrogaster* and *M. pennsylvanicus* in southern Indiana. Ecology 50:587-607.

Murray, K. F. 1965. Population changes during the 1957-1958 vole (*Microtus*) outbreak in California. Ecology 46:163-71.

Myers, J. H., and C. J. Krebs. 1971. Genetic, behavioral, and reproductive attributes of dispersing field voles, *Microtus pennsylvanicus* and *Microtus ochrogaster*. Ecol. Monogr. 41:53-78.

Snyder, D. B. 1963. The effects of endrin on vole (*Microtus pennsylvanicus*) reproduction in bluegrass meadows. Ph. D. Dissertation, Ohio State Univ., Columbus. 118 pp.

West, G. C. 1965. A population study of the structure of a *Microtus pennsylvanicus pennsylvanicus* Ord. population in northeastern Ohio. M. S. Thesis, Kent State Univ., Kent, Ohio.

Prairie Vole
Microtus ochrogaster (Wagner)

The prairie vole, an inhabitant of the northern Great Plains, has a more restricted distribution in Ohio than the meadow vole, possibly because of Ohio's location on the eastern edge of its range. Except for some animals taken in West Virginia, specimens caught in Jefferson County, Ohio, represent the easternmost records for the species in the United States (Henninger, 1921). Since the prairie vole probably moved into Ohio from the West much later than the meadow vole, its distribution is still limited in eastern and northern counties. Henninger (1921) says that the species had not been recorded in Ohio until 1921, when the first specimens were taken. In subsequent years, however, the prairie vole will probably extend its range in Ohio into suitable habitats to the north and east.

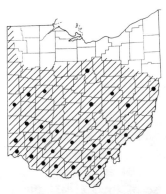

Microtus ochrogaster (Wagner)

Microtus ochrogaster and *Microtus pennsylvanicus* are so similar that they can be positively identified only by careful study of the enamel pattern of the cheek teeth (fig. 18). *Microtus ochrogaster* usually has a salt-and-pepper or grizzled appearance on the back because many dorsal hairs have yellow or straw-colored tips, whereas *M. pennsylvanicus* has uniform-colored, smooth fur. Prairie voles often have a dark mid-dorsal stripe not found in meadow voles. *Microtus ochrogaster* in Ohio normally has an all-gray belly, but there is often buff coloring also; *M. pennsylvanicus* has a clear gray or silver belly. Both species may have five plantar tubercles on the hind feet, but *M. pennsylvanicus* most often has six. Female priaire voles have three pairs of teats on the venter; meadow voles have four pairs. Most authors (among them, DeCoursey, 1957, in Greene County) state that *M. orhrogaster* usually frequents dry upland fields, whereas *M. pennsylvanicus* is more commonly found in more moist or even wet areas. However, I have caught both species in the same fields and runways.

The dental formula is as follows:

$$i\frac{1\text{-}1}{1\text{-}1},\ c\frac{0\text{-}0}{0\text{-}0},\ pm\frac{0\text{-}0}{0\text{-}0},\ m\frac{3\text{-}3}{3\text{-}3} = 16$$

Following are measurements of 26 adult prairie voles (*Microtus ochrogaster*) from Ohio:

	Total Length (mm)	Tail (mm)	Hind Foot (mm)
Extremes	128–53	21–36	16–19
Mean	138	28	17

Prairie voles have smaller litter sizes on the average than do meadow voles (DeCoursey, 1957; Jameson, 1947). Otherwise, their reproductive habits, growth rate, and life histories are similar.

Bole and Moulthrop (1942), working with subspecies of *M. ochrogaster,* found populations of the western subspecies, *M. o. ochrogaster,* limited to Hamilton and Clermont counties in extreme southwestern Ohio. They also described a new subspecies in Ohio, *M. o. ohionensis.* Subsequent workers (Hall and Cockrum, 1953; Hall and Kelson, 1959; and others) have accepted Bole and Moulthrop's treatment of the species. On the basis of more recent data, however, including many more locality records and specimens for comparison, I have reason to doubt the validity of the distinct subspecies, *ohionensis.* My findings suggest that all the prairie voles in Ohio belong to the subspecies, *M. o. ochrogaster,* though the more eastern specimens tend to be darker, less buff-bellied, and less grizzled on the back than those in the central and western parts of their range.

Prairie voles are serious pests in orchards (especially apple orchards) of southern Ohio, where they may actually do damage that is blamed on the meadow vole.

SELECTED REFERENCES

Bole, B. P., Jr., and P. N. Moulthrop. 1942. The Ohio recent mammal collection in the Cleveland Museum of Natural History. Sci. Publ. Cleveland Mus. Nat. Hist. 5:83–181, esp. 155–61.

DeCoursey, G. E., Jr. 1957. Identification, ecology, and reproduction of *Microtus* in Ohio. J. Mammal. 38:44–52.

Fitch, H. S. 1957. Aspects of reproduction and development in the prairie vole *(Microtus ochrogaster).* Univ. of Kans. Publ., Mus. Nat. Hist. 10(4):129–61.

Hall, E. R., and E. L. Cockrum. 1953. A synopsis of the North American microtine rodents. Univ. of Kans. Publ., Mus. Nat. Hist. 5(27):373–498.

Hall, E. R., and K. R. Kelson. 1959. The mammals of North America. Ronald Press Co., New York. 2 vols., 1162 pp.

Henninger, W. F. 1921. Two mammals new for Ohio. J. Mammal. 2:239.

Jameson, E. W., Jr. 1947. Natural history of the prairie vole. Univ. of Kans. Publ., Mus. Nat. Hist. 1:125–51.

Krebs, C. J. 1970. *Microtus* population biology: behavioral changes associated with the population cycle in *M. ochrogaster* and *M. pennsylvanicus.* Ecology 51:34–52.

Krebs, C. J., B. L. Keller, and R. H. Tamarin. 1969. *Microtus* population biology: demographic changes in fluctuating populations of *M. Ochrogaster* and *M. pennsylvanicus* in southern Indiana. Ecology 50:587–607.

Martin, E. P. 1956. A population study of the prairie vole *(Microtus ochrogaster)* in northeastern Kansas. Univ. of Kans. Publ., Mus. Nat. Hist. 8(6):361–416.

Myers, J. H., and C. J. Krebs. 1971. Genetic, behavioral, and reproductive attributes of dispersing field voles, *Microtus pennsylvanicus* and *Microtus ochrogaster.* Ecol. Monogr. 41:53–78.

Woodland Vole or Pine Vole or Pine Mouse
Microtus pinetorum (LeConte)

The old name "pine vole" is something of a misnomer, and the name "woodland vole" is more accurate, for the species is certainly not found exclusively in pine woods. In fact, I have never been successful in catching one in a stand of pine. Although seldom seen and not well known, this short-tailed, auburn-colored mouse (formerly *Pitymys pinetorum* LeConte) is probably present in many areas of the state, even though it has been reported from only 22 counties. Bailey (1900) recorded 1 specimen of the woodland vole from Ohio in 1900. Bole

(1938) commented that approximately 150 additional specimens had been taken since 1900, and I was able to examine 225 to 250 museum skins for this study. The woodland vole is found in all the states bordering Ohio and throughout the entire eastern and northeastern sections of the United States. It has been found in a wide range of habitats, from the preferred deciduous woods to grassy fields, old fields, rocky, brush-covered hillsides, orchards, berry thickets, and buildings. The largest concentration that I have encountered occupied a 4.6-meter (15-foot) strip of grassy terrain along a small stream that passed through an abandoned farm. I am at a loss to explain the complete absence of records from central and northwestern Ohio since seemingly prime habitat is readily available there. Stotler (1976, personal communication), however, indicates that extensive trapping in northwestern Ohio by researchers from Bowling Green State University and Defiance College has thus far failed to locate this species. Further research is required to explain its apparent absence from sizable portions of the state.

Microtus pinetorum (LeConte)

The woodland vole is easily recognized among the microtine rodents by its very short tail, which is only slightly longer than the hind foot, and by its short, dense, rich chestnut, brown, or dark red mole-like fur. The feet are gray. The belly may be gray, silver-gray, or even buff, depending upon the season and the age of the mouse. The bright, dark eyes are proportionally smaller than in other microtines. The ears are small, rounded, and completely hidden by fur. The head is round with the nose short and blunt. The four clawed toes on the front feet and the five on the hind are adapted to digging. Woodland voles are nervous, quick-moving, very alert animals. They have a special ability for escaping from cages; the slightest crack or loose piece of screening will soon be found and used as an escape route.

The skull of the woodland vole is similar to, but shorter than, that of the other microtines. In addition both the second and the third upper molar teeth have four irregular closed loops.

The dental formula is as follows:

$$i\frac{1-1}{1-1}, c\frac{0-0}{0-0}, pm\frac{0-0}{0-0}, m\frac{3-3}{3-3} = 16$$

Following are measurements of 37 adult woodland voles (*Microtus pinetorum*) from Ohio:

	Total Length (mm)	Tail (mm)	Hind Foot (mm)
Extremes	109–34	16–25	15–18
Mean	119	20	16

The woodland vole has been a difficult mouse to study because its populations are cyclic and are concentrated in certain restricted areas. In 1958–59 it was rather common at two localities near Cincinnati, but in 1960 the voles had all but disappeared. During the winter of 1962–63, I caught 6 woodland voles in 18,000 trap nights. Hamilton (1938) found 500 to 700 voles per hectare (200 to 300 per acre) in a New York orchard. Such populations, however, are probably rare and occur only in orchards or other small pockets where environmental conditions are ideal. Benton (1955) found that populations of these voles were very local, highly variable, and usually not evenly distributed throughout the habitat.

Woodland voles usually produce small litters of one to eight, with two to four per litter being average. The only litter that I have been successful in observing in the laboratory contained three young. It is possible that some individuals breed throughout the year, but my limited field data from Ohio agree with the findings of Benton (1955) that the woodland vole breeds from January through October. The earliest date on which I have caught a pregnant female is 25 February, and the latest date, 17 October. Males with enlarged testes have been taken in early January and in September in Ohio. The gestation period is estimated to be 21 to 22 days. Each female probably produces four to six litters each season, but this too is conjecture. Raynor (1960) discovered a pregnant female woodland vole

BEAVER
Castor canadensis

EASTERN HARVEST MOUSE
Reithrodontomys humulis

SOUTHERN FLYING SQUIRREL
Glaucomys volans

WHITE-FOOTED MOUSE
Peromyscus leucopus

WOODLAND VOLE
Microtus pinetorum

MEADOW VOLE
Microtus pennsylvanicus

attending a nest containing ten young that were obviously of three different age groups. He concluded that they represented three different litters. This is especially interesting since Hamilton (1938) notes that Bachman reported seeing nine young woodland voles taken from a nest, "but it seems highly probable that this was a litter of [another species of] Microtus." Perhaps Bachman was correct in his observation and had found a nest being used by more than one female.

Only Hamilton (1938) and Benton (1955) have reported on the appearance and development of the young woodland vole. The newborn closely resemble newborn meadow voles. Hair appears on day 5 or 6; the ear pinnae unfold on day 8; and the eyes open between days 9 and 12. The young are weaned early in the third week. Hamilton (1938) followed the weight increment for the first 24 days after birth.

Woodland voles are good burrowers and construct a maze of runways similar to those of meadow voles, but sometimes deeper. In wooded areas the runs are usually found just beneath the top layer of leaves, in the soft, easily worked leaf mold. They are also commonly used by other kinds of mice and by shrews. Woodland vole runs are not easily distinguished from meadow vole runs, except that woodland voles' runways often intermittently open on the surface for a short distance before disappearing belowground again. Woodland vole burrows in orchards may be located several feet or more below the surface, following the roots of trees that are being used as food (Benton, 1955). Woodland voles either completely avoid areas where the soil is packed, coarse, or lacking in sufficient humus, or they use runways made by other fossorial mammals such as moles. These animals are not completely subterranean and are occasionally caught some distance from any obvious runways. They have been seen briefly aboveground hunting nervously about the forest floor. Individuals apparently remain in the same area for a considerable time; their home range is estimated to include 0.1 hectare (0.25 acre) or less (Burt, 1940; Benton, 1955).

The woodland vole often builds its nests and makes its food caches beneath decaying logs and stumps in woodlots. The 20 cm-long oblong or globular nests that I have found have been made of oak, maple, tulip, and hickory leaves that have been cut into small pieces. Several runways may lead to the nest from different directions. Similar nests in fields and grasslands are constructed of dry grass, rootlets, and other available plant materials.

Benton (1955) says that woodland voles store food only in areas where food is not always abundant. He found very little food storage in the orchards that he studied. I once observed woodland voles transporting and storing small tubers of the Jerusalem artichoke (*Helianthus tuberosus*) beneath a fallen barn door. Numerous runways led to two separate caches beneath the door; each cache contained a large cupful of fresh tubers. I have seen stores in woodlots that contained several handfuls of beech nuts, hickory nuts, and acorns mixed together in a depression beneath an old stump or log. Roots of young green bull thistles (*Cirsium vulgare*) were obviously being eaten by these voles in a field where I trapped in the spring of 1959. Grasses of all kinds, seeds, berries, fruits, roots, and the bark of trees and shrubs were also eaten. Although principally herbivorous, woodland voles make use of many different kinds of food. This probably accounts in part for their wide range of habitats. Some insects and other small invertebrates are regularly eaten, as are trapped or fallen peers. Woodland voles kept in our laboratory remained healthy only when a small amount of raw liver and vitamins were included in their diet.

Damage done by woodland voles is comparable to that done by meadow voles except that the former is primarily a problem for the orchard grower and the nurseryman. Woodland voles eat young tree roots and attack underground portions of the trunks, sometimes completely stripping off the bark for 30 cm or more. Because of their subterranean habits, the orchardist may not suspect that these voles are in his orchard until the leaves start to yellow and the tree fails to grow. The nurseryman may discover that his young trees and shrubs are being eaten from below and that tubers and bulbs are being destroyed. Moles are often considered the culprits because their runways are so evident. Moles, however, feed on worms and insect larvae, and the woodland voles use the large mole runs for easy avenues of access to the nursery stock. Control measures used for meadow voles are also effective in controlling woodland voles.

The taxonomy of *M. pinetorum* in Ohio is uncertain. In his revision of the woodland vole, Bailey (1900) included only one subspecies, *scalopsoides*, in Ohio. Bole (1938), however, and Bole and Moulthrop (1942) extended the range of the southern subspecies *auricularis* to include extreme southern Ohio. These same authors were of the opinion that a number of specimens from Belmont County in eastern Ohio belonged to the subspecies *pinetorum.* Handley (1952) described the new subspecies *carbonarius* from Pulaski County, Kentucky. The range of this subspecies was said to include Lawrence County in south central Ohio. Hall and Kelson (1959) consider three subspecies as occurring in Ohio: *M. p. auricularis* in extreme southern and southwestern Ohio; *M. p. carbonarius* in south central Ohio (only Lawrence County at present); and *M. p. scalopsoides* throughout the rest of the state. After studying most of the known specimens from Ohio, I am of the opinion that *M. p. auricularis* does not occur here. Until we have many more specimens from the southern and central parts of Ohio for comparison, however, the taxonomy of the group will remain confused. The species *M. pinetorum* is badly in need of revision, but this will have to wait until (1) the molt pattern and color variations are better understood, and (2) we have a better means of determining the approximate age of individual mice so that only animals the same age group can be used for comparison.

SELECTED REFERENCES

Arata, A. A. 1965. Taxonomic status of the pine vole in Florida. J. Mammal. 46:87-94.

Bailey, V. O. 1900. Revision of American voles of the genus *Microtus.* North Amer. Fauna 17:1-88.

Benton, A. H. 1955. Observations on the life history of the northern pine mouse. J. Mammal. 36:52-62.

Bole, B. P., Jr. 1938. The pine mouse of southern Ohio. J. Mammal. 19:377.

Bole, B. P., Jr., and P. N. Moulthrop. 1942. The Ohio recent mammal collection in the Cleveland Museum of Natural History. Sci. Publ. Cleveland Mus. Nat. Hist. 5:83-181, esp. 161-62.

Burt, W. H. 1940. Territorial behavior and populations of some small animals in southern Michigan. Misc. Publ., Univ. of Mich. Mus. Zool. 45:1-58.

Collins, R. L. 1974. Ecology and appraisal of pine vole damage in Ohio apple orchards. M. S. Thesis, Ohio State Univ., Columbus. 123 pp.

Hall, E. R., and K. R. Kelson. 1959. The mammals of North America. Ronald Press Co., New York. 2 vols., 1162 pp.

Hamilton, W. J., Jr. 1938. Life history notes on the northern pine mouse. J. Mammal. 19:163-70.

Handley, C. O., Jr. 1952. A new pine mouse (*Pitymys pinetorum carbonarius*) from southern Appalachian Mountains. J. Washington Acad. Sci. 42:152-53.

_____. 1953. Abnormal coloration in the pine mouse (*Pitymys pinetorum*). J. Mammal. 34:262-63.

Neill, W. T., and J. M. Boyles. 1955. Notes on the Florida pine mouse, *Pitymys parvulus* Howell. J. Mammal. 36:138-39.

Paul, J. R. 1964. Second record of an albino pine vole. J. Mammal. 45:485.

Raynor, G. S. 1960. Three litters in a pine mouse nest. J. Mammal. 41:275.

Schantz, V. S. 1960. Record of an albino pine vole. J. Mammal. 41:129.

Winters, R. O. 1972. Problems in controlling pine mice in Ohio. Pp. 24-26 *in* James E. Forbes, ed. Proceedings of the New York pine mouse symposium. U. S. Dept. Interior. Bur. Sport Fisheries & Wildlife, Div. Wildlife Serv., Kingston, N.Y.

Muskrat
Ondatra zibethicus (Linnaeus)

This rodent is by far the largest member of the family Cricetidae and is perhaps best identified by the long, scaly, sparsely haired, black tail that is flattened laterally. No other North American mammal has a laterally flattened tail. The muskrat is semiaquatic and well adapted for living in water. The hind feet are much larger than the front feet and are partially webbed. They have stiff bristles that form swimming fringes between and along the lower sides of the toes. The underfur is thick, woolly, and waterproof, and is an effective insulator against the cold. Ohio muskrats have rich golden brown, chocolate, reddish-brown, or very dark brown upper parts. The sides are lighter brown fading to silver or silver-brown on the belly. The eyes and ears are small and mostly concealed in the soft fur. The skull is easily recognized by its flattened outline and large size, the large orange incisors with smooth surfaces, and the broad zygomatic arch. The cheek teeth are high-crowned and flat on top with conspicuous sharp triangles formed by the enamel ridges. The dental formula is as follows:

$$i\frac{1-1}{1-1}, c\frac{0-0}{0-0}, pm\frac{0-0}{0-0}, m\frac{3-3}{3-3} = 16$$

Following are measurements of 39 adult muskrats *(Ondatra zibethicus)* from Ohio:

	Total Length (mm)	Tail (mm)	Hind Foot (mm)
Extremes	473-673	216-80	73-89
Mean	559	251	81

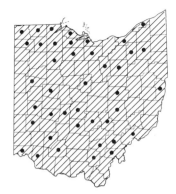

Ondatra zibethicus (Linnaeus)

Muskrats are found throughout the United States except on the Atlantic Coast south of South Carolina and along the Pacific Coast south of Oregon. They have adapted rather well to civilization; wherever water conditions are suitable in Ohio, there are muskrats. Rivers, lakes, streams, farm ponds, and marshes serve as home sites for these rodents. The maintenance staff of a cemetery completely surrounded by residential areas in Cincinnati has a problem controlling the size of the muskrat population in one of its large ponds. Because of their secretive and nocturnal habits, however, muskrats are not commonly seen even when present in sizable numbers. They construct large conspicuous houses in marshes and swamps, but in farm ponds and fish hatcheries and along streams and ditches they are more apt to dig burrows into the banks. The openings to the burrows are below the water level. Favorable habitats often support large populations of muskrats. From 1,134 hectares (2,800 acres) of marshland in the southwestern tip of Sandusky Bay, Ohio, 11 trappers caught 10,191 muskrats in a three-month trapping period (Anderson, 1947). Enders (1932) said that muskrats in Ohio were about equally divided between the "pond-rats" and the "stream-rats." He concluded: "The concentration in marshes is more noticeable, but the large area of stream banks, as compared with marshes, equalizes the numbers."

Characteristic muskrat lodges are similar to beaver lodges except for their smaller size. Also, they are made of whatever vegetable material is available rather than from branches and parts of trees. The house is built in water that is seldom more than 0.5 m deep and rests on a platform of vegetable matter and muck that has been piled there by the muskrats. The floor of the house is from 6 to 10 cm above the water level, and it may be from 1 to 3 m in diameter. The nest inside the house is usually lined with finer material than that used on the outside. Smaller houses, those 1 m or less in diameter and 0.5 m high, contain a single nest, but houses 2 m in diameter and 1 m or more high may enclose several nests. The dome-shaped houses are entered and exited by one or more plunge holes that open directly down into the water. Several muskrats may share a single nest. The house is completely watertight, and even in subfreezing weather the temperature within the nest is several degrees above freezing. In central and southern Ohio, where there are no extensive marshes, muskrats tunnel into dams and stream banks and make their nests in a chamber at the end of the tunnel. The opening to the tunnel is below the water level, and the tunnel slopes upward to the living chamber. There may also be an escape tunnel leading from the chamber onto dry land. Plant materials are transported to the living chamber, and a nest similar to that found in the pond houses is constructed there.

The breeding season of the muskrat in Ohio extends from March through November, although on occasion pregnant females have been caught in December, January, and February. Most females produce two litters annually, each containing about 6 young. Litter sizes of 1 to 11 are known. Errington (1939) and McCann (1944), working in Iowa and Minnesota respectively, found that under ideal conditions female muskrats may rear 19 or more young per year. During the breeding season male and female muskrats pair up, but Svihla and Svihla (1931) saw individual males mate with as many as three different females. Errington (1939) suggests that they are "loosely monogamous"! Muskrats have well-developd musk glands that enlarge, notably in the male, during the breeding season. The paired glands located just below the skin open onto the glans penis. The yellow, musky, sweet-smelling product of the glands is transferred to the feeding platforms, houses, logs, and all other places visited by the muskrat. Presumably the odor attracts members of the opposite sex, or perhaps, as with shrews, it serves as a warning to other muskrats of the same sex that the area is already occupied. The males initiate sexual activity, but only receptive females will allow males to mate with them. Mating takes place in the water. Errington (1939) has summarized his data on the growth of young muskrats as follows:

At birth blind, and almost helpless, weighing 21 grams and measuring 100 mm in total length, the young commonly opened their eyes between the fourteenth and sixteenth days. . . . Weaning took place mostly in the fourth week, and the young (about 180 grams and 285 mm) were living independently by the end of their first month. By this time their mothers, if pregnant, were generally ready to give birth to subsequent litters. . . . The majority of the young of the year taken by trappers in November and December were from five to eight months old and very adult-like except for their imperfect sexual development. Limited data indicate that, by the end of the first year, muskrats reach average weights close to 1100 grams, and lengths close to 550 mm. The larger young males showed signs of approaching sexual maturity by mid-winter, or at eight to nine months.

Enders (1932) studied the summer food habits of the muskrat in Ohio. As expected, the main items eaten were plant materials, but freshwater mussels and muskrat meat were also ingested. Of the 39 species of plants eaten, the following were among the most common: cattail *(Typha),* 5 or 6 species of common grasses, sedges, curly dock *(Rumex),* smart weed *(Polygonum),* pond lily *(Nymphaea),* water lotus *(Nelumbo),* clover *(Trifolium), Aster,* and Jerusalem artichoke *(Helianthus).* Muskrats are fond of corn; they may travel several hundred meters from a stream and eat the stems, leaves, and ears. Oats and wheat are eaten when available. Christian (1964) observed muskrats leaving a pond and feeding on dandelion *(Taraxacum)* roots even though cattails and sedges and several other usually preferred foods were closer. Numerous studies have indicated that muskrats eat the stems, roots, leaves, and flowers of almost any kind of plant. What is not eaten is used for building materials. Freshwater mussels are apparently an important item in the diet, especially in winter when green plants may not be available (Svihla and Svihla, 1931; Ender, 1932). Muskrats dive to the bottom of streams and bring mussels to the surface, often to a regular feeding station. Then, while holding the mollusk in the front feet, they pry open the shells with the incisors. Hundreds and sometimes even thousands of shells may accumulate at the feeding stations. Muskrats killed on the roads by automobiles may be eaten by other muskrats.

Food storage is practiced to some extent, especially by bank-dwelling muskrats. "Duck potatoes *(Sagittaria)* and ears of corn may be packed by the bushel in the chambers and ramifying blind alleys of some bank burrows. Duck potatoes may fill most of the chamber space and extensions thereof in certain marsh lodges." (Errington, 1963, p. 22.)

Although there is much information on home range and territoriality of the muskrat, I find it difficult to synthesize. Some workers have caught animals 8 km (5 mi) from where they were originally marked, but others say that the muskrat remains within 50 to 60 m of its house. The latter statement is probably most often correct, although in the fall of the year, kits (70 to 90 days of age) wander considerable distances to find new and unoccupied home sites. Individual home ranges also increase or decrease with changing population pressures and changes in water conditions. A female nursing a litter or one about to give birth may display territoriality, driving away any other muskrat that approaches too close to her house. On the other hand, three or four muskrats plus a nursing female and her young have been found together in a single nest. In general, it appears that aggressive territorial defense is employed only against muskrats that are new to a neighborhood; regular residents appear to be highly tolerant of one another.

For many years fur dealers in Ohio paid more money for muskrat fur than for any other species. From 1968 to 1978 considerably more than 500,000 muskrat pelts were sold each year, and the average price paid for a single pelt varied from $0.84 to $4.77 (Appendix Table 2). Muskrat still ranks first in total number of furs sold in the state each year, but raccoon, which brings approximately four times the price per pelt, has taken over as the chief money-maker.

The greatest natural enemy of the muskrat is the mink. There are authentic accounts, however, of muskrat and mink occupying adjoining burrows and of adult muskrats attacking and driving off mink. Other predators of the muskrat are owls, predacious fish (pike), snakes, large snapping turtles, Norway rats, and foxes. Even extended droughts, floods, and early freezes take their toll.

SELECTED REFERENCES

Anderson, J. M. 1947. Sex ratio and weights of southwestern Lake Erie muskrats. J. Mammal. 28:391–95.

Arata, A. A. 1959. Ecology of muskrats in strip-mine ponds in southern Illinois. J. Wildl. Manage. 23:177–86.

Beckett, J. V., and V. Gallicchio. 1967. A survey of helminths of the muskrat, *Ondatra z. Zibethica* Miller, 1912, in Portage County, Ohio. J. Parasitol. 53:1169–72.

Bednarik, K. E. 1953. An ecological and food habits study of the muskrat in the Lake Erie marshes. M. S. Thesis, Ohio State Univ., Columbus. 123 p.

Beer, J. R., and R. K. Meyer. 1951. Seasonal changes in the endocrine organs and behavior patterns of the muskrat. J. Mammal. 32:173-91.

Christian, J. J. 1964. Selective feeding on dandelion roots by muskrats. J. Mammal. 45:147.

Donohoe, R. W. 1961. Muskrat reproduction in areas of controlled and uncontrolled water-level units. M. Sc. Thesis, Ohio State Univ., Columbus. 280 pp.

———. 1966. Muskrat reproduction in areas of controlled and uncontrolled water-level units. J. Wildl. Manage. 30:320-26.

Enders, R. K. 1932. Food of the muskrat in summer. Ohio J. Sci. 32:21-30.

Errington, P. L. 1937. The breeding season of the muskrat in northwest Iowa. J. Mammal. 18:333-37.

———. 1939. Observations on young muskrats in Iowa. J. Mammal. 20:465-78.

———. 1963. Muskrat populations. Iowa State Univ. Press, Ames. 665 pp.

Leedy, D. L. 1950. Ohio's status as a game and fur-producing state. Ohio J. Sci. 50:88-94.

McCann, L. J. 1944. Notes on growth, sex, and age ratios, and suggested management of Minnesota muskrats. J. Mammal. 25:59-63.

Ohio Division of Wildlife. 1975. Fur harvest in Ohio, 1964-1974. Ohio Dept. Nat. Resources Publ. 178. 2 pp.

Rausch, R. L. 1946. Parasites of Ohio muskrats. J. Wildl. Manage. 10:70

Rice, E. W., and O. B. Heck. 1975. A survey of the gastrointestinal helminths of the muskrat, *Ondatra zibethicus*, collected from two localities in Ohio. Ohio J. Sci. 75:263-64.

Svihla, A., and R. D. Snihla. 1931. The Louisiana muskrat. J. Mammal. 12:12-28.

Southern Bog Lemming
Synaptomys cooperi Baird

This interesting, short-tailed mouse has a wide range that includes much of southeastern Canada and extends south to South Carolina and west to Kansas and Arkansas. Nowhere is it common, and it is unusual to find large numbers in a single collection or from any one locality. As the common name implies, the southern bog lemming frequently lives in bogs and marshes, but not exclusively. In Ohio it is found most often in grassy fields, but the distribution pattern is spotty; only certain areas in any one county will harbor a population. The reason for this distribution is not clear.

Although the southern bog lemming was described in 1857 by Baird, information concerning it has accumulated very slowly. In 1955 Wetzel examined 658 specimens and finally clarified the tax-

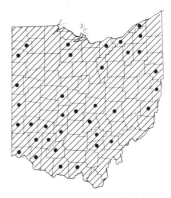

Synaptomys cooperi Baird

onomy. Connor (1959) presented an excellent report on the life history and ecology of the species, and Getz (1961) investigated the physical and biotic factors influencing the distribution of the bog lemming.

In Ohio, Hine (1910) stated that the bog lemming was widely distributed over the state and in some places common. Enders (1930) collected 1 specimen in Ashland County. Bole (1939) discussed the cyclic nature of bog lemming population levels in the Cleveland area. Preble (1942) found these animals common in the tall grass of wet pastureland in Morrow County. Bole and Moulthrop (1942) examined more than 70 specimens, primarily from northeastern Ohio. Oehler (1942) reported on 32 specimens that he caught on grassy hillsides around Cincinnati, and Gottschang (1965) commented on specimens taken in old fields in southwestern Ohio during the winter of 1962-63. It is apparent from these reports that the bog lemming occurs throughout most of Ohio.

The bog lemming has a short, broad, rounded head. The eyes are small and beady and hidden by fur; the ears are short, barely projecting above the head. *Synaptomys* is similar in appearance to *Microtus* but should not be confused with it. The skull of the bog lemming is heavier and broader than other small microtines in Ohio and is easily recognized by the distinctive shallow groove along the side of each upper incisor and the enamel pattern of the cheek teeth. The shallow groove can be readily seen by pushing back the upper lips and looking carefully along the side of the incisor. The dental formula is as follows:

$$i\frac{1-1}{1-1},\ c\frac{0-0}{0-0},\ pm\frac{0-0}{0-0},\ m\frac{3-3}{3-3}\ =\ 16$$

Following are measurements of 31 adult southern bog lemmings (Synaptomys cooperi) from Ohio:

	Total Length (mm)	Tail (mm)	Hind Foot (mm)
Extremes	107–24	11–20	16–19
Mean	114	15	17.5

The tail is equal to, or shorter than, the hind foot (rarely 1 mm longer than the foot). The woodland vole *(Microtus pinetorum)* also has a short tail, but there are no grooves on the incisors and the fur is red or reddish-brown and smooth. Adult bog lemmings are grizzled in appearance with gray, green-gray, or grayish-brown colors predominating. The ends of many of the hairs on the back and sides are straw-colored or yellow, giving the animals their grizzled appearance. The belly is uniform gray or silver; occasionally a buff patch may be present. The young are dark, dull gray.

In the University of Cincinnati collection, there is a series of bog lemmings taken in February and March in Hamilton and Clermont counties that have long, shaggy, thick, light-gray fur, and another series from Scioto County caught in November and December have much shorter, thinner, brownish-gray coats. These differences may have something to do with the molt pattern. Bole and Moulthrop (1942) described an autumn molt (October to December) and a summer molt (June to mid-August), but Connor (1959) observed only an autumn molt. Only the November, December, and January specimens that I have collected have conspicuous areas of new fur on them. Summer specimens certainly have shorter hair and lighter-colored coats, possibly due to molting.

In nests 10 to 15 cm in diameter made of grass or shredded leaves constructed above or below the ground, litters of 1 to 6 (average, 4) are born during any month of the year. Just how many litters a single female produces each year in the field is not known. One captive female gave birth to 6 litters, totaling 22 young, in a period of approximately 26 weeks (Connor, 1959). This is probably a higher rate of reproduction than in the field. Young may be produced throughout the year in Ohio, but there is positive evidence of reproductive activity only for the months of April through October.

Connor (1959) has been the only mammalogist to breed bog lemmings successfully in captivity. He reports that females breed within four to six hours after giving birth to their litters. The gestation period is 23 to 26 days. Weights of newborn young ranged from 3.1 to 4.3 g and averaged 3.9 g (Connor, 1959). Stegeman (1930) found that two newborn bog lemmings weighed 2.4 and 2.7 g, respectively. Growth rate is rapid for the first three weeks, after which it slows and gradually levels off. The newborn are naked and pink, the pinnae are folded against the head, and the eyes are closed. Hair is evident in 4 to 7 days; the eyes open between days 10 and 12; the pinnae unfold between days 2 and 3. The young are weaned at about 16 days of age, but they remain with the female until at least 20 days of age. In a litter that was born in captivity on 25 February in my laboratory, two of the three mice were eaten by the mother on the 4th day. The remaining male was found dead in the cage when he was 41 days old. His development paralleled that reported by Stegeman (1930) and Connor (1959).

The home range of the bog lemming has been variously estimated to be from 0.04 hectare (0.11 acre) (Getz, 1960) to more than 0.2 hectare (0.5 acre) (Connor, 1959). Population density fluctuates with the season as well as from year to year, but whether this species has populations that are truly cyclic is not known. Stegeman (1930) found that there were 35 bog lemmings per hectare (14 per acre) in his study plots, but Bole (1939) and Connor (1959) and others have given more conservative estimates of 5 to 15 mice per hectare (2 to 6 per acre).

Bog lemmings are hardy animals that are active both day and night. I have caught them on days when the temperature was well below freezing and the fields were partially covered with snow and ice. They are not readily attracted to traps by bait; most individuals are caught by placing the treadle of the trap directly across a runway, and it makes little difference whether the trap is baited or not. Bog lemmings construct small tunnel-like runways usually just below the thick cover of grass in fields or in the upper one to two inches of soft leaf mold in bogs and hedgerows. Active runs are easily recognized because the ground is worn bare of grass by the many passing feet, and the bright, light-green droppings of the animals are scattered along the runways. Fresh grass cuttings, neatly piled like miniature cords of wood, are also evident along active runways. When bog lemmings and voles are present in the same field, they make free use of each other's runways. There have been reports of bog lemmings living in deciduous woods (Hamilton, 1941), but I have not found them there. Runways are often found, however, along the borders between grassy fields and woodlots that are frequently used by bog lemmings.

Besides finely chopped leaves, grasses, and sedges, bog lemmings are known to include in their diet some insects and other small invertebrates. In the laboratory captive animals have eaten oats, apples, carrots, lettuce, seeds, and berries, but those that I have raised required fresh green grass daily for survival.

Because so little is known about this mouse in its natural habitat, it is impossible to assess accurately its economic importance, but it seems unlikely that the bog lemming, with its widely scattered, small populations will ever become an economic burden to man.

SELECTED REFERENCES

Bole, B. P., Jr. 1939. The quadrat method of studying small mammal populations. Sci. Publ. Cleveland Mus. Nat. Hist. 5:15-77.

Bole, B. P., Jr., and P. N. Moulthrop. 1942. The Ohio recent mammal collection in the Cleveland Museum of Natural History. Sci. Publ. Cleveland Mus. Nat. Hist. 5:83-181, esp. 146-51.

Connor, P. F. 1959. The bog lemming, *Synaptomys cooperi,* in southern New Jersey. Publ. Mus. Mich. State Univ., Biol. Series 1:165-248.

Enders, R. K. 1930. Some factors influencing the distribution of mammals in Ohio. Occ. Papers, Univ. of Mich. Mus. Zool. 212:1-27.

Getz, L. L. 1960. Home ranges of the bog lemming. J. Mammal. 41:404-5.

———. 1961. Factors influencing the local distribution of *Microtus* and *Synaptomys* in southern Michigan. Ecology 42:110-19.

Gottschang, J. L. 1965. Winter populations of small mammals in old fields of southwestern Ohio. J. Mammal. 46:44-52.

Hamilton, W. J., Jr. 1941. On the occurrence of *Synaptomys cooperi* in forested regions. J. Mammal. 22:195.

Hine, J. S. 1910. Ohio species of mice. Ohio Nat. 10:65-72.

Howell, A. B. 1927. Revision of the American lemming mice (genus *Synaptomys*). North Amer. Fauna 50:1-38.

Oehler, C. 1942. Notes on lemming mice at Cincinnati, Ohio. J. Mammal. 23:341-42.

Preble, N. A. 1942. Notes on the mammals of Morrow County, Ohio. J. Mammal. 23:82-88.

Stegeman, L. C. 1930. Notes on *Synaptomys cooperi cooperi* in Washtenaw County, Michigan. J. Mammal. 11:460-66.

Wetzel, R. M. 1955. Speciation and dispersal of the southern bog lemming, *Synaptomys cooperi* (Baird). J. Mammal. 36:1-20.

FAMILY MURIDAE
Old World Rats and Mice

All of the animals in this family are members of the Old World fauna. The 100± genera in the family are all mouse-like or rat-like animals of medium to small size, with long, sparsely-haired, scaly tails. Two genera, *Rattus* and *Mus,* have been transported abroad and now are distributed worldwide. The skull of the Muridae is similar to that of the Cricetidae except that (1) the cheek teeth are tuberculate and the tubercles (cusps) arranged in three rather than two longitudinal rows; and (2) the posterior edge of the palatine bones extends beyond the posterior edge of the last upper molar.

KEY TO THE SPECIES OF MURIDAE IN OHIO

1. Total length greater than 250 mm; skull more than 25 mm long; first upper molar not as long as second and third molars combined . . . . *Rattus norvegicus* (Norway Rat), p. 108
1'. Total length less than 250 mm; skull less than 25 mm long; first upper molar longer than second and third molars combined (fig. 20) . . . . *Mus musculus* (House Mouse), p. 110

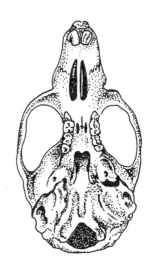

Fig. 20. Ventral view of the skull of the house mouse *(Mus musculus),* showing the first molar larger than the second and third combined. (Drawing by Elizabeth Dalvé.)

Norway Rat or Common Rat or Brown Rat
Rattus norvegicus (Berkenhout)

Mammalogists generally agree that the Norway rat was introduced along the eastern seaboard of the United States in about 1775 from England. It is altogether possible that this pest was subsequently introduced into Ohio as early as 1788 at the founding of Marietta, the first settlement in the Northwest Territory. History has shown that rats have followed man and civilization by hitchhiking rides on various means of transportation, particularly ships and boats. Certainly, boats navigating the Ohio River in the late 1700s and early 1800s carried rats as well as men as they moved inland. The earliest reference to the Norway rat in Ohio apears to be in Kirtland's report of 1838 on the zoology of Ohio. He says: "The Norway rat is not a native of our country, but is now extending itself in every direction, especially along the shores of our canals and navigable streams." By 1850 the rat was well established in all the eastern counties, and by 1880 it was probably present throughout Ohio, as it is today.

Rattus norvegicus (Berkenhout)

Norway rats have coarse brown or grayish-brown fur. The belly is only slightly lighter in color than the back. The ears are prominent and naked. The tail is scaly and sparsely haired and is usually shorter than the combined head and body length. Skull characteristics are the same as those given for the family Muridae. The dental formula is as follows:

$$i\frac{1\text{-}1}{1\text{-}1}, c\frac{0\text{-}0}{0\text{-}0}, pm\frac{0\text{-}0}{0\text{-}0}, m\frac{3\text{-}3}{3\text{-}3} = 16$$

Following are measurements of 20 adult Norway rats (*Rattus norvegicus*) from Ohio:

	Total Length (mm)	Tail (mm)	Hind Foot (mm)
Extremes	277–458	125–203	31–48
Mean	367	167	39

Man has done almost everything possible to exterminate rats, yet these animals have always defied the challenge and are probably present in as great numbers today as at any time in the past. Rats are as much at home in sewers and garbage dumps as they are in homes or open fields. They are common along many waterways thoughout Ohio. Their basic requirements are adequate cover and sufficient food. Since they are omnivorous, they can and do live almost everywhere. The greatest concentrations occur on farms, where barns and storage buildings offer protection from the environment and provide abundant, safe nesting sites. Food in the form of stored wheat, oats, and corn is usually plentiful, and the chicken house offers chicken feed, fresh eggs, and young poultry. Large cities harbor the next largest concentrations of rats. In classic studies in 1949 that have not been repeated, Davis and Fales (1950) estimated 1 rat for every 15 citizens in Baltimore, and Davis (1950) estimated a rat population of 1 rat per 36 persons in New York. Garbage dumps, trash heaps, and slum areas have the highest populations of rats within cities, but even the highest-class residential areas may have their rat problem.

Rats are prolific animals that reproduce throughout the year, although few litters are born during the winter. Females are capable of reproducing when they are 11 to 12 weeks old. Davis (1949b) concluded that age and size at sexual maturity differ in different localities. The female is multiparous and polyestrous, coming into heat about every 10 days until she is bred. The gestation period is 20 to 21 days, but this time is extended when a female is nursing young. Litters of one to twelve are produced, but six or seven is average. Feral rats normally produce five or six litters each year. The blind, naked young develop rapidly. The eyes open on about day 15. Weaning occurs at three weeks. Feral Norway rats have a short life expectancy; fewer than 5 percent live longer than one year (Davis, 1953).

Rats are secretive animals that are most active after dark, especially during the first one to two hours after sunset. In open fields a series of well-worn runways leads from the entrance of the underground burrow to the feeding stations. In buildings a careful inspection of the floors along the walls will usually disclose numerous footprints or grease marks on the walls, indicating that rats are using the walls as they travel to and from their feeding places in the dark. Rats are surprisingly sedentary; if food and shelter are adequate, they will not stray more than 60 meters from their nest. Davis et al. (1948) found very little movement by the rats from one building to another on the several farms that they studied. Colonies generally repopulate from within when voids are created as animals die.

The Norway rat is the most destructive of all mammals, causing more property damage and spreading more diseases than any other species known. Eadie (1954) says: "The total annual monetary loss due to rats in this country [USA] has been estimated by various observers at approximately $200,000,000, although many believe this figure to be conservative. The average annual loss to individual farms has been [estimated] ... at ... from ten to eighty dollars per farm. This loss results from such activities as eating and contaminating livestock feeds, stored fruits and vegetables, and crops in gardens and fields; destroying poultry and eggs, baby pigs, and lambs; and effecting a great variety of property damage."

At least ten human diseases and several livestock diseases are known to be transmitted by rats and/or their parasites. Bubonic plague, sylvatic plague, and murine typhus fever are three well-known rat-borne diseases, and diseases such as leptospirosis and salmonellosis may be contracted by eating food or drinking water contaminated by rat urine or feces.

Once established, rats are difficult to eradicate. However, they cannot live where there is not enough food. The first step in eliminating rats, therefore, is to make absolutely certain that all sources of food are either removed or kept in rat-proof containers. Since rats are excellent climbers and jumpers, foods stored high on open shelves are not necessarily safe from them. Whole neighborhoods must participate in any effective rat eradication program. On the farm the rat control problems are obviously intensified. Many excellent government bulletins that outline specific rat control programs are available to the farmer.

Some of the recently developed poisons are highly effective against rats and with care are relatively safe to use in the home. Rat traps baited with oats, apple, bacon, or peanut butter and carefully placed across runways may prove effective in the initial phases of control, but rats soon become suspicious of traps and avoid them. The more toxic poisons and gases can be used only by licensed applicators of pesticides. Jackson et al. (1973) have evidence that some rats have already developed resistance to the anticoagulant rodenticides!

The white rat, an albino strain of the Norway rat, is of great value in laboratory experiments and is also a popular pet. The feral Norway rat, however, is a serious threat to man's welfare.

SELECTED REFERENCES

Benedict, F. G., and J. M. Petrik,. 1930. Metabolism studies on the wild rat. Amer. J. Physiol. 94:662-85.

Bole, B. P., Jr., and P. N. Moulthrop. 1942. The Ohio recent mammal collection in the Cleveland Museum of Natural History. Sci. Publ. Cleveland Mus. Nat. Hist. 5:83-181, esp. 163-64.

Brooks, J. E. 1973. A review of commensal rodents and their control. New York State Dept. of Health. Critical Reviews in Environmental Control 3(4):405-53.

Calhoun, J. B. 1962. The ecology and sociology of the Norway rat. U. S. Dept. Health, Education & Welfare, Public Health Serv. Publ. No. 1008. 288 pp.

Chapman, F. B. 1938. Exodus of Norway rats from flooded areas. J. Mammal. 19:376-77.

Christian, J. J., and D. E. Davis. 1956. The relationship between adrenal weight and population status of urban Norway rats. J. Mammal. 37:475-86.

Davis, D. E. 1949a. A phenotypical difference in growth of wild rats. Growth 13:1-6.

———. 1949b. The weight of wild brown rats at sexual maturity. J. Mammal. 30:125-30.

———. 1950. The rat population of New York, 1949. Amer. J. Hygiene 52(2):147-52.

———. 1953. The characteristics of rat populations. Quart. Rev. Biol. 28:373-401.

Davis, D. E., and W. T. Fales. 1950. The rat population of Baltimore, 1949. Amer. J Hygiene 52(2):143-46.

Davis, D. E., J. T. Emlen, and A. W. Stokes. 1948. Studies on home range in the brown rat. J. Mammal. 29:207-25.

Drummond, D.C., and K.D. Taylor. 1970. Practical rodent control. Pp. 742-68 in World Food Program Storage Manual, Part 3. FAO. Rome.

Eadie, W. R. 1954. Animal control in field, farm, and forest. Macmillan Co., New York. 257 pp.

Emlen, J. T., and D. E. Davis. 1948. Determination of reproductive rates in rat populations by examination of carcasses. Physiol. Zool. 21:59-65.

Emlen, J. T., A. W. Stokes, and D. E. Davis. 1949. Methods for estimating populations of brown rats in urban habitats. Ecology 30:430–42.

Emlen, J. T., A. W. Stokes, and C. P. Winsor. 1948. The rate of recovery of decimated populations of brown rats in nature. Ecology 29:133–45.

Hamilton, W. J., Jr. 1936. Rats and their control. Cornell Extension Bull. 353:1–32. 11 figs.

Hartman, C. 1922. A brown rat kills a rattler. J. Mammal. 3:116–17.

Jackson, W. B. 1977. Evaluation of rodent depredations to crops and stored products. European and Mediterranean Plant Protection Organization Bull. 7(2):439–58.

Jackson, W. B., J. E. Brooks, A. M. Bowerman, and D. Kaukeinen. 1973. Anticoagulant resistance in Norway rats in U. S. cities. Pest Control 41(4):56–64, 81.

Kirtland, J. P. 1838. Report on mammals. Geol. Surv. Ohio, Second Annu. Rep. Pp. 160–61, 175–77.

Perry, J. S. 1946. The reproduction of the wild brown rat (*Rattus norvegicus* Erxleben). Proc. Zool. Soc. London 115:19–46.

Price, E. O. 1977. Burrowing in wild and domestic Norway rats. J. Mammal. 58:239–40.

Rowe, F. P. 1975. Control of rodents in stored products and urban environments. Pp. 339–48 *in* F. B. Golley, K. Petrusewicz, and L. Ryszkowski. Small mammals: their productivity and population dynamics. Cambridge Univ. Press, 451 pp.

Silver, J. 1927. The introduction and spread of house rats in the United States. J. Mammal. 8:58–60.

Silver, J., W. E. Crouch, and M. C. Betts. 1942. Rat proofing buildings and premises. U. S. Dept. Interior, Fish & Wildlife Serv., Conserv. Bull. No. 19. 26 pp.

Silver, J., and F. E. Garlough. 1941. Rat control. U. S. Dept. Interior, Fish & Wildlife Serv., Conserv. Bull. No. 8. 27 pp.

Storer, T. I. 1948. Control of rats and mice. Univ. of Calif. Agric. Extension Serv. Circ. 142. 37 pp.

House Mouse
Mus musculus Linnaeus

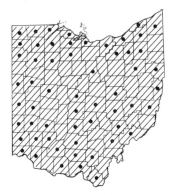

Mus musculus Linnaeus

No one is certain exactly when the house mouse was introduced into America from Europe; presumably the house mouse and the Norway rat arrived at about the same time. The house mouse was known to the earliest naturalists in Ohio. Today, of course, it is resident throughout the entire state. It is probably the most abundant species in the world.

Hall (1927) reports outbreaks of house mice in the United States in which 80,000 or more animals occupied a single acre (0.4 hectare) of ground (not in Ohio, however). I have never found feral house mice common here, although they are likely to be found in a wide variety of habitats. Goodpaster (1941) reported catching them occasionally far from human habitation in traps set for other species. Bole and Moulthrop (1942) stated that outside cities house mice are found in sedge meadows and damp grasslands. Enders (1930) caught house mice in a small marsh and in deep woods. More commonly, house mice occur abundantly around and in buildings, where they live commensally with man. Preble (1942) found them common in houses in Morrow County, and Metzger (1955) found them very common around buildings, granaries, and so on in Perry County. Cornfields and soybean fields have yielded specimens. They are commonly found in weedy fencerows and fields with a heavy cover of weeds and grasses.

The house mouse is a small mouse, easily recognized by its brown, tan, or grayish-brown coat. The belly is tan, light brown, or gray; animals from southern Ohio often have an orange or buff cast to the underparts. Considerable variation in coloration is common in this species. The tail is naked, scaly, annulated, and slightly bicolored. The eyes are small and the ears prominent and naked or nearly naked. House mice are also easily recognized by the musky odor of secretions from three glands that surround the anal opening. Descriptions of the skull are included in the key for the family Muridae. When viewed from the side, a subapical notch is apparent in the upper incisors (fig. 21). The dental formula is the same as for the Norway rat. Following are measurements of 50 adult house mice (*Mus musculus*) from Ohio:

	Total Length (mm)	Tail (mm)	Hind Foot (mm)
Extremes	132–209	64–105	15–20
Mean	158	76	18

Fig. 21. Lateral view of the house mouse *(Mus musculus)*, upper incisor, showing subapical notch. (Drawing by Elizabeth Dalvé.)

Considerable variation in external measurements as well as in coloration is apparently the result of rather rapid evolutionary change in commensal populations. Schwarz and Schwarz (1943) have shown that wild and commensal populations within the same region may be quite different.

House mice raised in the laboratory reproduce throughout the year as do commensal populations living in buildings. However, fewer litters are produced in the winter and midsummer than during the remainder of the year (Brown, 1953). Feral house mice in Ohio produce few litters from November through March. I have one record of an adult female caught 28 November that contained 5 fully developed embryos. As many as 12 young may be produced in a single litter, but 5 to 7 is average. The young are born in a nest made of shredded paper, leaves, grass, hay, or any soft and readily available material. The young, weighing less than 1 g at birth, are naked, blind, and helpless. The eyes open on or about day 14, and the young are weaned when about 20 days old. Males and females develop at the same rate and may breed when only 35 days of age. Most individuals are fully mature at about 60 days of age. The gestation period is 19 to 20 days, but this period is increased by approximately 1 day for each young being nursed by a pregnant female (Asdell, 1943). Females exhibit a postpartum heat and may mate with several males within a short period of time. Adult females may have as few as 5 or as many as 10 litters each year.

House mice in the field may use runways made by other mammals, but do not make runs of their own. These primarily nocturnal animals are most active during the one to two hours following sunset. Indoors they have a tendency to be active both day and night.

Caldwell (1964) stated that wild house mice are migratory or "seminomadic," moving from one home range to another. He reported that only 1 of 22 mice captured in any one grid remained as long as two months. The average home range for resident mice was 0.16 hectare (0.41 acre) (Caldwell, 1964). The movements of house mice living in buildings are usually determined by the construction of the building and food availability. Mice living in a storage room where food is always available may not move even 30 m away from the nest, but a mouse living on the floor above may have to travel more than 30 m to get its food. In a study conducted by Brown (1953), the greatest distance between points of capture for mice living in a barn was 24.1 m (79 ft) for males and 13.7 (45 ft) for females. The average distance was 6.1 m (19.9 ft) for males and 3.8 m (12.4 ft) for females. Brown did not detect any immigration or emigration between the buildings and the surrounding fields. It has been suggested that house mice may move from the fields into buildings with the onset of cold weather. Wild house mice sometimes congregate in large numbers under haystacks or beneath corn shocks during cold winters, and house mice living in buildings tend to be colonial.

Whether house mice display territoriality is debatable. Females about to give birth sometimes defend their nests against other adult mice. A social group living in a nest drives away an "outsider." Crowcroft (1955), observing males defending a certain area around nest boxes, considered this behavior a form of territorial defense.

Estimates of the total amount of damage done by house mice in the United States each year are not available, but experts agree that it is considerable. Only the Norway rat is more destructive. House mice are serious pests in any building where grain and feed are stored because of both the amount of material consumed and the dirt, urine, and black fecal pellets scattered through the remains. Furniture can be ruined when house mice rip holes in chairs and mattresses in search of nesting materials. Access holes are frequently eaten in woodwork and floors. Grain and corn crops left standing in the field over winter may be largely destroyed by house mice and other species of small mammals. Even frozen meat kept in cold rooms is subject to depredation by house mice.

Fortunately, house mice are easier to trap than rats. Most populations can be successfully controlled by using large numbers of common snapback mousetraps. These can be baited with a wide variety of food, but a mixture of peanut butter and rolled oats is very effective. If possible, the natural food source should be removed during the trapping period. Most poison baits used for killing rats are also effective against house mice, but precautions must be taken in using them.

Cats, weasels, foxes, skunks, hawks, owls, and snakes are natural predators of mice.

Among the many color mutations of the house mouse is the albino strain, or common laboratory "white mouse," which is raised by the millions for research purposes. There is also a sizable trade in white mice through pet shops.

Hundreds of studies have been reported on the physiology, genetics, psychology, economic importance, and social habits of the house mouse. However, there is still much to be learned concerning the life and social habits of feral populations of this mouse. Comparatively few detailed field studies have been made of this species, none of them in Ohio.

SELECTED REFERENCES

Asdell, S. A. 1943. Patterns of mammalian reproduction. Comstock Publ. Assoc. Ithaca, N. Y. 437 pp. 2d ed. 1964, 670 pp.

Bole, B. P., Jr., and P. N. Moulthrop. 1942. The Ohio recent mammal collection in the Cleveland Museum of Natural History. Sci. Publ. Cleveland Mus. Nat. Hist. 5:83-181, esp. 164-65.

Breakey, D. R. 1963. The breeding season and age structure of feral house mouse populations near San Francisco Bay, California. J. Mammal. 44:153-68.

Brown, R. Z. 1953. Social behavior, reproduction, and population changes in the house mouse *(Mus musculus* L.). Ecol. Monogr. 23:17-40.

Caldwell, L. D. 1964. An investigation of competition in natural populations of mice. J. Mammal. 45:12-30.

Crowcroft, P. 1955. Territoriality in wild house mice, *Mus musculus* L. J. Mammal. 36:299-301.

Enders, R. K. 1930. Some factors influencing the distribution of mammals in Ohio. Occ. Papers, Univ. of Mich. Mus. Zool. 212:1-27.

Goodpaster, W. W. 1941. A list of the birds and mammals of southwestern Ohio. J. Cincinnati Soc. Nat. Hist. 22:1-47.

Hall, E. R. 1927. An outbreak of house mice in Kern County, California. Univ. of Calif. Publ. Zool. 30: 189-203.

Laurie, E. M. O. 1946. The reproduction of the housemouse *(Mus musculus)* living in different environments. Proc. Royal Soc. London 133:248-81.

Metzger, B. 1955. Notes on mammals of Perry County, Ohio. J. Mammal. 36:101-5.

Moulthrop, P. N. 1942. Description of a new house mouse from Cuba. Sci. Publ. Cleveland Mus. Nat. Hist. 5:79-82.

Preble, N. A. 1942. Notes on the mammals of Morrow County, Ohio. J. Mammal. 23:82-86.

Schwarz, E., and H. K. Schwarz. 1943. The wild and commensal stocks of the house mouse, *Mus musculus* Linnaeus. J. Mammal. 24:59-72.

Smith, W. W. 1954. Reproduction in the house mouse, *Mus musculus* L., in Mississippi. J. Mammal. 35: 509-15.

Snell, G. D., ed. 1941. Biology of the laboratory mouse. Dover Publ., New York. 497 pp.

Southern, H. N., and E. M. O. Laurie. 1946. The house mouse *(Mus musculus)* in corn ricks. J. Anim. Ecol. 15:134-49.

Strecker, R. L. 1954. Regulatory mechanisms in housemouse populations: the effect of limited food supply on an unconfined population. Ecology 35:249-53.

Strecker, R. L., and J. T. Emlen. 1953. Regulatory mechanisms in house-mouse populations: the effect of limited food supply on a confined population. Ecology 34:375-85.

FAMILY ZAPODIDAE
Jumping Mice

Four living genera are recognized in this family, two of which, *Zapus* and *Napaeozapus,* are found in North America. The North American species are easily recognized by their mouse-like form, the greatly elongated hind legs and feet, and the long, whip-like tail. The closest relative is found in China and the remaining genus in northern Europe and Asia.

KEY TO THE SPECIES OF ZAPODIDAE IN OHIO

1. Tail not tipped with white; general color of upper parts straw-colored, yellowish-brown or with an olive tinge; thin orange-yellow line separating back from belly; four cheek teeth on each side of upper jaw (premolar small and peg-like, easily overlooked); an open field species found throughout Ohio .... *Zapus hudsonius* (Meadow Jumping Mouse) p. 112

1'. Tail usually tipped with white; general color of upper parts bright fulvous or red; without a thin orange-yellow line separating back from belly; three cheek teeth on each side of upper jaw; a woodland species known only from eastern and northeastern Ohio........ .... *Napaeozapus insignis* (Woodland Jumping Mouse) p. 115

Meadow Jumping Mouse
Zapus hudsonius (Zimmermann)

The meadow jumping mouse is a rather common species, generally distributed throughout Ohio. Its

Zapus hudsonius (Zimmermann)

$i\frac{1-1}{1-1}, c\frac{0-0}{0-0}, pm\frac{1-1}{0-0}, m\frac{3-3}{3-3} = 18$

Following are measurements of 10 adult meadow jumping mice (*Zapus hudsonius*) from Ohio (also see Rybak et al., 1975):

	Total Length (mm)	Tail (mm)	Hind Foot (mm)
Extremes	179–203	106–21	26–30
Mean	192	114	29

range in North America extends as far south as Alabama in the east and central Colorado in the west. It is also common throughout Canada and southern Alaska. The presence of this animal in Ohio has been noted by the earliest biologists (among them, Kirtland, 1838; Brayton, 1882; Vickers, 1894; and Preble, 1899). As early as 1923, Gossard reported the known records of the meadow jumping mouse from Ohio. Dexter (1954b) better defined the range for the species in Ohio. More recently, Gottschang (1961), Rybak et al. (1975), and McLean (1976) have added to this range distribution. More than 300 specimens have been reported from a large number of areas over the state.

These saltatorial animals are easily recognized by their long hind legs and feet, and by the tail, which is longer than the body. When startled from their nest or hiding place, jumping mice make a series of prodigious jumps. Actual measurements indicate routine leaps of up to 1 m (3.3 ft). Jumping mice have coarse (hispid) coats of rather short, stiff, yellow or orange black-tipped hairs. The black-tipped hairs are most numerous on the back, where they form a broad, dark band down the center. The sides are predominantly orange or straw-colored. There is a narrow band of pure orange or orange-yellow hair that extends along either side, separating the dorsal fur from the snow-white belly fur. The undersurface of the legs and tops of the feet are white. The long, scaly, sparsely haired tail is bicolored throughout its length, white below and gray above. There is a short pencil of hairs on the end of the tail. The prominent ears are rimmed with orange or yellow. The skull is distinctive, its characteristic features being: large, conspicuous infraorbital foramen; zygomatic arches depressed and low on the skull; palatine foramen short and excessively bowed; upper incisors deeply grooved; peg-like premolar in front of the three molars in the upper tooth row. The dental formula is as follows:

Although the meadow jumping mouse is generally considered to be a field species, it actually lives in a variety of habitats. Dexter's list (1954b) of habitats, in order of apparent preference, include "damp meadows; grain and hay fields; margins of streams and glacial swamps and bogs; shrubby second growth; borders of woodlands; and beech-maple or beech-maple-hemlock woods and ravines." Quimby (1951) found this jumping mouse most frequently in willow-alder thickets and grass and/or sedge meadows. Getz (1961) found them only in moist situations and slightly more abundant where there was standing water. Since moisture is an important factor in the ecology of the meadow jumping mouse, it may be that the lack of sufficient wet areas in the Till Plains of Ohio accounts for the scarcity of the mouse in this area. Upland coniferous forests also yield very few specimens. However, this mouse is moderately common in northeastern Ohio.

Jumping mice are our most profound hibernators. Some individuals enter the hibernal chamber in September and do not emerge until the following May. Not all meadow jumping mice enter into hibernation as early as September, however; there are numerous records of active animals in October and November. A female caught by my son in Warren County was trapped on 1 October on a night with light frost and the temperature at 7°C (44°F).

A heavy deposit of stored fat is developed before the jumping mouse enters hibernation. Morrison and Ryser (1962) found individuals gaining weight at the rate of 2 g per day over a three-day period during the month preceding hibernation. Some mice increased their normal weight by 95 percent. Unlike the woodchuck, jumping mice use their stored fat during the hibernation period; when they awake in spring they may weight 25 to 35 percent less than the previous fall. Jumping mice also emerge much later in the spring than other hibernators. Emergence dates vary from year to year, but they always occur after all traces of winter have disappeared and spring is well under way. Quimby (1951) states that most of the animals that he studied in Minnesota appeared on the study plots during the first two weeks in May. One or two individuals appeared during the last week of April. Males always emerged a week or two before the females. On 25 April 1930 R. M. Goslin found a nest 27.5 cm (11 in) belowground at Independence, Cuyahoga County, that contained a hibernating meadow jumping mouse (Dexter, 1954a). Since jumping mice have been found out of hibernation and active as early as April in Wisconsin (Quimby, 1951), New York (Hamilton, 1935), and Vermont (Sheldon, 1934), it is highly probably that some individuals in Ohio also become active in April.

Most observations indicate that the meadow jumping mouse hibernates in well-drained areas. Dry hillsides, mounds of earth, old trash heaps, woodchuck dens, and upland fields have all been used. The winter sleeping quarters, a nest made of shredded leaves and/or grass, may be located from a few centimeters to as much as a meter below the surface. Apparently, numerous jumping mice may be attracted to the same general area when looking for places to hibernate. Several workers (Schwartz, 1951; Waters and Stockley, 1965) have found three or more nests within the same mound of earth or in the same general vicinity. Dilger (1948) reported a male and a female occupying the same hibernation nest in New York, and B. Raithel located five females in a single nest near Rootstown, Portage County, Ohio (Dexter, 1954a). Most hibernation nests, however, contain only a single mouse. The classic hibernation position is with the head and nose curled under and tightly tucked into the inguinal region and with the long tail coiled repeatedly around the body like a large spring.

Most females breed soon after emerging from hibernation and produce their first litters in June. The gestation period varies from 17 to 21 days. Individual females are known to have produced two litters in a season; Quimby (1951) suggests that some probably produce three litters each year. Some females taken in October in Ohio have shown obvious signs of nursing young, but none have been pregnant. The last litters are probably born in August and possibly early September. Litter sizes of one to eight have been reported; the average is five or six. The limited number of litters reported for Ohio have contained either three or four young. The newborn young are pink and helpless, weighing between 0.7 and 0.8 g each. Hair appears on or about day 9. The pinnae are open about day 19, and the eyes open between days 22 and 25, at which time weaning starts. The adult pelage is acquired and the young are ready to leave the nest by 30 days of age.

Home range has been estimated to be from 0.1 hectare (0.25 acre) to 0.8 hectare (2.0 acres) per mouse. Males have larger ranges than females. Population size is difficult to estimate because the mice tend to group together in restricted areas of favorable habitat. Hamilton (1935) states that the meadow jumping mouse is probably one of the most common mammals in certain localities, but most researchers have been more conservative in their estimates. Densities of 5 to 25 per hectare (2 to 10 per acre) are probably common.

Meadow jumping mice are primarily seed eaters; the dry seeds produced by most grasses are the chief component of their diet. In obtaining these seed heads, the mice chop the stems, which they seldom eat, into three or four 2.5-cm (1-in) pieces in much the same manner as do lemmings. They are "messy" feeders, leaving behind pieces of seed coats and husks in little piles. Meadow jumping mice are also insectivorous. Water is drunk by captive mice in the laboratory, but their water requirements in the wild are not known. I once sat a short distance from a jumping mouse that was busy cleaning itself in a small, rapidly flowing stream. After throwing water over itself with its front feet, it then rubbed and shook its body like a bird taking a bath. It was not observed drinking water, however.

Bole and Moulthrop (1942) were enthusiastic in their treatment of this genus in Ohio, naming two new subspecies and including a third within its boundaries. Krutzsch (1954) revised the genus and concluded that only one subspecies, *Zapus hudsonius americanus,* occurs in Ohio.

Meadow jumping mice are not abundant enough in Ohio to have adverse economic significance to man. It is interesting that, in addition to falling prey to all of the common terrestrial predators, jumping mice, because of their frequent association with water, are occasionally eaten by game fish and even large frogs.

SELECTED REFERENCES

Bole, B. P., Jr., and P. N. Moulthrop. 1942. The Ohio recent mammal collection in the Cleveland Museum of Natural History. Sci. Publ. Cleveland Mus. Nat. Hist. 5:83-181, esp. 165-72.

Brayton, A. W. 1882. Report on the Mammalia of Ohio. Rep. Geol. Surv. Ohio 4:1–185.

Dexter, R. W. 1943. A record of *Zapus hudsonius* in northeastern Ohio. J. Mammal. 24:267.

———. 1954a. Notes on nests of the meadow jumping mouse, *Zapus hudsonius.* J. Mammal. 35:121.

———. 1954b. Distribution of the meadow jumping mouse *Zapus hudsonius* in Ohio. J. Mammal. 35:233–39.

Dilger, W. C. 1948. Hibernation site of the meadow jumping mouse. J. Mammal. 29:299–300.

Getz, L. L. 1961. Notes on the local distribution of *Peromyscus leucopus* and *Zapus hudsonius.* Amer. Midl. Nat. 65:486–500.

Gossard, H. A. 1923. Ohio records of the jumping mouse. Ohio J. Sci. 23:284–86.

Gottschang, J. L. 1961. *Zapus* from Warren County, Ohio. Ohio J. Sci. 61:125.

Hamilton, W. J., Jr. 1935. Habits of jumping mice. Amer. Midl. Nat. 16:187–200.

Jones, J. K., Jr. 1950. Another record of a swimming jumping mouse. J. Mammal. 31:453–54.

Kirtland, J. P. 1838. Report on zoology of Ohio. Second Annu. Rep., Geol. Surv. Ohio, pp. 157–200.

Krutzsch, P. H. 1954. North American jumping mice (genus *Zapus*). Univ. of Kans. Publ. Mus. Nat. Hist. 7(4):349–472.

McLean, E. B. 1976. Additional records of the jumping mouse, *Zapus hudsonius,* in Ohio. Ohio J. Sci. 76:32.

Metzger, B. 1955. Notes on mammals of Perry County, Ohio. J. Mammal. 36:101–5.

Morrison, P., and F. A. Ryser. 1962. Metabolism and body temperature in a small hibernator, the meadow jumping mouse, *Zapus hudsonius.* J. Cell. & Comp. Physiol. 60:169–80.

Preble, E. A. 1899. Revision of the jumping mice of the genus *Zapus.* North Amer. Fauna 15:1–43.

Preble, N. A. 1944. A swimming jumping mouse. J. Mammal. 25:200–201.

Quimby, D. C. 1951. The life history and ecology of the jumping mouse, *Zapus hudsonius.* Ecol. Monogr. 21:61–95.

Rybak, E. J., E. J. Neufarth, and S. H. Vessey. 1975. Distribution of the jumping mouse, *Zapus hudsonius,* in Ohio: a twenty-year update. Ohio J. Sci. 75:184–87.

Schwartz, C. W. 1951. A new record of *Zapus hudsonius* in Missouri and notes on its hibernation. J. Mammal. 32:227–28.

Sheldon, C. 1934. Studies on the life histories of *Zapus* and *Napaeozapus* in Nova Scotia. J. Mammal. 15:290–300.

Vickers, E. W. 1894. Notes on *Zapus hudsonicus, Condylura cristata,* and *Sorex platyrhinus.* Second Annu. Rep., Ohio Acad. Sci., p. 15.

Waters, J. H., and B. H. Stockley. 1965. Hibernating meadow jumping mouse on Nantucket Island, Massachusetts. J. Mammal. 46:67–76.

Whitaker, J. O., Jr. 1963. A study of the meadow jumping mouse, *Zapus hudsonius* (Zimmerman), in central New York. Ecol. Monogr. 33:215–54.

———. 1972. *Zapus hudsonius.* Amer. Soc. Mammal., Mammalian Species No. 11:1–7.

Woodland Jumping Mouse
Napaeozapus insignis (Miller)

The woodland jumping mouse has a more restricted range in Ohio than the meadow jumping mouse and is much less known. It was not reported until 1928 when Enders called attention to a specimen taken by Arthur B. Fuller at Mentor, Lake County, on 30 August 1925. Subsequent captures have been made in Ashtabula, Cuyahoga, Geauga, Jefferson, Harrison, Noble, and Belmont counties. To my knowledge a total of fewer than sixty specimens has been collected. The woodland jumping mouse is a boreal animal that inhabits cool ravines and wooded gorges. Although its range extends as far south as the Appalachian Mountains of northern Georgia, it is primarily restricted to latitudes north of Ohio. It is difficult to explain why this mouse inhabits counties bordering Lake Erie in the northeastern portion of Ohio and Belmont, Noble, Harrison, and Jefferson counties 160 km (100 mi) south, but has not been found in the counties in between. Perhaps greater trapping pressure and/or closer observation will reveal a wider range in Ohio.

Napaeozapus insignis (Miller)

The woodland jumping mouse is comparable in size to the meadow jumping mouse but is darker and brighter in color. It has a reddish-brown or buff-yellow coat; the broad mid-dorsal band in darker and, therefore, more prominent than in the meadow jumping mouse. Usually the end of the tail is white, although many specimens do not show this characteristic. Foster (1947) claims that there is considerably less white on the tails of Ohio specimens than on the tails of New York and New Hampshire skins. The belly and tops of the feet are white, and the tail is prominently bicolored. The skull is similar to that of the meadow jumping mouse in every respect except for absence of the peg-like upper premolar. The dental formula is as follows:

$$i\frac{1-1}{1-1}, c\frac{0-0}{0-0}, pm\frac{0-0}{0-0}, m\frac{3-3}{3-3} = 16$$

Following are measurements of 4 adult woodland jumping mice (*Napaeozapus insignis*) from Ohio:

	Total Length (mm)	Tail (mm)	Hind Foot (mm)
Extremes	203–30	120–30	28–30
Mean	219	125	30

The woodland jumping mouse is primarily a woodland species. The meadow jumping mouse may wander into the territory of the woodland jumping mouse, but the reverse is not true. Woodland jumping mice are most frequently found along borders of small streams or in moist, grassy areas bordering woodland bogs and swamps. They are nocturnal animals that seldom appear during the day.

Exact data are not available from Ohio, but I believe that the woodland jumping mouse begins to hibernate at the same time as the meadow jumping mouse and that it emerges at about the same time in the spring. Preble (1956) found these mice entering hibernation as early as the first week in October and emerging the last week in April in New Hampshire. Almost nothing is known about where woodland jumping mice hibernate. They are known to build nests both above- and belowground in the summer, and it is presumed that they hibernate in underground nests similar to those of meadow jumping mice. Sheldon (1938) observed her captive mice burrowing below the surface and building nests of materials provided in the cages; they also usually plugged the holes leading to the nests from the inside. This is probably why few nests have been found in the field. Mice caught in August and September have more fat on them than animals taken at any other time of year, indicating again that formation of a layer of fat is necessary before the animals enter hibernation.

Litter sizes range from two to nine, but four or five is average. Most females proably produce two litters each year, one in June and another in August. The gestation period is 23 to 25 days. The female will eat her own young if the nest is disturbed in any way. The newborn young resemble the young of meadow jumping mice and average about 1 g in weight. Hair appears on the seventh to ninth day. At the end of three weeks the eyes and ears are open and the adult pelage is evident. The young probably leave the home nest permanently when four to five weeks old. Details of the development of the young from birth are presented by Layne and Hamilton (1954).

Where there is limited habitat, as in Ohio, there are concentrations of individuals inhabiting certain areas of favored habitat. Sheldon (1938) found this to be true in her studies in Vermont, but Preble (1956) states that in the area he studied in New Hampshire, where there was sufficient favorable habitat, the mice spread out and were not found in concentrated numbers.

Home ranges are difficult to assess for this species. Blair (1941), live-trapping a 7.36-hectare (18.18-acre) hardwood forest in Michigan, found that the females had ranges of 0.4 to 2.6 hectares (1.0 to 6.5 acres), and the males ranges of 0.4 to 3.6 hectares (1.0 to 9.0 acres). There were 27 females and 25 males living in the 7.36 hectares (18.18 acres). Well-developed trails made and used by woodland jumping mice may be found winding among grass stems and ferns, but more commonly the mice appear to wander freely over their entire territory.

Woodland jumping mice feed on a variety of foods such as seeds of all the grasses found in their natural habitat, berries, small fleshy fruits, alder cones, green vegetation, numerous kinds of insects and insect larvae, and other small invertebrates. The stomachs of 103 mice caught by Whitaker (1963) in New York contained, by volume, 37 percent subterranean fungi, 24 percent various seeds, 20 percent fleshy fruits and miscellaneous vegetation, and 19 percent insect and other animal material. Sheldon (1938) discovered that her captive mice preferred sunflower seeds and powdered milk.

Many predators are not quick enough to catch these agile mice. Since they live in out-of-the-way areas and occur in small numbers, they are not known to cause any problems for man. There are few mammals, however, with more aesthetic appeal than these colorful jumpers.

HOUSE MOUSE
Mus musculus

SOUTHERN BOG LEMMING
Synaptomys cooperi

MEADOW JUMPING MOUSE
Zapus hudsonius

RED FOX
Vulpes vulpes

LEAST WEASEL (Summer)
Mustela nivalis

LEAST WEASEL (Winter)
Mustela nivalis

SELECTED REFERENCES

Blair, W. F. 1941. Some data on the home ranges and general life history of the short-tailed shrew, red backed vole, and woodland jumping mouse in northern Michigan. Amer. Midl. Nat. 25:681-85.

Bole, B. P., Jr. 1935. *Napaeozapus insignis insignis* in Ohio. J. Mammal. 16:153-54.

Enders, R. K. 1928. Two new records for Ohio. J. Mammal. 9:155.

Foster, C. A. 1947. On tail color in *Napaeozapus*. J. Mammal. 28:62.

Hamilton, W. J., Jr. 1935. Habits of jumping mice. Amer. Midl. Nat. 16:187-200.

Hine, J. S. 1929. Distribution of Ohio mammals. Annu. Rep., Proc. Ohio Acad. Sci. 8(6):268.

Layne, J. N., and W. J. Hamilton, Jr. 1954. The young of the woodland jumping mouse, *Napaeozapus insignis insignis* (Miller). Amer. Midl. Nat. 52:242-47.

Newmann, R., and T. J. Cade. 1964. Photoperiodic influence on the hibernation of jumping mice. Ecology 45:382-84.

Preble, N. A. 1956. Notes on the life history of *Napaeozapus*. J. Mammal. 37:196-200.

Saunders, W. E. 1921. Notes on *Napaeozapus*. J. Mammal. 2:237-38.

Sheldon, C. 1934. Studies on the life histories of *Zapus* and *Napaeozapus* in Nova Scotia. J. Mammal. 15:290-300.

———. 1938. Vermont jumping mice of the genus *Napaeozapus*. J. Mammal. 19:444-53.

Snyder, L. L. 1924. Some details on the life history and behavior of *Napaeozapus insignis abietorum* (Preble). J. Mammal. 5:233-37.

Tolin, W. A. 1973. Bioaccumulation of mercury, lead, and cadmium in wildlife inhabiting coal strip-mined areas in Perry, Noble, and Harrison Counties, Ohio. Ohio Coop. Wildl. Research Unit, Quart. Progress Rep. 27(2):45-48.

Whitaker, J. O., Jr. 1963. Food, habitat and parasites of the woodland jumping mouse in central New York. J. Mammal. 44:316-21.

———. 1972. *Napaeozapus insignis*. Amer. Soc. Mammal., Mammalian Species No. 14:1-6.

ORDER CARNIVORA
Flesh Eaters

This order of medium- to large-sized, mammals has a worldwide distribution, although Australia has only one representative, the dingo, and some oceanic islands have no carnivores. They have evolved as primarily meat eaters along lines that are best adapted for pursuing, killing, and eating their prey. The well-developed digitigrade or semiplantigrade limbs have either four or five sharp-clawed toes. The heavy skull possesses powerful jaws equipped with a full heterodont dentition, the main charactristic common to all members of this diverse order. The incisors are small and sharp and modified for holding food. The canines are long, curved, and pointed, well suited for piercing and tearing. The fourth or last premolar in the upper jaw and the first molar in the lower jaw, the carnassial or sectorial teeth, are greatly developed and modified for shearing, cutting, and/or crushing. Three (and possibly five) families of carnivores are present in Ohio.

KEY TO THE FAMILIES OF CARNIVORA IN OHIO

1. Weight of adults more than 45 kg (100 lb); tail inconspicuous; 42 teeth; molar teeth flattened with crushing surfaces ....
.....................URSIDAE (Bear) p. 153
1'. Weight of adults less than 45 kg (100 lb); tail conspicuous; 42 teeth or fewer; if 42, molar teeth modified at least in part with cutting and shearing surfaces 2
2. Digitigrade; four toes on forefeet 3
2'. Plantigrade or semiplantigrade; five toes on forefeet 4
3. Cheek teeth crowns of cutting and shearing type with conspicuous cutting edges; claws completely retractile; tail shorter than, or only slightly longer than, hind foot; form cat-like................................
.............FELIDAE (Bobcat, Lynx) p. 155
3'. Cheek teeth with evident crushing and grinding surfaces; claws not retractile or only partly so; tail longer than hind foot; form dog-like ....
......CANIDAE (Dogs, Foxes, Wolves) p. 119
4. Plantigrade; tail bushy with six or seven dark rings; black mask through face; adults weighing more than 3.6 kg (8 lb); skull with 40 teeth, carnassial teeth modified with crushing surfaces ....
...........PROCYONIDAE (Raccoon) p. 128
4'. Semiplantigrade; tail not bushy or ringed; no mask through face; adults weighing less than 3.6 kg (8 lb); skull with 38 or fewer teeth, carnassial teeth modified with cutting and shearing surface..........................
......MUSTELIDAE (Weasels, Skunk) p. 131

FAMILY CANIDAE
Dogs, Foxes, and Wolves

Members of this family are easily recognized by their dog-like features. They generally have long, pointed muzzles, long legs, and a prominent long bushy tail. There are four digits on the front feet (a fifth being located higher on the leg) and four on the hind feet. The Canidae are swift runners with more than average endurance. They are omnivorous; the carnassial teeth are well developed and prominent, with crushing and grinding surfaces as well as cutting edges. The Canidae, native to all continents except Australia and Antarctica, have always been an interesting and important group to man. Certain of its members are important predators of livestock and poultry. Furs from some members of the group have great economic value, and some members provide sport for hunters and trappers.

There are four genera and eight species of Canidae in the United States. Three genera, each with a single species, are still present in Ohio. They are *Canis lantrans,* the coyote; *Vulpes vulpes,* the red fox; and *Urocyon cinereoargenteus,* the gray fox. A fourth species, *Canis lupus,* the timber wolf (see page 153) was a common resident in Ohio before 1800 and was probably still present until at least 1850. Unlike the coyote and the fox, the timber wolf does not adapt well to civilization, and as the forests were cut and the land became more densely populated, it was eliminated from the state's fauna.

KEY TO THE SPECIES OF CANIDAE IN OHIO

1. Weight of adults more than 7 kg (15 lb); tail less than one-half as long as head and body; color predominantly light-brown or cream; postorbital processes convex; temporal ridges united for most of their length, forming a prominent sagittal crest............
   ....*Canis latrans* (Coyote) p. 120
1'. Weight of adults less than 7 kg (15 lb); tail more than one-half as long as head and body; color red or salt-and-pepper gray; postorbital processes concave; temporal ridges separated for most of their length, not forming a prominent sagittal crest...................... 2
2. General color red, especially on the back; feet black; end of tail white; temporal ridges indistinct or gradually converging posteriorly, forming a V with a short sagittal crest; no conspicuous notch on bottom of lower jaw; upper incisors usually lobed (fig. 22) ....
   ....*Vulpes vulpes* (Red fox) p. 123
2'. General color grizzled or salt-and-pepper gray; feet yellowish or brown; tail tipped with black; temporal ridges prominent, widely separated, and lyrate (U-shaped), not forming a sagittal crest; conspicuous notch on bottom of lower jaw posteriorly; upper incisors not lobed (fig. 22).................
   ....*Urocyon cinereoargenteus* (Gray fox) p. 126

Coyote or Brush Wolf
Canis latrans Say

The history and present status of the coyote in Ohio are somewhat confused. There have been many reports of coyotes in the state, but few of these have actually been verified because the skins and skulls have rarely been preserved. Coyotes are easily confused with domestic dogs *(Canis familiaris),* and it is usually not possible to make positive identification from a photograph or pelt of an animal unless accompanied by the skull.

Canis latrans Say

The first confirmed report of coyote in Ohio was that of a male killed in Preble County in 1947 (Negus, 1948). The skin and skull are in the mammal collection of Miami University. Also in 1948 several notes appeared in the Ohio Conservation Bulletin concerning the presence of coyote in Ohio. Whitacre (1948), in the March issue, states that in 1942 a group of five coyotes, a male, a female, and three pups, "were finally brought to justice after raiding and killing chickens, ducks, and sheep in Lucas, Wood, and Henry counties." In the May issue (p. 31), there is a picture of a coyote (?) pelt taken by a hunter near Mosquito Reservoir, Trumbull County. Also in the May 1956 issue (p. 15) is a photograph of a Lorain County specimen (Miller, 1956).

It is puzzling that in their very thorough treatment of the coyote, Young and Jackson (1951) include Negus's paper and Whitacre's article in their bibliography but in the text do not include Ohio in their locality records. On the range map they show the coyote occurring only in the extreme northwestern corner of the state. Further, Hall and Kelson (1959, p. 845) do not list any locality records for Ohio, although they too include the extreme northwestern corner of Ohio within the range of the coyote. Burt (1957) does not cite specific locality records, but his range map for the coyote shows it occupying most of Ohio, northwestern Pennsylvania, and western New York.

There are few museum records in recent years of coyotes in Ohio. Four skulls are part of the Ohio University mammal collection: two from Harrison County, dated 19 April 1954 and 20 January 1961, a third from Franklin Township, Jackson County, and a fourth from Guernsey County. The Dayton

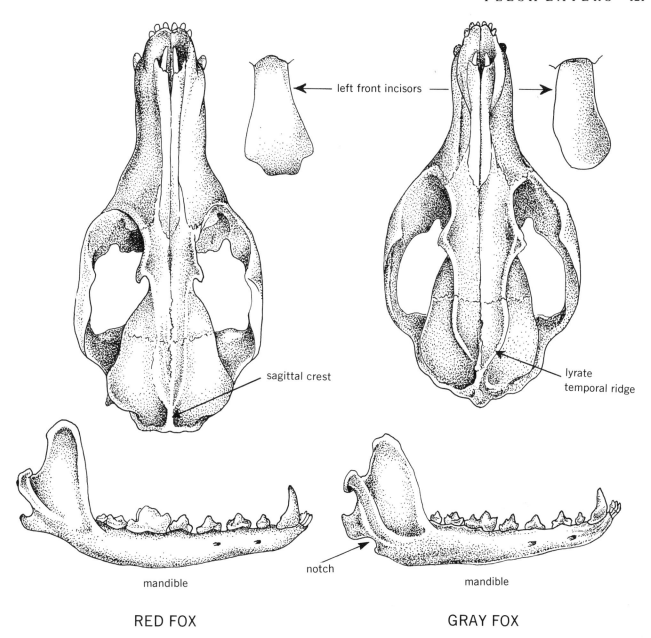

Fig. 22. Dorsal view of skulls and lateral view of mandibles of red fox *(Vulpes vulpes)* and gray fox *(Urocyon cinereoargenteus),* showing distinguishing characteristics. (Drawing by Elizabeth Dalvé.)

Museum of Natural History and Bowling Green State University also have specimens from Montgomery and Erie (Kelley's Island) counties, respectively. A Muskingum County specimen taken from Powelson Wildlife Area in 1976 was deposited in the Ohio State University Department of Zoology Teaching Collection. Sporadic reports have appeared in recent years in various newspapers over the state about the killing, trapping, or sighting of coyotes. These and the Ohio Conservation Bulletin reports are not presented in the distribution map because the skins and skulls have not been preserved and are not available for verification.

Coyotes closely resemble small German police dogs or border collies. When they run, the tail is held down rather than parallel or curled over the back as is characteristic of the dog or wolf. The muzzle is long and pointed, the ears are erect and pointed, and the tail is round and fluffy. The overall color is grayish-white or yellow-white. The top of the nose and back of the ears, legs, and feet are orange, rufous, or cinnamon-colored. There is a thin but prominent dark line extending down the front leg to the foot. The long guard hairs on the back are broadly tipped with black and form a decided mid-dorsal dark band that extends to the base of the tail, the tip of which is black. The white or cream-colored fur on the belly is long and silky. The hair on the back is pinkish-gray at the base; when this is visible, it imparts a pink color to the animal.

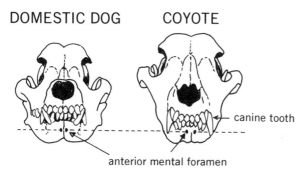

Fig. 23. Contrasting front views of the skulls of the domestic dog *(Canis familiaris)* and the coyote *(Canis latrans)*. The upper canines of the coyote usually extend below the anterior mental foramina (Schwartz and Schwartz, 1959). (Drawing by Elizabeth Dalvé.)

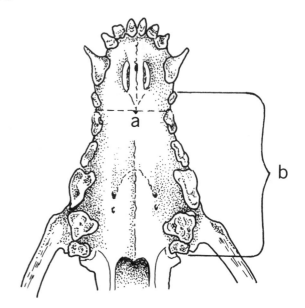

Fig. 24. Anterior end of the coyote *(Canis latrans)* skull, ventral view. If the length of the upper tooth row *(b)* is 3.1 or more times the palatal width *(a)*, the specimen is a coyote (Howard, 1949). (Drawing by Elizabeth Dalvé.)

The skull is larger than that of the fox, and the supraorbital bones are convex rather than concave. There is a prominent sagittal crest. When viewed head-on, the upper canines usually extend below the anterior mental foramina (fig. 23). In the dog the canines do not usually extend below the foramina. Howard (1949) used a ratio between the length of the upper-tooth row and palatal width between the upper first premolars for distinguishing between coyote and dog skulls: "If the molar tooth-row is 3.1 or more times that of the palatal width, the specimen is a coyote, but if it less than 2.7 times, it is a dog" (fig. 24). Howard's method accurately separates 80 to 90 percent of the coyote and dog skulls. Coyotes and dogs have been successfully mated many times; the resulting "coydog" offspring are usually fertile. Wild coydogs have led to much confusion in identification, since on occasion they have been reported as wolves, coyotes, and wild-dogs. The dental formula for the coyote is as follows:

$$i\frac{3-3}{3-3}, c\frac{1-1}{1-1}, pm\frac{4-4}{4-4}, m\frac{2-2}{3-3} = 42$$

Since no one has made a detailed study of the coyote in Ohio, we must depend on studies made elsewhere for information concerning its life history and habits. There is little question that the coyote is more destructive to domesticated animals than is either the gray fox or the red fox. Most of the early reports of coyotes in Ohio mention their killing poultry and livestock. Even today, however, the principal food items in the coyote diet are rabbits and small rodents (ground squirrels, chipmunks, gophers, mice). Sperry (1941) reported the percentage of occurrence of prey species from his excellent food habit study based on an analysis of stomach contents from 8,263 coyotes from throughout their western and midwestern range: rabbits, 34 percent; rodents and carrion, 32.5 percent; domestic livestock, 14 percent; deer, 3.5 percent; poultry, 1.0 percent; game birds, 1.0 percent; nongame birds, 1.0 percent; other miscellaneous vertebrates, 1.0 percent; invertebrates, 10 percent; vegetable food, 2.0 percent. Coyotes readily eat the dead of their own species; they are also said to kill and eat foxes readily. Coyotes may be omnivorous, but most studies indicate that usually less than 5 percent of their food is vegetable matter.

Female coyotes are monestrous with a 4- or 5-day estrus period. Breeding occurs in late winter (January and February) in our latitude. The young are born after a 60- to 65-day gestation period in a well-concealed den. Litter sizes may vary from 4 to 10, but Young and Jackson (1951, p. 81) record litters of 14, 17, and 19. However, 5 to 7 is the average litter size. Two or three females may share the same den while raising their young. The young coyotes emerge from the den when they are about three weeks old. When they are eight to ten weeks old, the entire family abandons the den and become nomadic until fall when the group disbands. Some females may breed when only one year old, but most are probably two years old when they have their first litter.

In reviewing the available data, there is little doubt that the coyote is present in Ohio, but it is not well established. Ohio has suitable habitat for the coyote in large areas of mixed farmland and woods, and although these animals normally inhabit the brushy areas of the western states, they have been gradually moving farther east. Individuals undoubtedly wander into Ohio on occasion, and perhaps some of them will establish themselves here.

For an exciting discussion on the importance and control of the coyote, the reader is referred to the "Symposium on Predatory Animal Control" (Stone, 1930, pp. 325-89). Young and Jackson (1951, pp. 115-69) present a more recent treatment of the subject. At present, at least, the coyote is not well enough established in Ohio to be economically important. However, coyote pups should not be brought into the state as pets because many of the introductions in the eastern states have been made in this way. Coyotes are clever and resourceful animals, and if numbers increase substantially, they could become a serious economic pest in Ohio. For the most inclusive, up-to-date treatment of the coyote (basic biology, behavior, ecology, systematics, and management) see Bekoff (1978).

SELECTED REFERENCES

Alcorn, J. R. 1946. On the decoying of coyotes. J. Mammal. 27:122-26.

Bekoff, Marc. 1977. *Canis latrans.* Amer. Soc. Mammal., Mammalian Species No. 79:1-9.

———. 1978. Coyotes: biology, behavior, and management. Academic Press, N. Y. 384 pp.

Burt, W. H. 1957. Mammals of the Great Lakes region. Univ. of Mich. Press, Ann Arbor. 246 pp.

Cain, S. A., J. A. Kadlec, D. L. Allen, R. A. Cooley, M. G. Hornocker, A. S. Leopold, and F. H. Wagner. 1972. Pest control—1971. Report to Council on Environmental Quality and U.S. Dept. Interior. 207 pp. Review: Howard, W. E., 1972 *in* J. Wildl. Manage. 36:1373-75.

Dalnik, E. H., R. L. Medford, and R. J. Schied. 1976. Bibliography on the control and management of the coyote and related canids with selected references on animal physiology, behavior, control methods, and reproduction. Agric. Res. Serv., Beltsville, Md.

Dobie, J. F. 1961. The voice of the coyote. Univ. of Nebr. Press. 386 pp.

Goodpaster, W. W., and D. F. Hoffmeister. 1968. Notes on Ohioan mammals. Ohio J. Sci. 68:116-17.

Hall, E. R., and K. R. Kelson. 1959. The mammals of North America. Ronald Press Co., New York. 2 vols., 1162 pp.

Hamilton, W. J., Jr. 1943. The mammals of eastern United States: an account of recent land mammals occurring east of the Mississippi. Comstock Publ. Co., Ithaca, N. Y. 438 pp.

Howard, W. E. 1949. A means to distinguish skulls of coyotes and domestic dogs. J. Mammal. 30:169-71.

———. 1974. Predator control: whose responsibility? BioScience 24(6):359-63.

Korschgen, L. J. 1957. Food habits of the coyote in Missouri. J. Wildl. Manage. 21:424-35.

McGinnis, Helen. 1979. Pennsylvania coyotes and their relationship to other wild *Canis* populations in the Great Lakes region in the northeastern United States. M. Sc. Thesis. Pa. State Univ., University Park. 227 pp.

Miller, D. 1956. Coyote coming east? Ohio Conserv. Bull. 20(5):15.

Murie, A. 1940. Ecology of the coyote in the Yellowstone. Fauna of the National Parks of the United States, U.S. Govt. Print. Office, Washington, D.C. Conserv. Bull. 4:1-206. 56 figs.

Negus, N. C. 1948. A coyote, *Canis latrans,* from Preble County, Ohio. J. Mammal. 29:295.

Schwartz, C. W., and E. R. Schwartz. 1959. The wild mammals of Missouri. Univ. of Mo. Press and Mo. Conserv. Comm. 341 pp.

Sperry, C. C. 1941. Food habits of the coyote. U.S. Dept. Interior, Fish & Wildlife Research Bull. 4:1-70.

Stone, W., Pres. 1930. Symposium on predatory animal control. Twelfth Annual Stated Meeting of the American Society of Mammals. J. Mammal. 11:325-89.

Whitacre, D. 1948. The mysterious coyote pack of Ohio. Ohio Conserv. Bull. 12(3):29.

Whiteman, E. E. 1940. Habits and pelage changes in captive coyotes. J. Mammal. 21:435-38.

Young, S. P., and H. Jackson. 1951. The clever coyote. Stackpole Co., Harrisburg, Pa., and Wildl. Manage. Inst., Washington, D. C. 411 pp.

Red Fox
Vulpes vulpes Linnaeus

It is interesting that although the scientific name used since 1920 for the common and familiar red fox has been *Vulpes fulva* (Desmarest), it now appears that *V. vulpes* Linnaeus is indeed its correct name. The problem is that no one is certain whether the red fox in eastern North America is a native species or whether it was introduced from Europe. Bone piles of prehistoric origin unearthed in Ohio, New York, and Pennsylvania, for example, have failed to produce red fox bones; Indians from these areas relate that the red fox was unknown to them before the coming of the white man. It is also known that European foxes had been introduced between 1700 and 1750 into New England, Pennsylvania, and Virginia for sporting purposes. It is believed that many of these foxes moved rapidly into the lands that were being cleared in the West. Churcher (1959) indicates, however, that red fox bones dated prior to the time Europeans first arrived in this country have been recovered from Indian mounds in southern Ontario. He states:

Thus the conclusion to be drawn is that a red fox was native to North America north of Lat. 40°N or 45°N, but was either scarce or absent from most of the unbroken mixed hardwood forests, where the gray fox was paramount. The European red fox was introduced into the eastern seaboard area about 1750, and either partially displaced the gray fox in the southern portion of the continent, or interbred with the scarce population of indigenous red fox to produce a hybrid population.

Churcher presents further evidence to show that the red foxes of North America and Eurasia belong to a single species. Since the European red fox is *V. vulpes* and since it was named before the North American red fox, the latter would also be *V. vulpes.*

The red fox is now found throughout the eastern United States and Canada. Thomas (1951), discussing the distribution of animals in Ohio, wrote, "*fulva* also had a northerly range in pre-Columbian times, from which it spread southward and southeastward with the clearing of the forests." Red foxes probably came into Ohio between 1750 and 1800, some native animals coming from the North and other introduced animals entering from Pennsylvania and Virginia. This species has been reported from every county in Ohio.

Vulpes vulpes Linnaeus

The red fox is a small- to medium-sized, dog-like mammal with erect, pointed ears and a long, round, white-tipped, bushy tail. The coat is most often red or yellowish red; the black feet and black on the back of the ears contrast sharply with the rest of the body. The belly, throat, and sides of the cheeks are white. Color variations are fairly common; pure black, silver, and other color variations may occur in the same litter with normal reds. To my knowledge we have no information on the frequency of occurrence of these genetic types in Ohio red foxes. The skull characteristics are those given in the key and shown in figure 22. The dental formula is as follows:

$$i\frac{3-3}{3-3}, c\frac{1-1}{1-1}, pm\frac{4-4}{4-4}, m\frac{2-2}{3-3} = 42$$

Following are measurements of 16 adult red foxes (*Vulpes vulpes*) from Ohio:

	Total Length (mm)	Tail (mm)	Hind Foot (mm)
Extremes	900–1,054	325–419	150–72
Mean	979	370	160

The red fox is an animal of open areas rather than woods. Its primary habitat is semicultivated farmlands with scattered deciduous woods. Although secretive and shy, it sometimes lives surprisingly close to civilization. A number of wooded areas within the city limits of Cincinnati are inhabited by red foxes. In the spring of 1968, a mother and her two pups regularly visited and removed the peanut butter-oats baits from the squirrel traps we had set in one of the local cemeteries. The animals became so accustomed to us that we often approached within 30 meters before they were disposed to run. The red fox is more active during the night, but it may be seen hunting at any hour of the day or night.

Mating takes place in the late winter (December through mid-February) in Ohio, and the young are born during February, March, and April, with the peak of whelping occurring in March. The gestation period is 51 to 53 days. Numerous studies on the reproductive habits of the red fox (Seagears, 1944; Gier, 1947; Sheldon, 1949; Hoffman and Kirkpatrick, 1954; Schofield, 1948; and others) indicate that the average litter size is 5 or 6. As many as 11 pups have been born in a litter, and Hoffman and Kirkpatrick (1954) found the remarkable number of 13 fetuses in one female. Litter sizes vary in different regions and from year to year, presumably because of yearly variations in environmental conditions and population densities. The various factors regulating litter sizes, however, are not well understood.

Females are monestrous, and most produce their first litter when they are yearlings. The young are born and raised in a burrow dug by the parents. The den is often an abandoned woodchuck hole that has been enlarged and renovated by the foxes. It may be located within a woods, in an open field, or on the border of a woodlot. Usually more than one den is prepared; if one is discovered or disturbed, the entire family will be moved into new quarters. Several families may share the same den.

Red foxes are at least semimonogamous. A male and a female remain together for one entire breeding season (midwinter through summer in Ohio), and some pairs may remain together for more than one breeding season. Sheldon (1950), however, believes that some males may mate with several females during the breeding period. Males are known to be good providers and usually help the female care for and raise the young. Sheldon (1949) states that the average weight of the young is about 100 g at birth. Their eyes open when they are seven or eight days old. By the end of the fourth or fifth week, they are playing around the den entrance. They catch and eat mice and other small rodents when eight weeks old but remain near the den and with the family until they are about three months old (usually August or September in Ohio).

The home range of an adult red fox normally encompasses 2.5 to 5.0 sq km (1 to 2 sq mi) at the most. In the fall, however, when the young leave the den and disperse, they may travel as much as 24 to 32 km (15 to 20 mi) before settling down. Ables (1965) marked a five-month-old male at Madison, Wisconsin, on 29 August 1962; the animal was later shot on 20 May 1963 in Montgomery County, Indiana, a distance of 394 km (245 mi).

The food habits of the red fox have been investigated by a number of workers. Much has been written concerning the effect of fox predation on wild game species and domestic poultry. However, the studies summarized by Korschgen (1959) and others have shown that the red fox's effect upon game species is minor, and under ordinary conditions it does little damage to poultry. Foxes are omnivores, eating whatever is available to them; thus their food habits change with the seasons. In summer and fall when fruits and berries are abundant, they may feed almost exclusively on these vegetative items. In winter and spring they eat principally rabbits, rodents (mostly mice), and an occasional bird, or if these preferred items are scarce or absent, they may turn to frozen apples, winter wheat, or garbage. The stomach of a crippled red fox examined by Dexter (1951) was filled with a mass of timothy grass *(Phleum pratense)* and a few miscellaneous herbs. Obviously, this fox was eating what it could "catch."

The economic status of the red fox in Ohio has varied considerably since about 1940, principally because of the changes in prices paid for the pelt. In summarizing the fox bounty and fur harvest reports of 1944–45 to 1956–57, Bednarik (1959) states: "In 1945–46, Ohio records show a harvest of 25,857 foxes valued at $73,857.00 or an average pelt value of $2.83, while in 1956–57, 834 foxes valued at $190.00 or an average pelt value of $0.22 were harvested." In the 1970s, with a significant rise in fur prices, an average fox pelt brought over $40.00 (Appendix Table 2) with prime pelts bringing up to $80.00 each! With the change in economic status, foxes are now protected as furbearers.

Many counties (and some private hunting clubs) in Ohio used to pay bounties of from $1.50 to $5.00 for a dead fox. In 1962, seventy counties paid fox bounties. By 1972, however, only seventeen counties continued the practice (Ohio Division of Wildlife, 1974), and effective 28 November 1975, bounties were eliminated by an act of the Ohio General Assembly. Bounties were originally levied to encourage the killing of foxes because it was believed that by killing the foxes that prey on game animals and birds, more game animals would be present and hunting would improve. Not only was the bounty system costly to the citizenry of the counties employing it (Bednarik [1959] reported $530,000 paid between 1944 and 1957), but it seldom, if ever, accomplished its purpose. The fact is that, despite the bounty system, the gray fox has steadily increased its numbers in Ohio since 1940, and the red fox is certainly holding its own or better. Ohio is not unique in this regard. The bounty system in some form has been used throughout this country since the late 1600s and in Europe since about 1550. Results are usually the same; corrupt practices in collecting the bounty, waste of taxpayers' money, and a failure to accomplish its purpose.

Because of their secretive habits, suspicious nature, speed, and ability to take advantage of available cover, red foxes may be difficult to trap or hunt. Trapping is especially difficult, requiring a great deal of time and patience and a high initial outlay of money for traps and other equipment. With increases in pelt value and the challenge this species offers the sportsman, the red fox has become an important "game animal," quite in contrast to the label "varmint" assigned in the past.

SELECTED REFERENCES

Ables, E. D. 1965. An exceptional fox movement. J. Mammal. 46:102.

Allen, S. H. 1974. Modified techniques for aging red fox using canine teeth. J. Wildl. Manage. 38:152–54.

Bednarik, K. E. 1959. The fox: thirteen years of bounty and fur harvest in Ohio. Ohio Dept. Nat. Resources, Div. Wildl. Publ. 177. 7 pp.

Bezdek, H. 1942. Encephalitis (?) in red fox in southwestern Ohio. J. Mammal. 23:98.

Churcher, C. S. 1959. The specific status of the New World red fox. J. Mammal. 40:513–20.

Dexter, R. W. 1951. Food of a crippled red fox. J. Mammal. 32:464.

Gale, L. R. 1947. Foods of foxes in southeastern Ohio. M. S. Thesis, Ohio Univ., Athens.

Gier, H. T. 1947. Populations and reproductive potentials of foxes in Ohio. Ninth Midwest Wildlife Conference. Mimeo. 6 pp.

Hendon, E. V. 1942. The status of the red fox in eastern Ohio. M. S. Thesis, Ohio State Univ., Columbus. 127 pp.

Hoffman, R. A., and C. M. Kirkpatrick. 1954. Red fox weights and reproduction in Tippecanoe County, Indiana. J. Mammal. 35:504–9.

Korschgen, L. J. 1959. Food habits of the red fox in Missouri. J. Wildl. Manage. 23:168–76.

Layne, J. N., and W. H. McKeon. 1956. Some aspects of red fox and gray fox reproduction in New York. N. Y. Fish & Game J. 3:44–74.

Mitchell, K. A. 1940. The status of the red fox in west-central Ohio. M. S. Thesis, Ohio State Univ., Columbus. 82 pp.

Morse, M. A., and D. S. Balser. 1961. Fox calling as a hunting technique. J. Wildl. Manage. 25:148–54.

Ohio Division of Wildlife. 1974. Fox bounty payments in Ohio, 1944–1972. Wildlife In-Service Note 253. Game Management Section, Ohio Dept. of Nat. Resources. 5 pp.

Russell, K. R. 1968. Fox bounty payments in Ohio, 1944–1967. Wildlife In-Service Note 97, Ohio Dept. Nat. Resources, Div. Wildl. 10 pp.

Schofield, R. D. 1958. Litter size and age ratios of Michigan red foxes. J. Wildl. Manage. 22:313–15.

Seagears, C. B. 1944. The red fox in New York. N. Y. Conserv. Dept., Education Bull. 83 pp.

Sheldon, W. G. 1949. Reproductive behavior of foxes in New York State. J. Mammal. 30:236–46.

———. 1950. Denning habits and home range of red foxes in New York State. J. Wildl. Manage. 14:33–42.

———. 1953. Returns on banded red and gray foxes in New York State. J. Mammal. 34:125–26.

Thomas, E. S. 1951. Distribution of Ohio animals. Ohio J. Sci. 51:153–67.

Vogtsberger, L. M., and G. W. Barrett. 1973. Bioenergetics of captive red foxes. J. Wildl. Manage. 37:495–500.

Gray Fox
Urocyon cinereoargenteus (Schreber)

The gray fox is a native North American species that was well established in the forests of Ohio long before the white man arrived. Since it is a woodland form, the clearing of land selected against it, and apparently the gray fox population was drastically reduced in Ohio during the latter part of the 1800s. Hicks (1941), however, reported that in 1941 the gray fox was "extremely abundant" in south central Ohio, with scattered or no records from counties in the northern part of the state. Today the gray fox is found in every county in Ohio, but it is still uncommon in northwestern Ohio, particularly in the western tier of counties bordering Indiana. Generally speaking, if a straight line is drawn diagonally across the state from Hamilton County in the southwest to Ashtabula County in the northeast, in the southeastern area gray foxes are extremely abundant to common, and in the northwestern area gray foxes are common to uncommon.

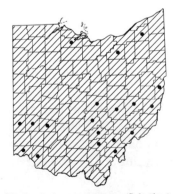

Urocyon cinereoargenteus (Schreber)

The gray fox has a more southern distribution than the red fox, rarely occurring in Canada in extreme southern Ontario and Quebec. It ranges westward to the Pacific Coast but is not found in the high plains, northern Rockies, or Pacific Northwest. Various subspecies extend through Mexico and Central America into South America.

Although somewhat smaller, the gray fox is similar in general appearance to the red fox. The ears and muzzle are shorter, and the fur is not as long and silky. The body color is predominantly salt-and-pepper gray or black instead of red. A black stripe starting on the rump extends along the dorsal surface of the tail and ends in a black tip. A black line extends from the nose through the eye and onto the side of the head. The cheek, lower jaw, and throat are white. A rich rufous patch of fur is present on the side of the head just below the ear. The backs of the ears are dark red or rufous. A broad band of light orange fur separates the gray fur on the back from the long white fur on the belly. Some of the orange fur may extend onto the belly, where it mixes with the white hair. The legs are rufous or light orange, and the feet are of the same color as the back.

Skull characteristics are presented in the key (p. 120) and figure 22. The dental formula is the same as for the red fox. Following are measurements of 10 adult gray foxes *(Urocyon cinereoargenteus)* from Ohio:

	Total Length (mm)	Tail (mm)	Hind Foot (mm)
Extremes	831–990	285–435	124–50
Mean	923	341	135

The gray fox is found more commonly in woodlands than in open fields, and for this reason it is probably seen less often than the red fox. The largest populations are in rural areas of the state where open fields and woodlots are still present. Like the red fox, however, it is adapting to civilization.

Gray fox dens are not often found because they are frequently located in dense thickets or woods. Wood (1958) states that old slab piles and sawdust heaps around abandoned sawmills are the most common denning sites. Hollow logs and stumps are also used, and the animals may burrow beneath an old barn or rock pile. The reproductive habits are similar to those of the red fox with one or two exceptions. Sheldon (1949), Layne (1958), and others agree that gray fox breeding activity occurs approximately one month later than the red fox breeding period. Mating in Ohio probably takes place from January through April, with the height of activity sometime in March. This agrees with the breeding season determined for the gray fox in other areas within the same general latitude as Ohio (Layne and McKeon, 1956; Layne, 1958). Litters are born in March, April, and May. The gestation period is believed to be 53 days. Litter sizes are smaller than those of the red fox, with a range of 1 to 9 (average, 4.35), as determined in eight different studies (see Selected References below). Adult size is reached within five to six months, and sexual maturity during the first year. Most gray foxes travel from 1.6 to 16 km (1 to 10 mi) when they disperse from the home den in the fall. However, one female marked by Sheldon (1953) in New York was trapped two and one-half years later 83 km (52 miles) away, a record travel distance for banded gray foxes.

Food habits of the gray fox are similar to those of the red fox. Small mammals and plant material are the two most important categories of food throughout the year, with birds, carrion, insects, and miscellaneous items filling out the diet.

Economically, the gray fox is less important than the red fox because the short, rather coarse fur is worth much less. Fur dealers offered a top price of only $0.15 for a prime pelt during the late 1950s, although the same pelt could bring over $30.00 in the late 1970s. Nevertheless, sales of approximately 20,000 gray fox pelts a year have been reported by Ohio fur buyers (Appendix Table 2). Most hunters are not interested in hunting this animal because of its habit of immediately climbing a tree or "holing-up" when pursued. The gray fox probably does less damage to livestock and poultry than does the red fox. The species carries rabies, however, and any individual that appears sick or that fails to run when approached should be avoided or killed.

SELECTED REFERENCES

Gale, L. R. 1947. Foods of foxes in southeastern Ohio. M. S. Thesis, Ohio Univ., Athens.

Gier, H. T. 1947. Populations and reproductive potentials of foxes in Ohio. Ninth Midwest Wildlife Conference. Mimeo. 6 pp.

Hicks, L. 1941. Official letter to Game Management section, Ohio Dept. Conserv.

Layne, J. N. 1958. Reproductive characteristics of the gray fox in southern Illinois. J. Wildl. Manage. 22:157-63.

Layne, J. N., and W. H. McKeon. 1956. Some aspects of red fox and gray fox reproduction in New York. N. Y. Fish & Game J. 3:44-74.

Lord, R. D., Jr. 1961. The lens as an indicator of age in the gray fox. J. Mammal. 42:109-11.

Richards, S. H., and R. L. Hine. 1953. Wisconsin fox populations. Wisc. Conserv. Dept., Tech. Wildl. Bull. No. 6. 78 pp.

Sheldon, W. G. 1949. Reproductive behavior of foxes in New York State. J. Mammal. 30:236-47.

———. 1953. Returns on banded red and gray foxes in New York State. J. Mammal. 34:125-26.

Sullivan, E. G. 1956. Gray fox reproduction, denning, range, and weights in Alabama. J. Mammal. 37:346-51.

Wood, J. E. 1958. Age structure and productivity of a gray fox population. J. Mammal. 39:74-86.

———. 1959. Relative estimates of fox population levels. J. Wildl. Manage. 23:53-63.

Wood, J. E., D. E. Davis, and E. V. Komarek. 1958. The distribution of fox populations in relation to vegetation in southeastern Georgia. Ecology 39:160-62.

FAMILY PROCYONIDAE
Raccoons and Allies

Members of this family reside primarily in the New World: North America, Central America, and the northern two-thirds of South America. Two forms of doubtful affinity, the pandas, inhabit western China, and in 1941 were successfully introduced into Russia. Procyonids are medium-sized mammals with plantigrade or semiplantigrade feet with five toes, each with a nonretractile claw. The tail is moderately long in most species and prehensile in some of the tree-dwelling forms. The skull is short and broad, and the teeth are better suited for an omnivorous diet than a carnivorous one. The carnassials are not strongly developed, and the cheek teeth have flat crushing surfaces.

Seven genera are recognized in the New World and two in Asia. *Procyon,* which ranges from Brazil northward into Canada, is the most common genus found in the United States. There are two other genera found only in Texas and along the Mexican border. Only one species of *Procyon* occurs in the United States: *P. lotor,* the common and well-known raccoon.

Raccoon
Procyon lotor (Linnaeus)

Raccoons inhabit every county in Ohio. During the 1930s they were scarce, but since that time they have steadily increased in numbers.

Raccoons are medium-sized mammals, adults usually weighing 4.5-16.0 kg (10-35 lb). Their most distinguishing external features are the black mask across the face and the moderately long, black-

Procyon lotor (Linnaeus)

ringed tail. The hair is coarse and long, and the usual color is salt-and-pepper gray. White, black, brown, blond, and a number of other color phases are known but are not common in Ohio specimens. The tips of the ears and tops of the feet are light brown. The skull is broad and heavy, and the hard palate extends well beyond the last molars. The canine teeth are well developed, but the cheek teeth are flat-crowned with crushing and grinding surfaces. There are two upper molars on each side rather than one as in the mustelids. The raccoon is the only mammal in Ohio with a total of 40 teeth. The dental formula is as follows:

$$i\frac{3-3}{3-3}, c\frac{1-1}{1-1}, pm\frac{4-4}{4-4}, m\frac{2-2}{2-2} = 40$$

Following are measurements of 50 adult raccoons *(Procyon lotor)* from Ohio:

	Total Length (mm)	Tail (mm)	Hind Foot (mm)
Extremes	690-840	190-295	96-115
Mean	757	247	105

Although active throughout most of the year, raccoons caught in the fall in Ohio have a thick subcutaneous layer of fat covering most of the body as well as large deposits of internal mesenteric fat. Williams (1909) determined that one-sixth of the total carcass weight of a raccoon taken in Butler County, Ohio, was stored fat. From December through February or March, raccoons are relatively quiet, remaining in the dens and sleeping most of the time. During this time the excess fat deposits are used as a fuel supply, and an adult raccoon may weigh less than one-half as much in March as it did in early December. When normal activity and feeding is resumed in the spring, the animals gradually gain weight until the following fall when there is again a sudden increase in feeding and weight gain just before the winter rest period.

Raccoons are flexible in their choice of den sites. The majority choose natural hollows in trees, but they may use also hollow logs, caves, crevices in rocks, or even holes in the ground. They are almost unique in that they cannot make their own dens, and they do not make nests in their dens. In Cincinnati, as elsewhere in Ohio, some raccoons use storm sewers as den sites (Schinner, 1969; Cauley, 1971).

Water is an important commodity in the life of a raccoon. Contrary to popular belief, however, raccoons do not necessarily wash their food before eating it. Most of the raccoon's food comes from streams or banks of streams. Stuewer (1943) found that all the dens in his study area were within 0.4 km (0.25 mi) of a body of water. Both Stuewer (1943) and Butterfield (1954a) caught most of their animals in traps that were set along streams and ponds.

Raccoons mate in February and March in Ohio. Males are promiscuous, mating with several females each year. The gestation period is 63 days. Normally, only one litter is produced by a female each year. Most litters in Ohio are born in April and May, although a few females that are not successfully mated in the spring may mate later and produce a late summer brood. McKeever (1958), working in southwestern Georgia and northwestern Florida, found raccoons giving birth to young from April to early October, with approximately half of the births occurring in May. George and Stitt (1951) observed three litters near Ann Arbor, Michigan, that were born in mid-March. Litter sizes range from two to seven, with four being average. Llewellyn (1953) has described the fetal development and Hamilton (1936) the early stages of postnatal development of the raccoon. The newborn young are well furred. The eyes open on or about the nineteenth day, and the annulations on the tail are evident at that time. Weaning takes place during the sixth or seventh week after birth, when the young raccoon weighs about 690 g (1.5 lb). The young remain with the mother until September or October, although some families may stay together until the next breeding period. Stuewer (1943), Pope (1944), and Wood (1955) present evidence indicating that 50 percent or more of the yearling females produce litters but that very few yearling males have successful matings. The growth rate of raccoons is relatively slow; individuals do not reach adult size until they are at least two years old (Stuewer, 1943). Little is known concerning life expectancy of the raccoon in the wild, but Haugen (1954) reported a female raccoon over 12.5 years of age shot within 0.8 km (0.5 mi) of where it was originally live-trapped and tagged as a juvenile. Most raccoons probably live only 6 to 9 years under natural conditions.

Few mammals use as many different kinds of food as does the raccoon. Nuts, fruits, berries, grass, leaves, seeds, insects, crustaceans, worms, eggs, fish, and various small mammals are all eaten. Undoubtedly, this ability to make use of a great variety of foods has been an important factor in the success of the species. In the summer fruits and berries comprise most of the raccoon's diet, and acorns predominate in the fall. Steuwer (1943) in his excellent account of the raccoon in Michigan states: "The most striking fact recorded concerning food habits was the preference exhibited by raccoons for acorn mast whenever it was available." My findings concur; the stomachs of two raccoons that I examined in November contained 90 percent acorn remains and 10 percent undetermined material. During winter animal food in the form of insects, mice, crayfish, fish, and earthworms may become more prominent dietary items. The raccoon's fondness for corn, green or dried, is well known. In Cincinnati the "citified" raccoons, having become adept at removing garbage can lids, grow fat on a nutritious diet of kitchen scraps. Dexter (1951) found earthworms in 50 percent of the stomachs (averaging 12.6 percent of total stomach volume) he examined of raccoons caught in winter in northern Ohio.

Raccoons are primarily nocturnal animals, but they occasionally leave the den during the day to bask in the sun or look for food. Their nightly wandering may take them several miles from the den site, but ordinarily, when food and water are readily available, they seldom wander more than 3.2 km (2 mi). Stuewer (1943) calculated the theoretical circular ranges of male raccoons to be a maximum of 847 hectares (2,092 acres), a minimum of 18.2 hectares (45 acres), and an average of 204 hectares (504 acres). Females and juveniles have home ranges approximately one-half this size. Our studies in Cincinnati indicate that raccoons living in suburban areas seldom have home ranges that exceed 3.2 hectares (8 acres).

In the fall when the litters disperse, individuals

regularly move 16 to 32 km (10 to 20 mi) from the home den before choosing a new den of their own. A young male tagged in Marshall County in Minnesota was trapped 264 km (165 mi) away three years later (Priewert, 1961). Though probably atypical, this does indicate that at times raccoons travel great distances, especially when moving into new, unoccupied territory. Although raccoons are solitary animals, adult home ranges broadly overlap, and individuals do not attempt to defend a territory. Population density outside cities probably approximates about one raccoon for every 12 to 16 hectares (30 to 40 acres) of suitable habitat. In cities raccoons may be found living in 0.36 hectare (0.9 acre).

As a furbearer in Ohio, the raccoon has achieved top monetary values. In the 1930s raccoon pelts brought as much as $5.00 each. During the 1940s the price dropped to an average of $3.50 per pelt, and in the 1960s it was about $1.00. However, in the late 1970s prices averaged over $15.00 (Appendix Table 2), and it became the primary furbearing species in the state.

The raccoon is highly regarded as a game species in Ohio. Many hunters pursue raccoons after dark with dogs and lights just for sport. Young raccoons make interesting and lively pets but usually with age become too destructive to be tolerated. A young raccoon belonging to a friend of mine ripped the stuffing out of the back seat of his automobile. On another occasion this same animal playfully (?) bit completely through his owner's ear!

Raccoons have no large enemies in Ohio; in a one-on-one situation there are few dogs that can overcome a full-grown, mature raccoon. Most areas in Ohio are well suited to the needs of the raccoon, and it probably will continue to increase in numbers throughout the state. In some park areas raccoons present a nuisance by frequently raiding and overturning garbage cans and trash receptacles. An outbreak of rabies in Columbus and Franklin County in 1977 confirms that raccoons occasionally carry rabies and therefore should not be handled indiscriminately.

SELECTED REFERENCES

Berard, E. V. 1952. Evidence of a late birth for the raccoon. J. Mammal. 33:247-48.

Butterfield, R. T. 1944. Populations, hunting pressure, and movements of Ohio raccoons. Trans. Ninth North Amer. Wildl. Conf. 9:337-43.

———. 1954a. Traps, live-trapping, and marking of raccoons. J. Mammal. 35:440-42.

———. 1954b. Some raccoon and groundhog relationships. J. Wildl. Manage. 18:433-37.

Cauley, D. L. 1971. The effects of urbanization on raccoon (*Procyon lotor*) populations. M. S. Thesis, Univ. of Cinncinnati, Cincinnati, Ohio. 55 pp.

Dexter, R. W. 1951. Earthworms in the winter diet of the opossum and raccoon. J. Mammal. 32:464.

Dorney, R. S. 1954. Ecology of marsh raccoons. J. Wildl. Manage. 18:217-25.

George, J. L., and M. Stitt. 1951. March litters of raccoons (*Procyon lotor*) in Michigan. J. Mammal. 32:218.

Goldman, E. A. 1950. Raccoons of North and Middle America. North Amer. Fauna 60:1-153.

Hamilton, W. J., Jr. 1936. The food and breeding habits of the raccoon. Ohio J. Sci. 36:131-40.

———. 1940. The summer food of minks and raccoons on the Montezuma Marsh, New York. J. Wildl. Manage. 4:80-84.

Haugen, O. L. 1954. Longevity of the raccoon in the wild. J. Mammal. 35:439.

Hoffmann, C. O., and J. L. Gottschang. 1977. Numbers, distribution, and movements of a raccoon population in a suburban residential community. J. Mammal. 58:623-36.

Llewellyn, L. M. 1953. Growth rate of the raccoon fetus. J. Wildl. Manage. 17:320-21.

McKeever, S. 1958. Reproduction in the raccoon in the southeastern United States. J. Wildl. Manage. 22:211.

Petrides, G. A. 1950. The determination of sex and age ratios in fur animals. Amer. Midl. Nat. 43:355-82.

Pope, C. H. 1944. Attainment of sexual maturity in raccoons. J. Mammal. 25:91.

Preble, N. A. 1940. The status and management of raccoon in central Ohio. M. S. Thesis, Ohio State Univ., Columbus. 149 pp.

———. 1941. Raccoon management in central Ohio. Ohio Wildlife Research Unit, Ohio State Univ., Release 161. 9 pp.

Priewert, F. W. 1961. Record of an extensive movement by a raccoon. J. Mammal. 42:113.

Sagar, R. G. 1956. A study of factors affecting raccoon reproduction in Ohio. M. S. Thesis, Ohio State Univ., Columbus. 78 pp.

Schinner, J. R. 1969. Ecology and life history of the raccoon (*Procyon lotor*) within the Clifton suburb of Cincinnati, Ohio. M. S. Thesis, Univ. of Cincinnati, Cincinnati, Ohio. 60 pp.

Stuewer, F. W. 1942. Studies of molting and priming of fur of the eastern raccoon. J. Mammal. 23:339-404.

———. 1943. Raccoons: their habits and management in Michigan. Ecol. Monogr. 13:203-58.

Urban, D. 1968. The ecology of racoons on a managed waterfowl marsh in southwestern Lake Erie. M. S. Thesis, Ohio State Univ., Columbus. 98 pp.

———. 1970. Raccoon populations, movement pat-terms, and predation on a managed waterfowl marsh. J. Wildl. Manage. 34:372-82.

Whitney, L. F. 1931. The raccoon and its hunting. J. Mammal. 12:29-38.

Williams, S. R. 1909. On hibernation in the raccoon. Ohio Nat. 9:495-96.

Wood, J. E. 1955. Notes on reproduction and rate of increase of raccoons in the post oak region of Texas. J. Wildl. Manage. 19:409-10.

FAMILY MUSTELIDAE
Weasels, Skunk, Mink, Otter, and Badger

Mustelids are medium- to small-sized carnivores that are more commonly found in the north temperate regions. They have a worldwide distribution except for Australia and Antarctica. Each foot of a mustelid has five toes with nonretractile claws. The limbs are digitigrade or semiplantigrade. The ears are short and rounded. Anal glands (musk glands) are present and, in some species, developed to a remarkable degree. The skull has a short rostrum and flattened cranium, and the tympanic bullae are flattened and only slightly inflated. The molar teeth have cutting surfaces and the carnassials are usually highly developed. There are never more than two pairs of molars in each jaw.

Twenty-six genera are recognized in the family (Cockrum, 1962). Nine genera and fifteen species are found in North America. In Ohio the family is represented by three genera and seven species.

SELECTED REFERENCES

Cockrun, E. L. 1962. Introduction to mammalogy. Ronald Press Co., New York. 455 pp.

KEY TO THE SPECIES OF MUSTELIDAE IN OHIO

1. Toes fully webbed; tail longer than 35 cm (14 in) and thickened at base .... *Lutra canadensis* (River Otter) p. 154
1'. Toes not webbed; tail less than 35 cm (14 in) long and not thickened at base 2
2. White median stripe on the forehead; color black, white, black with white stripes, or yellowish gray 3
2'. No white median stripe on forehead; color brown; (in winter may be white) 4
3. Tail longer than 17.5 cm (7 in) and bushy; color usually black with two white stripes originating on the shoulders and gradually separating as they pass down the back and over the rump; occasionally all black or all white; anal glands well developed; front claws less than 25 mm in length .... *Mephitis mephitis* (Striped Skunk) p. 140
3'. Tail shorter than 17.5 cm (7 in), not bushy; color salt-and-pepper or yellowish-gray, cheeks white; anal glands not well developed; front claws 25-32 mm in length and obviously adapted to digging .... *Taxidea taxus* (Badger) p. 138
4. Length 140 to 220 mm; tail 20-40 mm, not tipped with black .... *Mustela nivalis* (Least Weasel) p. 132
4'. Length greater than 220 mm; tail more than 40 mm and tipped with black 5
5. Length usually greater than 450 mm; overall color dark brown or chocolate; white or yellowish chin; white or yellowish spot sometimes on belly and throat .... *Mustela vison* (Mink) p. 137
5'. Length usually less than 450 mm; dorsal color brown; belly white or yellowish; in winter, coat may be white with black tip on tail 6
6. Tail shorter than ⅓ total length; white or yellow on belly extends to inside of legs and often onto feet and toes .... *Mustela erminea* (Ermine) p. 131
6'. Tail longer than ⅓ total length; white or yellow on belly does not extend to inside of legs or onto feet .... *Mustela frenata* (Long-tailed Weasel) p. 134

Ermine or Short-tailed Weasel
Mustela erminea Linnaeus

The ermine is primarily an animal of the cold north, although it is found throughout Pennsylvania and has been taken as far south as Bethesda, Maryland. It is probably not a regular resident of Ohio, but it should not be surprising to find that an occasional individual has wandered into the state from Michigan or Pennsylvania.

Henninger (1921) reported that "a fine male of *Mustela cicognanii*" (*M. erminea*) had been captured in Ohio. Prior to that, he had considered it unlikely that this species would ever be found in the state. The animal was deposited as specimen no. 925 in the collection of the Ohio State University Museum of Zoology. However, Hall (1937) discovered that the specimen had been misidentified and that it was actually a long-tailed weasel, *M. frenata*. Bole and Moulthrop (1942) reported that in November 1937 a female ermine was collected at Pepper Pike, Cuyahoga County, Ohio. This specimen was given

Mustela erminea Linnaeus

to the Cleveland Museum of Natural History. A male was caught in Trumbull Township, Ashtabula County, on 17 April 1964 (no. 2109 in the Ohio State University Museum of Zoology). Its measurements are as follows: total length, 252 mm; tail, 60 mm; hind foot, 36 mm; weight, 96.3 g. Another male specimen came into possession of the Ohio Division of Wildlife. It had been trapped in early 1978 near Mentor Marsh in Lake County, and its measurements are as follows: total length, 265 mm; tail, 70 mm; hind foot, 35 mm.

Hamilton (1933) took the following measurements of 46 adult ermines in New York. Male weasels are decidedly larger than females, and for this reason the measurements have been given separately for the sexes.

	Total Length (mm)	Tail (mm)	Hind Foot (mm)	Weight (g)
31 Males				
Extremes	251–95	65–80	32–38	66–105
Mean	272	71	35	81
15 Females				
Extremes	194–255	44–65	28–31	45–71
Mean	237	55	29	54

The ermine is easily mistaken for the long-tailed weasel, but differs from it in the following ways: (1) the ermine is a much smaller animal (see measurements of each); (2) the white or sulfur-white area on the venter is more extensive on the ermine, extending onto the feet, toes, and upper lip; (3) the tail of the ermine is black tipped in all seasons but proportionally shorter than that of the long-tailed weasel, seldom exceeding one-third of the total body length; (4) the ermine becomes completely white in winter except for the black tip on the end of the tail; (5) the skull of the ermine is smaller, and the postorbital processes are blunt rather than pointed.

On the whole, the general habits, life history, and reproductive habits of the ermine are similar to those of the long-tailed weasel (see p. 134).

SELECTED REFERENCES

Bissonnette, T. H. 1942. Anamalous seasonal coat-color changes in a small male Bonaparte's weasel (*Mustela cicognanii cicognanii* Bonaparte). Midl. Nat. 28:327–33.

Bole, B. P., Jr., and P. N. Moulthrop. 1942. The Ohio recent mammal collection in the Cleveland Museum of Natural History. Sci. Publ. Cleveland Mus. Nat. Hist. 5:83–181, esp. p. 122.

Deansloy, R. 1935. The reproductive processes of certain mammals. IX. Growth and reproduction in the stoat *(Mustela erminea)*. Phil. Trans. Royal Soc. London, Ser. B, 225:459–92.

Hall, E. R. 1937. *Mustela cicognanii*, the short-tailed weasel, incorrectly ascribed to Ohio. Amer. Midl. Nat. 18:304.

Hamilton, W. J., Jr. 1933. The weasels of New York: their natural history and economic status. Amer. Midl. Nat. 14:289–344.

Henninger, W. F. 1921. Two mammals new to Ohio. J. Mammal. 2:239.

Rust, C. C. 1962. Temperature as a modifying factor in the spring pelage change of short-tailed weasels. J. Mammal. 43:323–28.

(See also Selected References for *Mustela frenata*, p. 136.)

Least Weasel
Mustela nivalis Linnaeus

The least weasel, the smallest carnivore in the world, has a distribution from Alaska southward to Montana and Nebraska, through Minnesota, Wisconsin, Michigan, western Pennsylvania, and the mountains of Virginia and North Carolina (Hatt, 1940). Formerly *Mustela rixosa* (Bangs), it was first reported in Ohio in 1904 (Wright, 1905), in western Pennsylvania in 1901, and in Indiana in 1928. Addi-

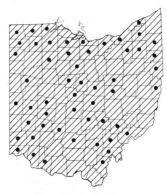

Mustela nivalis Linnaeus

tional Ohio records between 1904 and 1920 led Henninger (1923) to conclude that it occurs over most of the northern part of Ohio. More recent records have established its presence throughout central and southwestern Ohio, and I feel fairly certain that the species has a statewide distribution. Least weasels have not been found in large numbers anywhere within their range, but they may not be as uncommon as generally believed. An adult least weasel is not much bigger than a large field mouse. Hatt (1940) has suggested that least weasel abundance in any area may parallel microtine population cycles: when voles are abundant, weasels are abundant; and when voles are scarce, weasels are scarce.

The least weasel is easily recognized by its small size. It is similar in external appearance to other weasels except that the tail is not more than 40 mm long (approximately one-fifth the total body length) and does not have a black tip on the end in any season. The summer coat is rich chocolate brown above and white or brown spotted with white below. The amount of white fur and its distribution on the undersurface of this weasel, however, is highly variable in Ohio; seldom have I seen two animals with the same color pattern. Winter specimens may be completely white, completely brown, or any combination of white and brown. The skull is easily recognized by its small size (30–33 mm long). The dental formula is as follows:

$$i\frac{3-3}{3-3}, c\frac{1-1}{1-1}, pm\frac{3-3}{3-3}, m\frac{1-1}{2-2} = 34$$

Following are measurements of 24 adult least weasels (*Mustela nivalis*) from Ohio:

	Total Length (mm)	Tail (mm)	Hind Foot (mm)
11 Females			
Extremes	143–201	21–38	19–23
Mean	182	30	21
13 Males			
Extremes	180–215	23–36	21.0–23.5
Mean	193	33	22

The principal food of the least weasel in Ohio is probably *Microtus pennsylvanicus,* the meadow vole, for weasels are most often found in areas that support a high population of these voles. I once caught a female as she ran from beneath a shock of corn; another was caught in a trap set near a small hole in a weedy field. Least weasels have also been caught in buildings, along fencerows, and in forests in Ohio. When corn shocks were abundant, they appeared to be favorite home sites for this weasel during the winter months.

The reproductive habits of the least weasel have been described by Deanesly (1944), Hartman (1964), East and Lockie (1964, 1965), Heidt et al. (1968), and Heidt (1970). Litters of 4 to 10 are reported, but 5 appears to be average. Swenk (1926) found a nest on 1 July with 4 young. Allen (1940) reported finding 5 young in a nest that had been plowed up from 10 to 3 cm (4 to 5 in) below the surface of the ground. Females typically renovate gopher, mole, or mouse nests when rearing their young, often lining the nest with the fur from their latest victims (Polderboer, 1942). Young weasels have been recorded for almost every month of the year (Hall, 1951), suggesting that females may produce two or more litters each year. The gestation period is known to be 35 to 37 days (Heidt et al., 1968), indicating either a very short term or no delayed implantation for this species. Young weasels may reach their full growth in three months. Allen (1940) reports that 5 young weasels taken from their nest in July were fully grown by September.

The only food known to be used by the least weasel is small mammals, especially mice. The one specimen that I observed in captivity was kept alive for 30 days on a diet of deer mice and meadow voles. Allen (1940) reported that each of his five captive least weasels ate about one and one-half mice per day. Hatt (1940) reported that the least weasel feeds primarily on mice. Llewellyn (1942) kept an adult male in captivity for six days during which time it ate ten house mice having a total weight of 118 g. The weasel ate slightly more than one-half its own weight each day. Short (1961) says that in the field a least weasel must eat about half its own weight each day. He further concluded from this study that a maintenance requirement for a captive animal was about 1 g of suitable food per hour.

Observations that I have made suggest that at least one-half of the least weasel population in Ohio turns completely white each winter. Of 22 least weasels caught from December through March, 11 had coats that were predominantly white. Since 6 weasels taken during the last week in November had brown summer coats and 2 taken on 10 December had white winter coats (except for a few brown hairs on the back), it is suggested that in Ohio the molt may start during the first week of December. A female taken on 22 March retained her white coat except for brown spots on top of the head and on the nose. Weasels taken after March and before December all had brown summer coats.

Because least weasels are not abundant and because they are excellent "mousers," they should be considered beneficial. They are too small to be valuable as furbearers. Larger weasels, large birds of prey, snakes, cats, foxes, and other carnivorous mammals are all known to prey on the least weasel.

SELECTED REFERENCES

Allen, D. L. 1940. Two recent mammal records from Allegan County, Michigan. J. Mammal. 21:459-60.

Deanesly, R. 1944. The reproductive cycle of the female weasel (*Mustela nivalis*). Proc. Zool. Soc. London. 114:339-49.

East, K., and J. D. Lockie. 1964. Observations on a family of weasels (*Mustela nivalis*) bred in captivity. Proc. Zool. Soc., London. 143:359-63.

———. 1965. Further observations on weasels *(Mustela nivalis)* and stoats (*Mustela erminea*) born in captivity. Proc. Zool. Soc. London. 147:234-38.

Hall, E. R. 1951. American weasels. Univ. of Kans. Publ., Mus. Nat. Hist., 4:1-466.

Hartman, L. 1964. The behavior and breeding of captive weasels (*Mustela nivalis*). New Zealand J. Sci. 7: 147-56.

Hatt, R. T. 1940. The least weasel in Michigan. J. Mammal. 21:412-16.

Heidt, G. A. 1970. The least weasel, *Mustela nivalis* Linnaeus: developmental biology in comparison with other North American *Mustela*. Mich. State. Univ., Mus. Nat. Hist. Biol. Series Publ. 4(7):227-82.

Heidt, G. A., M. K. Petersen, and G. L. Kirtland, Jr. 1968. Mating behavior and development of least weasels (*Mustela nivalis*) in captivity. J. Mammal. 49: 413-19.

Henninger, W. F. 1923. On the status of *Mustela alleghениensis*. J. Mammal. 4:121.

Hoffmeister, D. F. 1956. Southern limits of the least weasel (*Mustela rixosa*) in central United States. Trans. Ill. Acad. Sci. 48:195-96.

Llewellyn, L. M. 1942. Notes on the Alleghenian least weasel in Virginia. J. Mammal. 23:439-41.

Lyon, M. W., Jr. 1933. Two new records of the least weasel in Indiana. Amer. Midl. Nat. 14:345-49.

Metzger, B. 1955. Notes on mammals of Perry County, Ohio. J. Mammal. 36:101-5.

Phillips, R. S. 1949. Strange behavior of a least weasel, *Mustela rixosa alleghениensis*. J. Mammal. 30:306.

Polderboer, E. B. 1942. Habits of the least weasel (*Mustela rixosa*) in northeastern Iowa. J. Mammal. 23: 145-47.

Short, H. L. 1961. Food habits of a captive least weasel. J. Mammal. 42:273-74.

Swanson, G., and P. O. Fryklund. 1935. The least weasel in Minnesota and its fluctuation in numbers. Amer. Midl. Nat. 16:120-26.

Swenk, M. H. 1926. Notes on *Mustela campestris* Jackson, and on the American forms of least weasels. J. Mammal. 7:313-30.

Whittaker, J. O., Jr. and E. G. Zimmerman. 1965. Additional *Mustela nivalis* records for Indiana. J. Mammal. 46:516.

Wright, A. A. 1905. Our smallest carnivore. Ohio Nat. 5:251-54.

Long-tailed Weasel
Mustela frenata Lichtenstein

Mustela frenata is the common weasel throughout Ohio. Hall and Kelson (1959, p. 911) include a total of 35 subspecies within this species, which occurs throughout the United States, southern Canada, Mexico, and Central America.

As long ago as 1933, Hamilton made the following comment concerning weasels: "Occurring in one form or another in every state of the Union, they have been almost entirely overlooked by mammalogists, except those bent on amassing a series of conventional skins for museums." The statement is as applicable today as it was in 1933 concerning our knowledge of weasels in Ohio, except that a large series of skins from Ohio is nonexistent. Other than the trapping reports issued each year by the Ohio Division of Wildlife, we know little about the weasel in Ohio.

The body of the long-tailed weasel is long and slender. The neck is long and the head small and triangular in outline. The legs are short. The tail exceeds one-third of the total body length, is furred but not bushy, and has a black tip in all seasons. The ears are short and rounded. Long-tailed wea-

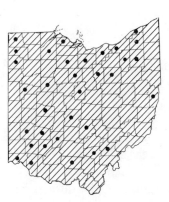

Mustela frenata Lichtenstein

sels are predominantly rich brown in color with a yellowish-white cast to the belly, throat, and chin. The white may continue onto the lower sides of the legs, but it seldom extends as far as the feet and toes. Except for the black on the end of the tail, Ohio specimens may turn pure white in winter.

As indicated in the key, the mustelid skull is easily recognized by the flat, noninflated, elongated bullae and short rostrum. The dental formula is as follows:

$$i\frac{3-3}{3-3}, c\frac{1-1}{1-1}, pm\frac{3-3}{3-3}, m\frac{1-1}{2-2} = 34$$

Male weasels are decidedly larger than females, and for this reason the measurements have been given separately for the sexes. Following are measurements of 16 adult long-tailed weasels (*Mustela frenata*) from Ohio:

	Total Length (mm)	Tail (mm)	Hind Foot (mm)
9 Males			
Extremes	325–445	106–58	37–51
Mean	351	130	43
7 Females			
Extremes	299–330	87–110	32–38
Mean	320	103	34

This weasel is a strictly carnivorous mammal that has been accused at times of being a wanton, uncontrolled, bloodthirsty killer of a wide variety of animal life. These energetic, aggressive animals center their attention primarily on small mammals. Mice, rats, shrews, chipmunks, squirrels, and rabbits are the principal items in their diet throughout the year. Some birds and a few amphibians and reptiles are also eaten, and insects are occasionally included in the diet. Long-tailed weasels appear to have a special liking for blood and will lap the blood flowing from a recently killed victim before eating the carcass. They have been seen attacking birds and mammals many times their own size, but Allen (1938) believes that rabbits are about the largest mammal the long-tailed weasel can kill. On the other hand, he reports several instances of a weasel being killed by a rabbit kept in the same cage with it. I once saw a long-tailed weasel killing a large field mouse (*Microtus pennsylvanicus*) near Ithaca, New York. The weasel held the mouse from behind with both front and hind legs wrapped tightly around the body of the mouse. The weasel's jaws were grasping the head of the mouse, which was bleeding profusely and at that point was unable to move. This apparently is the usual method of the weasel in subduing its prey.

Female long-tailed weasels produce one litter of young annually, usually in April or May. There may be 2 to 8 young in a litter, but 5 or 6 is average. Wright (1942a, 1947) states that breeding occurs during July or August, and the young are born the following April or May. The female long-tailed weasel, like most mustelids, exhibits delayed implantation. The eggs undergo only limited division following fertilization. The unimplanted blastocysts then lie dormant in the uterus until 21 to 28 days before birth when new divisions occur and the embryos complete their development. Young female weasels breed for the first time when they are 3 to 4 months old and produce litters the following spring. Males do not breed until they are 15 to 18 months old. The testes of adult males start to develop in late March, reach peak development in July, and regress during September and October.

Hamilton (1933) has studied the growth and development of young weasels in captivity and relates that even though the young are eating meat at 3 weeks and are capable of walking some distance at 4 weeks, the eyes do not open until day 36. By 7 weeks the young males weigh as much as the adult female; the young females weigh three-fourths as much. In the field the family probably disbands about this time or earlier. Seven weasels raised by Sanderson (1949) reached the peak of their growth at 10 weeks.

Kirtland (1838) and Brayton (1882) both refer to *Mustela frenata* as the "ermine." Strictly speaking, this is incorrect, since the true ermine is *M. erminea*. In common usage, however, any weasel that has turned white in the winter is referred to as ermine. In Ohio most long-tailed weasels remain brown all year. I have seen only two white long-tailed weasels from Ohio, one taken 9 November in Portage County and the other taken 10 December in Hamilton County. A local fur buyer states that not more than three or four long-tailed weasels with white coats have been included in the several thousand pelts that he has seen over several years. Hamilton (1933) states that the fall molt occurs during the latter part of October and the first part of November; the spring molt most often occurs in March. The black fur on the end of the tail is present in both winter and summer pelages.

These bundles of energy hunt mostly after dark. They bound rather than walk, and the back is often arched so that they appear shorter than they actually are. Their serpentine bodies enable them to enter openings that appear much too small to accommodate them. I once caught a full-grown long-tailed weasel in a Sherman live trap that measured 5 × 5 × 15 cm (2 × 2 × 6 in). Not only did the weasel fit in the trap, but he could turn around in it with ease. It made no difference which end of the trap I opened; he was always looking at me!

The long-tailed weasel has a well-developed pair of scent glands that open into the anus. The musk produced by these glands has a sweet, oily, nauseating quality that is quite different from, and, in my opinion, is even less tolerable than, that produced by the skunk. Fortunately, weasels do not spray their musk about.

Between 1938 and 1947 about 10,000 pelts were marketed each year (Leedy, 1950). By 1961–62 only 1,873 weasels were reported sold in Ohio, and in 1971–72 the number dropped to 188 (Appendix Table 2). Weasels account for a very low percentage of the fur pelts taken in Ohio each year. Also, fewer skins are being turned in by trappers, and the impression gained by fur buyers is that farmers are complaining less and less about weasels raiding their poultry yards. It is of further interest that the largest buyer and processor of raw furs in southern Ohio recently told me that Ohio weasels have decreased in size over the past ten years. He states that large, adult male pelts are especially uncommon in the raw fur trade today. Most of the weasels taken by professional trappers today are caught in sets made for skunk or raccoon. The low prices paid for the pelt (averaging about $0.50 each) do not make it worth the trapper's time to try to catch them.

Weasels occasionally enter poultry houses, where they may kill every bird in the flock and destroy every egg. On the other hand, weasels may live in the vicinity of poultry yards and never bother the birds. They are attracted to the poultry yard by the rats and mice that are living on the easily available grain and chicken feed. If each weasel eats one mouse per day (a conservative figure), and if there are 40,000 weasels in Ohio (four times the average number trapped each year by trappers, 1938–47), more than 14,500,000 mice are destroyed each year by long-tailed weasels! Possibly half this many rats are also killed. It is obvious that the long-tailed weasel is a valuable economic asset to the farmer and should be protected rather than killed wantonly. Weasel control measures away from farm buildings are not warranted; in fact, they should be discouraged.

Large hawks and owls and most predatory mammals prey on weasels. Large snakes sometimes catch and eat them, and domestic and feral cats living in rural areas regularly kill weasels.

SELECTED REFERENCES

Allen, D. L. 1938. Notes on the killing technique of the New York weasel. J. Mammal. 19:225–29.

Bissonnette, T. H. 1944. Experimental modification and control of molts and changes of coat-color in weasels by controlled lighting. N. Y. Acad. Sci. Ann. 45:221–60.

Brayton, A. W. 1882. Report on the Mammalia of Ohio. Rep. Geol. Surv. Ohio 4:1–185.

Green, C. V. 1936. Observations on the New York weasel, with remarks on its winter dichromatism. J. Mammal. 17:247–49.

Hall, E. R. 1951. American weasels. Univ. of Kans. Publ., Mus. Nat. Hist. 4:1–466.

Hall, E. R., and K. R. Kelson. 1959. The mammals of North America. Ronald Press Co., New York. 2 vols., 1162 pp.

Hamilton, W. J., Jr. 1933. The weasels of New York: their natural history and economic status. Amer. Midl. Nat. 14:289–344.

Hamlett, G. W. D. 1935. Delayed implantation and discontinuous development in the mammals. Quart. Rev. Biol. 10:432–47.

Kirtland, J. P. 1838. Report on the zoology of Ohio. Second Annu. Rep., Geol. Surv. Ohio. Pp. 160–61, 175–77.

Leedy, D. L. 1950. Ohio's status as a game and fur producing state. Ohio J. Sci. 50:88–94.

Leopold, A. 1937. Killing technique of the weasel J. Mammal. 18:98–99.

Miller, F. W. 1931. A feeding habit of the long-tailed weasel. J. Mammal. 12:164.

Petrides, G. A. 1950. The determination of sex and age ratios in fur animals. Amer. Midl. Nat. 43:355–82.

Sanderson, G. C. 1949. Growth and behavior of a litter of captive long-tailed weasels. J. Mammal. 30:412–15.

Warren, E. R. 1932. When do weasels mate? J. Mammal. 13:71–72.

Wright, P. L. 1942a. Delayed implantation in the long-tailed weasel (*Mustela frenata*), the short-tailed weasel (*Mustela cicognani*), and the marten (*Martes americana*). Anat. Rec. 83:341–53.

_____. 1942b. A correlation between the spring molt and spring changes in the sexual cycle in the weasel. J. Exp. Zool. 91:103–10.

_____. 1947. The sexual cycle of the male long-tailed weasel (*Mustela frenata*). J. Mammal. 28:343–52.

Mink
Mustela vison Schreber

The mink, probably the best-known furbearer, ranges throughout the United States and most of Canada. Fur records indicate that this mammal is found throughout Ohio, but to my knowledge there are fewer than 20 mink skins in Ohio that have been properly labeled, measured, and preserved for study. The live animal is seldom encountered in the field because of its nocturnal, secretive habits.

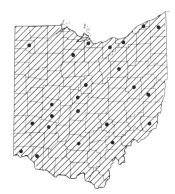

Mustela vison Schreber

Minks are large, weasel-like mammals that are rich chocolate brown in color. Often there are one or more irregular patches of white fur on the belly. The shape and distribution of these patches are so distinctive that they can be used in positive identification (McCabe, 1949). The tail, which is about one-third the total length of the animal, is definitely bushy and noticeably darker than the rest of the body. Mink do not turn white in the winter. The dentition and skull are similar to other weasels except for skull size and the dumbbell-shaped last upper molar. Following are measurements of 7 adult minks (*Mustela vison*) from Ohio:

	Total Length (mm)	Tail (mm)	Hind Foot (mm)
3 Males			
Extremes	540–620	190–212	55–75
Mean	570	199	66
4 Females			
Extremes	432–570	127–65	57–70
Mean	497	151	64

Minks are decidedly semiaquatic animals. Although they occasionally travel over land for some distance, they are most often found in the vicinity of streams or lakes. The thick underfur prevents water from penetrating to the skin, and the hind feet have some webbing between the toes; otherwise minks are not noticeably adapted to living in the water. However, they swim well enough to catch fish, and they can remain submerged for considerable periods of time.

Korschgen (1958) has reviewed the literature concerning the food habits of the mink. Most studies have revealed that muskrats, frogs, mice, crayfish, molluscs, and fish are the major items in the diet, along with birds, insects, and various other arthropods. A mink may kill more than it can eat at one time, storing the excess in its burrow. Schnell (1964) found the carcasses of 8 cotton rats (*Sigmodon hispidus,* medium-sized rats inhabiting damp areas in the southern United States) stuffed into the side chamber of a mink den. This same mink had killed 40 cotton rats on a 0.7-hectare (1.7 acre) island in less than two months!

Much of the information concerning reproductive habits has been gained from studying ranch minks. A restricted breeding period in late winter and early spring is indicated, with most matings occurring in February and March. Mitchell (1961) found that in Montana the height of the breeding season occurred in March. Litters are born from 40 to 75 days after copulation. Delayed implantation is exhibited, with parturition occurring about 30 days after implantation (Enders, 1952). Litters of more than ten are known, but four or five is probably the usual number. Females produce only one litter each year. Marshall (1936) found evidence that wild minks are promiscuous, a characteristic of ranch minks also. The newborn young, about 7.5 cm (3 in) long, have a light covering of silver hair. The hair becomes pale reddish-gray (Svihla, 1931) when the animals are about two weeks old. The eyes open when they are approximately five weeks old. They can move about and take care of themselves when seven or eight weeks of age. The family remains together until September or October before disbanding.

Home range studies indicate that males move greater distances and more frequently than do females. Marshall (1936) found that a female restricted her activities within an 8-hectare (20-acre) plot, whereas several males covered this area plus an additional undetermined area during the same period. Ritcey and Edwards (1956) found adult males moving more than juveniles, and adult females the most sedentary of all. Mitchell (1961) also discovered males moving farther than females, with the diameter of the movements of any male up to 4.8 km (3 mi). When not moving about, minks live in burrows, hollow logs, tree stumps, or old muskrat houses. The den usually contains a nest of leaves and/or other plant materials; sometimes it also contains the fur of recently eaten mammals.

Generally speaking, the mink, which has few natural enemies, is probably an asset rather than a liability in Ohio. There are still some ten thousand or more wild mink pelts sold to Ohio fur dealers each year (Appendix Table 2). For many years a single mink pelt was worth more than that of any other furbearer in Ohio. With the increased popularity of mink in the garment industry, however, and the subsequent establishment of commercial mink-raising farms, all mink furs have decreased in value. A fur dealer in Cincinnati remembered paying as much as $20.00 for a prime mink skin in the early 1900s, but since about 1965 the price has rarely risen above $10.00. All fur prices increased in 1972, and average mink pelts brought about $11.00 each. The market has since fluctuated considerably.

SELECTED REFERENCES

Allen, J. A. 1940. The principles of mink ranching. Can. Dept. Mines & Nat. Resources, Game & Fisheries Branch. pp. 1-148.

Bassett, C. F., and L. M. Llewellyn. 1949. The molting and fur growth pattern in the adult mink. Amer. Midl. Nat. 42:751-56.

Elder, W. H. 1951. The baculum as an age criterion in mink. J. Mammal. 32:43-50.

Enders, R. K. 1952. Reproduction in the mink *(Mustela vison)*. Proc. Amer. Phil. Soc. 96:691-755.

Hamilton, W. J., Jr. 1940. The summer food of minks and raccoons on the Montezuma Marsh, New York. J. Wildl. Manage. 4:80-84.

Hamlett, G. W. D. 1935. Delayed implantation and discontinuous development in the mammals. Quart. Rev. Biol. 10:432-47.

Hansson, A. 1947. The physiology of reproduction in mink *(Mustela vison* Schreb.) with special reference to delayed implantation. Acta Zoologica 28:1-136.

Korschgen, L. J. 1958. December food habits of mink in Missouri. J. Mammal. 39:521-27.

Marshall, W. H. 1936. A study of the winter activities of the mink. J. Mammal. 17:382-92.

McCabe, R. A. 1949. Notes on live-trapping mink. J. Mammal. 30:416-23.

Mitchell, J. L. 1961. Mink movements and populations on a Montana river. J. Wildl. Manage. 25:48-54.

Ritcey, R. W., and R. Y. Edwards. 1956. Live trapping mink in British Columbia. J. Mammal. 37:114-16.

Schnell, J. H. 1964. A mink exterminates an insular cotton rat population. J. Mammal. 45:305-6.

Svihla, A. 1931. Habits of the Louisiana mink *(Mustela vison vulgivagus)*. J. Mammal. 12:366-68.

Badger
Taxidea taxus Schreber

Although little known to the earlier writers, the badger has probably been a resident of Ohio in small numbers since at least the early or mid-1800s. Kirtland (1838) does not even mention it, and Brayton (1882) believed it to be extirpated from the state. The first accounts of the animal are those of Moseley (1934), who reports a badger taken in 1882 in a wolf trap in Wood County and another reported by a fur dealer in 1887. Moseley reported that he had data for 70 badgers caught or killed in ten counties in northwestern Ohio, mostly taken in the early 1900s. He attributed a subsequent increase in sightings to closer observation plus an actual increase in badger numbers. Bole and Moulthrop (1942) add little to our knowledge, but leave the impression that they too believed that badger numbers were increasing in Ohio. Hamilton (1943) states: "Dr. L. E. Hicks has written me that badgers have always occurred in Wood, Henry, and Fulton counties in northwestern Ohio." In 1947 Leedy reported badgers from Morrow and Knox counties and questioned whether the extended range and apparent increase in spermophiles *(Spermophilus tridecemlineatus)* had been a factor in the extension of the range of the badger. Reports coming to me concerning the presence of badgers in Ohio since 1950 have been sporadic, and generally I have not been able to confirm them. The one live specimen that I have seen was brought to the Cincinnati Zoo from Highland County, were it had been run down, clubbed, and captured by a school bus driver. My general im-

Taxidea taxus Schreber

pression is that badgers have been extending their range south and east in Ohio, but that their numbers have increased little, if any, over the past 50 to 75 years.

The badger, like the skunk, has a white or near-white stripe extending from the nose over the occiput and reaching at least onto the shoulders. Here the similarity between the skunk and badger ends. The badger is a heavy, powerfully built animal with broad shoulders, short muscular neck, short brush-like tail, prominent but fairly short ears, and small, dark, penetrating eyes. Because the legs are short and the fur long, especially on the back and sides, the animal appears "flat," its legs and feet usually hidden as it walks or runs. The feet have five toes, each bearing a heavy, long, sharp claw. The middle claw on the front foot may be as long as 30 mm (1.2 in). The otherwise dark face has two white or cream-colored cheek patches that end just behind the eyes. There is a distinct vertical black bar in front of each ear. The chin, throat, and belly are yellowish-white; the sides and back, grizzled gray. Older individuals have a decidedly frosted appearance. Badgers are said to molt once each year, probably in the fall. Jackson (1961) gives their weight as 6.4 to 11.8 kg (14 to 16 lb), with the male averaging about 5 percent larger than the female.

The skull is larger and heavier than any of our other mustelids. Also, the tympanic bullae are inflated, prominent, and elongated. The upper molars are triangular in shape, and the hard palate extends well beyond the posterior margin of the molar teeth. The dental formula is as follows:

$$i\frac{3\text{-}3}{3\text{-}3}, c\frac{1\text{-}1}{1\text{-}1}, pm\frac{3\text{-}3}{3\text{-}3}, m\frac{1\text{-}1}{2\text{-}2} = 34$$

Measurements for one badger (a male) from Geauga County are: total length, 788 mm; tail, 130 mm; and hind foot, 112 mm.

Since no one has critically studied the badger in Ohio, the life history and habits presented here are the results of studies made elsewhere. Badgers are ordinarily solitary animals except during the mating season. They breed in August and September in our latitude. Like other mustelids, the females exhibit delayed implantation. The fertilized eggs undergo a limited number of divisions to the blastula stage, then remain inactive as blastocysts in the uterine tract until approximately mid-February or early March when implantation occurs. One to five (usually two or three) young are born approximately six weeks later. The young are lightly furred and blind. The eyes are open at four weeks, and the animals are weaned at about eight weeks, or when they are approximately half grown. They remain with their mother until late fall when the family scatters. However, surprisingly little is known for certain about many of the reproductive habits of the badger.

Badgers are primarily carnivorous. A food-habit study (Snead and Hendrickson, 1942), made from the analysis of 239 badger droppings in Iowa, shows the frequency (percentage of occurrence) of major diet items as follows: ground squirrels, 68 percent; mice, 46 percent; cottontails, 25 percent; and insects, 27 percent. Also found were birds, pocket gophers, one snake, and traces of plant material.

Badgers may be seen at almost any time of day, although they are most active from sunset until sunrise. During the summer months they occasionally emerge during the day and take sun baths near the entrance to their burrows. The burrow is quite conspicuous. The entrance must be large enough to accommodate the owner, and surrounding the hole there is almost always a sizable mound of dirt thrown there by the badger during the excavating process. A badger, when challenged, can easily "out-dig" a man with a shovel! In the southern part of their range, badgers are active all year; but in Ohio and farther north, the animals enter a state of semihibernation when the temperatures approach freezing. Presumably they remain asleep in a nest belowground during these inclement periods.

In Ohio badgers are not abundant enough to be considered economically significant. Only about 140 pelts were reported by Ohio fur buyers between 1958 and 1979, and the average price per pelt did not exceed $4.00 until 1973 (Appendix Table 2). The fur, although durable, is coarse and little used in the clothing industry. The hair used in "badger hair" shaving brushes is taken from *Meles* sp., a close European relative. Badgers have no serious natural enemies in Ohio. Even a well-trained, good fighting dog usually comes out second best in an encounter with a full-grown badger!

SELECTED REFERENCES

Bole, B. P., Jr., and P. N. Moulthrop. 1942. The Ohio recent mammal collection in the Cleveland

Museum of Natural History. Sci. Publ. Cleveland Mus. Nat. Hist. 5:83-181, esp. 125-26.

Brayton, A. W. 1882. Report on the Mammalia of Ohio. Rep. Geol. Surv. Ohio 4:1-185.

Davis, W. B. 1946. Further notes on badgers. J. Mammal. 27:175.

Dexter, R. W. 1945. Another record of the badger as a highway casualty. J. Mammal. 26:89

Errington, P. L. 1937. Summer food habits of the badger in northwestern Iowa. J. Mammal. 18:213-16.

Hamilton, W. J., Jr. 1943. The mammals of the eastern United States: an account of recent land mammals occurring east of the Mississippi. Comstock Publ. Co., Ithaca, N. Y. 438 pp.

Hamlett, G. W. D. 1932. Observations on the embryology of the badger. Anat. Rec. 53:283-303.

Jackson, H. H. T. 1961. Mammals of Wisconsin. Univ. of Wisc. Press, Madison. 504 pp.

Kirtland, J. P. 1838. Report on the zoology of Ohio. Second Annu. Rep., Geol. Surv. Ohio. Pp. 160-61, 175-77.

Leedy, D. L. 1947. Spermophiles and badgers move eastward in Ohio. J. Mammal. 28:290-92.

Long, C. A. 1973. *Taxidea taxus*. Amer. Soc. Mammal., Mammalian Species No. 26:1-4.

Moseley, E. L. 1934. Increase of badgers in northwestern Ohio. J. Mammal. 15:156-58.

Schantz, V. S. 1953. Additional information on distribution and variation of eastern badgers. J. Mammal. 34:388-89.

Snead, E., and G. O. Hendrickson. 1942. Food habits of the badger in Iowa. J. Mammal. 23:380-91.

Striped Skunk
Mephitis mephitis Schreber

"Sweet William" is known to everyone, if not by sight, then by smell or reputation. The striped skunk is found throughout the United States, most of Canada, and the northern one-third of Mexico. Skunks have been reported from every part of Ohio.

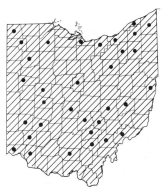

Mephitis mephitis Schreber

The large bushy tail and the two snow-white stripes extending along either side of the back are the trademarks of the skunk. Coloration, however, is highly variable within the species; Ohio specimens range from completely black to almost pure white. The two white stripes that are usually present on the back may be broad and continuous or narrow and incomplete. In most specimens a white line extends back from the nose between the eyes and onto the forehead. Even in the predominantly black animals, there is usually a white tip on the end of the tail. The belly, back, and legs are black. The feet are semiplantigrade with much longer and better-developed claws on the front feet than on the hind feet. The dental formula is as follows:

$$i\frac{3-3}{3-3}, c\frac{1-1}{1-1}, pm\frac{3-3}{3-3}, m\frac{1-1}{2-2} = 34$$

Following are measurements of 5 adult skunks (*Mephitis mephitis*) from Ohio:

	Total Length (mm)	Tail (mm)	Hind Foot (mm)
Extremes	528-654	188-218	62-68
Mean	588	196	65

The odor produced by the anal glands is the most distinctive feature of the skunk. Two large anal glands lie just below the surface of the skin, one on either side of the vent (anus). A duct leading from each gland opens by means of a nipple-like papilla into the anus. When the animal becomes excited, the tail is elevated, the muscles surrounding each gland are suddenly and forcefully contracted, and a fine, spray-like stream of musk is ejected toward the target. Chemically, the musk is a sulfur-alcohol (butyl-mercaptan) mixed with other sulfur-containing compounds. In addition to being highly malodorous, the substance is strongly acid, and produces an acute burning sensation when directed into the eyes. Some have claimed that the musk can cause permanent blindness, but Cuyler (1924) stated that in past centuries trappers and hunters used a drop or two of musk in their wash-water to keep their eyes clear and bright! Cuyler and his dogs were often squirted directly in the face and eyes without sustaining any permanent or harmful effects.

Although not true hibernators, skunks store quantities of body fat in the fall and, with the advent of cold weather, retire to underground nests where they may remain for several weeks or a month at a time. Their metabolism does not drop appreciably, however, and they emerge from the den during mild spells in the winter. Hamilton (1937) observed one female that remained in the den for 44 days, but this was termed "exceptional." Trapping results and direct observations indicate that more females than males "den up" and for longer periods of time. Most pelts handled by fur buyers are males. Few, if any, immature animals are taken in winter, indicating that first-year animals remain in the dens throughout most, if not all, of the winter (Hamilton, 1937). Several skunks may use the same wintering den; these groups often consist of old females plus young males and females.

Breeding activity starts in late Feburary in Ohio and continues through March. Females are in heat for approximately four to five days and will mate several times during this period. Males are probably promiscuous, and move from one den to another mating with females that are receptive to their advances. The gestation period is 60 to 65 days (Wight, 1931). Each of the 3 to 10 (average, 5) pink-skinned, blind, newborn skunks in a litter weighs between 30 and 35 g. A coat of fine hair is present, and the color pattern is evident even at this early age. The eyes open during the third week, and by the time the young are six weeks old they are accompanying their mother on her nightly hunting excursions. The entire family may remain together until the following spring. Stegeman (1937) has described the development of the young in captivity.

Skunks are primarily nocturnal animals, and seldom wander from the den during daylight hours. Under ordinary circumstances they remain within 1.6 to 3.2 km (1 to 2 mi) of the den during their nightly forages. Skunks are omnivorous. Hamilton (1936), Kelker (1937), and others have demonstrated that insects are the predominant item in the diet during the spring and summer months. As fall approaches, small mammals and fruits become the main dietary items. Berries, nuts, eggs, carrion, grasses, garbage, leaves, and small invertebrates of all kinds are eaten. Skunks are apparently fond of bees and thus are responsible for heavy depredations to apiculturists. Not only do they eat the dead bees found lying around the hive, but they also eat the live bees as the excited insects emerge from the hive (Storer and Vansell, 1935). Apparently skunks are completely impervious to bee stings. Skunks occasionally invade chicken houses, where they eat the eggs and occasionally the chickens. Bird nests, especially those of waterfowl, may be plundered for the eggs and/or young birds.

Judging from the number of carcasses seen on roads in all parts of Ohio, it would appear that skunks live in many habitats. However, they are more common in the brushy, rocky, semi-open areas of the state. They do not necessarily shy away from densely inhabited areas, and often live beneath an old barn or outbuilding and utilize urban garbage as a ready source of food. Skunks live in dens that they have dug or in dens that have previously been used by woodchucks or foxes. These dens may be located beneath buildings, in open fields, on hillsides, under logs, or in the woods. Allen and Shapton (1942) found that most skunk dens in Michigan were in renovated woodchuck burrows and that most of the active ground holes were located on the slopes around marshes, kettleholes, and stream bottoms, or in the edge of woodlots and patches of brush. One or more nests of grass in the den system are used as sleeping and wintering quarters.

Skunks are subject to many different kinds of diseases, and this in part accounts for their scarcity during certain years. For example, the skunk population was so drastically reduced in Ohio between 1940 and 1946 that it was given protection from hunters and trappers in 1947 and 1948. By the early 1950s skunks were present in good numbers, and they have generally maintained their populations since that time. One fur buyer told me that trappers in Ohio believe that highway fatalities help keep the skunk population low. Skunks are poor pedestrians; they are slow-moving and are blinded and confused by, but not fearful of, the headlights of oncoming automobiles. This accounts for their high mortality rate on the highways.

Skunks are economically important for several reasons. Skunk fur is attractive and durable, and for many years skunk rated as the 7th or 8th most important fur producer in Ohio. Between 1938 and 1947 the average number of pelts reported taken in Ohio was 61,760 (Leedy, 1950). However, with declining fur prices, only about 700 per year were sold for the period 1963-73. Since 1973, there has been an increase in skunk pelts reported by Ohio fur buyers, but the average price is generally less than $2.00 per pelt (Appendix Table 2).

Although skunks do occasionally raid chicken houses and apiaries and cause much damage, this is the exception rather than the rule. They can also be a serious problem in duck and goose nesting areas. The amount of damage done in these few instances, however, is more than compensated for by the myriads of harmful insects and small rodents destroyed by this animal. Skunks are noted carriers of rabies; a rabid skunk will not hesitate to bite another animal or man. Skunks make attractive and interesting pets, however, if "deodorized" and immunized.

Skunks have few natural enemies. Hawks and owls may take one on occasion, but only a desperately hungry carnivore is willing to risk the certain bath of musk for a meal.

SELECTED REFERENCES

Allen, D. L., and W. W. Shapton. 1942. An ecological study of winter dens, with special reference to the eastern skunk. Ecology 23:59-68.

Bailey, V. 1937. Deodorizing skunks. J. Mammal. 18: 481-82.

Chapman, F. B. 1946. An interesting feeding habit of skunks. J. Mammal. 27:397.

Cuyler, W. K. 1924. Observations on the habits of the striped skunk *(Mephitis mesomelas varians)*. J. Mammal. 5:180-89.

Hamilton, W. J., Jr. 1936. Seasonal food of skunks in New York. J. Mammal. 17:240-46.

———. 1937. Winter activity of the skunk. Ecology 18: 326-27.

———. 1963. Reproduction of the striped skunk in New York. J. Mammal. 44:123-24.

Kelker, G. H. 1937. Insect food of skunks. J. Mammal. 18:164-70.

Leedy, D. L. 1950. Ohio's status as a game and fur producing state. Ohio J. Sci. 50:88-94.

Mutch, G. R. P., and M. Aleksiuk. 1977. Ecological aspects of winter dormancy in the striped skunk *Mephitis mephitis)*. Can. J. Zool. 55:607-15.

Selko, L. F. 1938. Notes on the den ecology of the striped skunk in Iowa. Amer. Midl. Nat. 20:455-63.

Stegeman, L. C. 1937. Notes on young skunks in captivity. J. Mammal. 18:194-202.

Storer, T. I., and G. H. Vansell. 1935. Bee-eating proclivities of the striped skunk. J. Mammal. 16:118-21.

Storm, G. L. 1972. Daytime retreats and movements of skunks on farmlands in Illinois. J. Wildl. Manage. 36:31-45.

Verts, B. J. 1960. A device for anesthetizing skunks. J. Wildl. Manage. 24:335-36.

———. 1963. Equipment and techniques for radio-tracking striped skunks. J. Wildl. Manage. 27:325-39.

Wight, H. M. 1931. Reproduction in the eastern skunk *(Mephitis mephitis nigra)*. J. Mammal. 12:42-47.

ORDER ARTIODACTYLA
Even-toed Hoofed Ungulates

The ungulates, or hoofed mammals, are separated into two large orders, the Perissodactyla and the Artiodactyla. The Perissodactyla are the odd-toed ungulates, including the horse-like mammals, the tapirs and the rhinoceroses. The Artiodactyla are the even-toed hoofed mammals, including the peccaries and other pig-like forms, bison, antelopes, deer-like mammals, and both domestic and wild cattle, sheep, and goats. They have the third and fourth toes equally developed, whereas the other toes are greatly reduced or completely missing. The limbs are elongated and adapted for so-called unguligrade locomotion. From a prone position, artiodactyls characteristically get up rear end first rather than front end first. In the skull a diastema is present between the front teeth and the cheek teeth, and the premolars and molar teeth are not alike (the premolars are never molariform). In the group only pigs have incisors in the upper jaw. Horns or antlers may be present.

Artiodactyls have a worldwide distribution with the exception of Australia and New Zealand, into which they have been introduced, and Antarctica.

The skulls of domestic and/or prehistoric ungulate mammals are often found in the field or are uncovered from "digs" at prehistoric sites. Since these skulls are easily confused with the only living wild member of this group in Ohio, the white-tailed deer, *Odocoileus virginianus,* as well as among themselves, I have included the following skull identification key.

KEY TO THE SKULLS OF UNGULATES

1. Incisors present in upper jaw (3 in each half) . 2
1'. Incisors not present in upper jaw 3
2. Eye socket entirely enclosed by bony ring ... .Horse, Mule, Ass
2'. Eye socket not entirely enclosed by bony ring, open behind ....Pig
3. Large space or opening in skull bones just in front of eye sockets; (antlers may be present) ....Deer
3'. No large space or opening in skull bones in front of eye sockets 4
4. Cheek tooth row 108 mm (4.25 in) or more in length; skull length 305 mm (12 in) or more .. ..Cow
4'. Cheek tooth row less than 108 mm (4.25 in) in length; skull length less than 305 mm (12 in) .   5
5. Deep depression in skull bones (lachrymal bones) immediately in front of eye sockets; horns curved down and out ....Sheep
5'. No deep depression in skull bones immediately in front of eye sockets; horns parallel and directed back ....Goat

FAMILY CERVIDAE
Deer

Deer and their allies are widely distributed throughout the Old World and the New World. Four living genera are present in North America with the single species *Odocoileus virginianus* (formerly *Dama virginiana*), the white-tailed deer, being present in Ohio. The most striking characteristic of the entire family is the presence of antlers in the males (rarely in females).

White-tailed Deer
Odocoileus virginianus (Boddart)

The white-tailed deer (fig. 25) was abundant throughout Ohio at the time of settlement and was

Fig. 25. Mature white-tailed buck *(Odocoileus virginianus)* with eight points. (Photograph by Ronald A. Austing.)

an important food source for the Indians and pioneers. However, as settlers increased, the number of deer decreased; in 1882 Brayton wrote: "The Virginia Deer is rarely met with in Ohio at present, except as domesticated in parts." Although a closed season of 225 days was first established for deer in 1857, open seasons of lengths of more than a month and no bag limits continued throughout

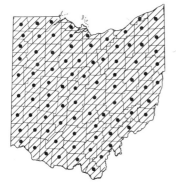

Odocoileus virginianus (Boddart)

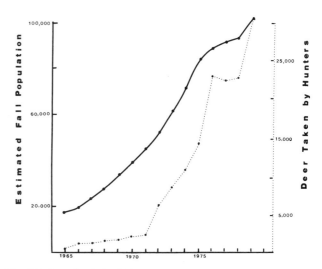

Fig. 26. Growth of the Ohio deer herd and deer taken by hunters, 1965-1979. *Sources:* Donohoe and Stoll (1976); Ohio Division of Wildlife (1979); Donohoe, Stoll, and M. W. McClain (1979, personal communication). (Courtesy of the Ohio Biological Survey.)

most of the late 1800s (Dambach, 1948) until, in 1904 according to Oney (1945), "southern Ohio saw its last wild deer." A few stragglers undoubtedly wandered into the state from Pennsylvania and Michigan, but from a practical standpoint the species was completely absent until 1922. Between 1922 and 1930, two hundred animals, purchased from various sources, were released by the Ohio Conservation Department on the Roosevelt Game Reserve in Scioto County. The herd, under strict protection, flourished and by 1932, when the restraining fences were removed, numbered about 1,000. The number of deer then gradually increased to a point where farmers in the area were complaining that the animals were damaging their crops. Consequently, the first modern open hunting season for deer in Ohio was held in 1943 in Adams, Scioto, and Pike counties, and 168 bucks were reported killed. Steady immigration from Pennsylvania and Michigan rapidly increased, and by 1960 all 88 counties had at least a few deer. By 1965 the Ohio deer herd numbered approximately 20,000 and was increasing rapidly in all regions of the state, especially in the unglaciated hill country (Donohoe and Stoll, 1976). Regulated hunting by zones during the 1960s and 1970s did not significantly retard this rate of increase. During the 1978 hunting season at least one deer was reported taken by either gun or bow and arrow from every county in the state (Ohio Division of Wildlife, 1979). The estimated deer population in 1979 was over 100,000 (Donohoe and Stoll, 1979, personal communication). In an effort to reduce the deer herd to minimize damage to orchards and grain crops and to reduce the number of deer-related highway accidents, the Ohio Division of Wildlife authorized a weeklong, statewide, open shotgun season in 1979, and also increased the allotment of antlerless permits. This resulted in a record 30,053 deer taken by hunters (figure 26). A summary of Ohio deer hunting seasons from 1943 to 1978 is presented in Appendix Table 3. Many deer are killed on Ohio highways; in 1974 more than 30 percent of the reported total deer mortality resulted from this cause (Donohoe and Stoll, 1976).

White-tailed deer are large, even-toed, hoofed animals with long legs. Adult males and an occasional female carry antlers. The coat in summer is predominantly light brown or chestnut in color. The throat and underparts are white. A white band extends across the muzzle. The tail is surprisingly large and fluffy; when the animal is alarmed, the tail is flicked vertically, revealing a bright white undersurface. In fall the summer coat is shed and replaced by a salt-and-pepper gray or darker brown coat. Young deer (fawns) have reddish coats with white spots.

The skull is easily recognized by its large size (more than 30 cm, or 12 in, long), absence of upper incisors (which all other native mammals in Ohio possess), and presence of a large opening between the lachrymal and nasal bones in front of each eye socket. The dental formula is as follows:

$$i\frac{0-0}{3-3}, c\frac{0-0}{1-1}, pm\frac{3-3}{3-3}, m\frac{3-3}{3-3} = 32$$

(Occasionally the upper canines are partially developed.)

Ohio white-tailed deer are among the healthiest and largest in the nation. The average Ohio adult buck is 175 to 200 cm (60 to 79 in) long with live weight of about 88 kg (196 lb). The live weight of the average adult doe is about 61 kg (135 lb) (Robert Stoll, 1978, personal communication). Deer from the western part of the state are usually larger than those of the plateau counties. The field-dressed weight (which is usually 20–25 percent less than live weight because of removal of lungs and intestines) of representative Ohio deer from three regions of the state as reported by Donohoe and Stoll (1976) is shown in the table below.

Field-dressed weight (including heart and liver)

Region of Ohio	Number of Deer		Mean Weight kg (lb)	
	Male	Female	Male	Female
Western				
Fawn	98	83	34.4 (75.7)	32.2 (70.9)
Yearling	51	47	58.7 (129.2)	47.2 (104.0)
Adult	45	46	73.3 (161.4)	52.6 (115.9)
Northeast				
Fawn	65	71	32.1 (70.6)	29.3 (64.6)
Yearling	50	47	50.4 (111.0)	44.1 (97.2)
Adult	70	39	65.4 (144.1)	50.7 (111.7)
Hill counties				
Fawn	58	62	31.2 (68.8)	29.1 (64.2)
Yearling	34	23	51.2 (112.7)	44.4 (97.8)
Adult	80	46	66.9 (147.3)	45.6 (100.5)

The large numbers of deer in eastern and southeastern Ohio are supported by the large amount of forested or semiforested, brushy habitat located there. With so much acreage of abandoned farmland in many stages of secondary succession, the deer herd there is probably the largest it has ever been. In agricultural areas of western and northern Ohio, larger woodlots offer necessary daytime cover and protection for deer, and fields and orchards provide food. Transformation of many woodlots into agricultural fields in this area of the state since 1970 has limited the growth of the deer herd in many of these counties. Deer are remarkably adaptable animals, however, and have been able to move into all areas of the state, including even those that are highly urbanized.

Deer are grazers and browsers, constantly moving from one food source to another, nibbling a twig here, picking a few leaves there, or chewing forbs and grasses somewhere else (Nixon et al., 1970). Although such unlikely items as snails and fish have been reported in their diet (Burgess, 1924), deer are primarily herbivores. Farm crops (winter wheat, corn, alfalfa, soy beans, and hay) are important items in the diet of many Ohio deer. The common occurrence of such nutrient-rich food items in their diet explains the excellent condition of Ohio deer. Hunters in neighboring states are envious of the average weight and condition of Ohio's "corn-fed" deer.

The diet of Ohio deer ideally consists of easily fermented foodstuffs capable of supplying their high energy requirements. Top food items are wild crabapple *(Pyrus coronaria)*, corn *(Zea mays)*, sumac *(Rhus* spp.), Japanese honeysuckle *(Lonicera japonica)*, grasses *(Gramineae)*, greenbrier *(Smilax* spp.), clover *(Trifolium* spp.), soybean *(Glycine max)*, jewelweed *(Impatiens* spp.), acorns *(Quercus* spp.), and dogwood *(Cornus* spp.) (Nixon et al., 1970). The combined results of many studies indicate that in forested areas young mixed-softwood-hardwood stands in the "brush" stage produce the most deer food per unit area (in some instances more than 227 kg per hectare (200 lbs per acre). It has been estimated that a 45-kg (100 lb) deer required about 2.25 kg (5 lb) of fresh browse daily.

White-tailed deer in our latitude breed from October through February. Most matings occur in Ohio between 3 and 16 November with populations north of latitude 40° reaching peak breeding activity about a week earlier than those in southern Ohio (Nixon, 1971). Active sperm have been found in males as early as September and as late as March (Cheatum and Morton, 1946). In the fall there is increased ovarian activity (Verme, 1961; Nixon, 1971), which results in the development of one or more mature ova. Ovulation occurs during a period of heat, and this estrus period lasts for 24 hours or less. Only during this period will the doe accept the service of a buck. If the doe is not successfully mated during her first estrus, new follicles develop in the ovaries and approximately 28 days later a second estrus occurs. As many as three successive estrus periods in one season have been recorded for a single doe (Severinghaus and Cheatum, 1956, p. 95–97). Virtually all yearling and adult does conceive each year, and in Ohio usually carry twins; but triplets and quadruplets also have been recorded (Nixon, 1971). The gestation period has been determined to be approximately 201 days.

Most fawns are born in May and early June. They have a full covering of hair when born. An hour or two after birth, they are standing for several minutes at a time on uncertain legs. The uncertainty and awkwardness are rapidly overcome, however, and by the time they are two to three weeks old they are able to outmaneuver a man trying to catch them. Although they start nibbling on young green shoots and leaves when only three to four weeks old, the weaning process is a slow one. There is general agreement that the fawns are not completely weaned until they are about four months old. Fawns usually shed their spotted coats in September. At this time a male may weigh 45 kg (100 lb) and is easily confused with a small, mature female. Approximately 75 percent of Ohio does breed during their first breeding season when they are only six to eight months old (Robert Stoll, 1978, personal communication). The remainder do not mate until they are approximately one and one-half years old, the age at which bucks attain sexual maturity and commence breeding activities.

The annual growth cycle of antlers in the male white-tailed buck has been described in detail by Wislocki (1942) and others. Antlers are among the most rapidly growing animal structures. Normally they start growing in May and are completed by September. They first appear when the buck is one year old as small "buttons" barely protruding above the hairline. Older bucks have branched antlers bearing several tines (points). The size and number of points on the antlers is not an indication of age but, rather, reflects in a general way the physical condition of the buck. As a result the number of points varies widely. Many one- and one-half-year-old bucks in Ohio have four to six points (Robert Donohoe, 1979, personal communication).

While the antlers are growing, they are covered by a protective layer of skin called "velvet." When they have reached their full growth, the blood supply at the base is cut off. The skin dies and is scraped away as the buck rubs the antlers against hard surfaces while polishing and sharpening individual tines. Antlers will be used in combat during the ensuing rutting season. Following the breeding season, necrosis weakens the antlers at their base, and they are dropped usually in late December or January. The cyclic development of the antlers is under hormonal control (Mirarchi et al., 1977) and is closely correlated with the breeding season and the reproductive cycle.

In regions where there are great climatic differences between summer and winter, or where food availability changes from season to season, deer sometimes make annual migrations of 16 to 32 km (10 to 20 mi). In Ohio, however, deer have rather small home ranges and appear reluctant to leave them. During the breeding season their home ranges probably do not cover more than 5 to 7 sq km (2 to 3 sq mi), and at other times it appears that the seasonal range of an individual deer does not greatly exceed 1.609 km (1 mi) (Severinghaus and Cheatum, 1956, p. 154). Since a deer usually spends its entire lifetime within a restricted area, it becomes very familiar with the area, a distinct advantage, especially during hunting season (Carhart, 1946). In the more northern states where deer "yard up" during the winter, they have been known to starve rather than move out of an area with which they are familiar.

The number of deer that any one area can support throughout the year is determined primarily by the amount of winter food available, a quantity that obviously is affected by many factors and one that changes from year to year. Nevertheless, Hosley (1956, p. 225) says that "there is surprisingly close agreement in the carrying capacity reported. Twenty deer per section or a deer to 32 acres [12.8 hectares] can be taken as an approximate average carrying capacity in the states represented." Although Ohio is included in the area referred to by Hosley, a personal note in 1978 from Donohoe states: "In Ohio's best range right now there are approximately seven to twelve deer per section. Because of people-deer conflicts (e.g., crop damage and deer-vehicle collisions), I doubt if Ohio could stand 20 per section as the author [Hosley] suggests."

SELECTED REFERENCES

Armstrong, R. A. 1950. Fetal development of the northern white-tailed deer *(Odocoileus virginianus borealis* Miller). Amer. Midl. Nat. 43:650-66.

Atwood, E. L. 1941. White-tailed deer foods of the United States. J. Wildl. Manage. 5:314-32.

Barbour, T., and G. M. Allen. 1922. The white-tailed deer of eastern United States. J. Mammal. 3:65-78.

Bellis, E. D., and H. B. Graves. 1971. Deer mortality on a Pennsylvania interstate highway. J. Wildl. Manage. 35:232-37.

Brayton, A. W. 1882. Report on the Mammalia of Ohio. Rep. Geol. Surv. Ohio 4:1-185.

Burgess, T. W. 1924. Fish-eating deer. J. Mammal. 5:64-65.

Carbaugh, B., J. P. Vaughan, E. D. Bellis, and H. B. Graves. 1975. Distribution and activity of white-tailed deer along an interstate highway. J. Wildl. Manage. 38:570-81.

Carhart, A. H. 1946. Hunting North American deer. Macmillan & Co., New York. 232 pp.

Chapman, F. B. 1938. The development and utilization of the wildlife resources of unglaciated Ohio. 2 vols. Ph. D. Dissertation, Ohio State Univ., Columbus. 791 pp.

_____. 1939. The whitetail deer and its management in southeastern Ohio. Trans. Fourth North Amer. Wildl. Conf. Pp. 257-67.

Chapman, F. B., and D. L. Leedy. 1948. The 1947 deer hunt at a glance. Ohio Conserv. Bull. 12(2):18-19.

Cheatum, E. L., and G. H. Morton. 1942. On the occurrence of pregnancy in white-tailed deer fawns. J. Mammal. 23:210-11.

_____. 1946. Breeding season of white-tailed deer in New York. J. Wildl. Manage. 10:249-63.

Dambach, C. A. 1948. The relative importance of hunting restrictions and land use in maintaining wildlife populations in Ohio. Ohio J. Sci. 48(6):209-29.

DeCapita, M. E. 1975. Evaluation of strip-mine reclamation for terrestrial wildlife restoration. M. S. Thesis, Ohio State Univ., Columbus. 134 pp.

Dexter, R. W., S. J. Cortese, and S. A. Reed. 1952. An analysis of food habits of the white-tailed deer. Ohio Wildl. Invest. 3(3):34-39.

Donohoe, R. W. 1962. Deer age and sex ratios. Game Research Ohio 1:48-51.

Donohoe, R. W., and R. Stoll, Jr. 1976. Forest game. pp. 2-1 to 2-17 in Ohio Division of Wildlife. Reasons for seasons, hunting and trapping. Ohio Dept. Nat. Resources Publ. No. 68 (R 1276).

Downs, A. A., and W. E. McQuilkin. 1944. Seed production of southern Appalachian oaks. J. Forestry 42:913-20.

Erickson, A. B. 1952. A late breeding record for the white-tailed deer. J. Wildl. Manage. 16:400.

Gilfillan, M. C. 1952. How fast do Ohio deer increase? Ohio Conserv. Bull. 16(8): 5, 32.

_____. 1957. Where to hunt deer in Ohio. Ohio Conserv. Bull. 21(12):4-5, 29.

Hamerstrom, F. N., Jr., and J. Blake. 1939. Winter movements and winter foods of white-tailed deer in central Wisconsin. J. Mammal. 20:206-15.

Haugen, A. O., and L. A. Davenport. 1950. Breeding records of white-tailed deer in the Upper Peninsula of Michigan. J. Wildl. Manage. 14:290-95.

Heet, G. C. 1977. Habitat utilization of white-tailed deer in southeastern Ohio determined by radiotelemetry. M.S. Thesis, Ohio State Univ., Columbus. 86 pp.

Hosley, N. W. 1956. Management of the white-tailed deer in its environment. Pp. 187-259 in W. P. Taylor, ed. The deer of North America *(which see)*.

Kellogg, R. 1956. What and where are the whitetails? Pp. 31-55 in W. P. Taylor, ed. The deer of North America *(which see)*.

Knowlton, F. F., and W. C. Glazener. 1965. Incidence of maxillary canine teeth in white-tailed deer from San Patricio County, Texas. J. Mammal. 46:352.

Mirarchi, R. E., P. F. Scanlon, R. L. Kirkpatrick, and C. B. Schreck. 1977. Androgen levels and antler development in captive and wild white-tailed deer. J. Wildl. Manage. 41:178-83.

Nixon, C. M. 1962. Leptospirosis in deer. Game research in Ohio 1:143:44.

_____. 1965. Some observations of deer range in southeastern Ohio. Game Research in Ohio 3:107-222.

_____. 1970. Deer range appraisal in the midwest. Pp. 11-18 in White-tailed deer in the Midwest: a Symposium. North Central Forest Experiment Station, St. Paul, Minn. USDA, Forest Service Research Paper NC-39. 34 pp.

_____. 1971. Productivity of white-tailed deer in Ohio. Ohio J. Sci. 71:217-25.

Nixon, C. M., M. W. McClain, and K. R. Russell. 1970. Deer food habits and range characteristics in Ohio. J. Wildl. Manage. 34:870-86.

Ohio Division of Wildlife. 1979. Deer season results, 1978. Ohio Dept. Nat. Resources, Div. Wildl. Publ. 166 (R379). 2 pp.

Olson, S. F. 1932. Fish-eating deer. J. Mammal. 13:80-81.

Oney, J. 1945. Ohio white-tails. Ohio Conserv. Bull. 9(12):27.

Rice, W. R., and J. D. Harder. 1977. Application of multiple aerial sampling to a mark-recapture census of white-tailed deer. J. Wildl. Manage. 41:197-206.

Riney, T. 1951. Standard terminology for deer teeth. J. Wildl. Manage. 15:99-101.

Ruhl, H. D. 1956. Hunting the whitetail. Pp. 261-331 in W. P. Taylor, ed. The deer of North America *(which see)*.

Schriver, A. D. 1976. Activity of white-tailed deer in southeastern Ohio. M. S. Thesis, Ohio State Univ., Columbus.

Severinghaus, C. W., and E. L. Cheatum. 1956. Life and times of the white-tailed deer. Pp. 57-186 in W. P. Taylor, ed., The deer of North America *(which see)*.

Stoll, R. J., Jr. and R. W. Donohoe. 1973. White-tailed deer harvest management in Ohio. Ohio Dept. Nat. Resources, Game Management Section, Div. Wildl., Inservice Document 73. 39 pp.

Stuber, J. W. 1943. Present deer situation in Ohio. Ohio Conserv. Bull. 7(12):4-5, 32.

Taylor, W. P., ed. 1956. The deer of North America: the white-tailed mule and black-tailed deer, genus *Odocoileus*, their history and management. Stackpole Co., Harrisburg, Pa., and Wildlife Management Inst., Washington, D. C. 668 pp.

Thomas, J. W., R. M. Robinson, and R. G. Marburger. 1965. Social behavior in a white-tailed deer herd containing hypogonadal males. J. Mammal. 46:314-27.

Trodd, L. L. 1962. Quadruplet fetuses in a white-

RACCOON
Procyon lotor

STRIPED SKUNK
Mephitis mephitis

WHITE-TAILED DEER
Odocoileus virginianus

tailed deer from Espanola, Ontario. J. Mammal. 43:414.

Verme, L. J. 1961. Late breeding in northern Michigan deer. J. Mammal. 42:426–27.

Wislocki, G. B. 1942. Studies on the growth of deer antlers. I. On the structure and histogenesis of the antler of the Virginia deer *(Odocoileus virginianus borealis)*. Amer. J. Anat. 71:371–415.

———. 1943. Seasonal changes in the male reproductive tract of the Virginia deer *(Odocoileus virginianus borealis)* with a discussion of the factors controlling the antler-gonad periodicity. Essays in Biology, Univ. of Calif. Press, Berkeley. Pp. 631–47.

EXTIRPATED SPECIES AND THOSE OF INCIDENTAL OR DOUBTFUL OCCURRENCE

In addition to the 54 species of mammals already described, there are others that have occurred or might occur in Ohio. Included are some species that at one time inhabited the state, but for one reason or another have been extirpated. On occasion individuals move back into Ohio; although they are usually only "visitors," it would not be too surprising to find some of these species reestablishing themselves in the state. Some other species that live in neighboring states may wander into Ohio from time to time; they too might become permanent residents. A third group includes species that are not native to the state, but that have been purposely released or have escaped from confinement.

Certain domestic animals, e.g., cat and dog, lacking human care, may successfully live a wild (feral) existence. Such species have not been considered in this discussion, although I have included characteristics for several of them in keys elsewhere to avoid confusion with the wild species. Unfortunately, the records for "doubtful" species are often established on hearsay evidence and/or secondhand stories that are impossible to verify. A house cat footprint made in soft mud spreads out, enlarges, and immediately becomes a "mountain lion print"; a bobtailed, feral house cat is identified as a "bobcat"; and so on. I would remind the reader that under field conditions correct identification of a species is often difficult even for the trained biologist. Only when the animals can be carefully examined in the laboratory by a trained mammalogist can positive identification be made. There is no substitute for a well-prepared, properly tagged and identified museum skin and skull.

Following are brief descriptions for 22 species that have occurred or might occur in Ohio.

GRAY MYOTIS. *Myotis grisescens* Howell. The gray myotis is similar to other species of *Myotis,* but generally larger. The wing membrane is attached to the ankle instead of to the side of the foot as in other *Myotis* species. This species ranges from eastern Kentucky west to Missouri and south to northern Florida. One specimen of this species was shot in flight over the Ohio River. I do not consider this southern species to be a regular member of the Ohio bat fauna. It is listed by the U. S. Fish and Wildlife Service as an endangered species (Tuttle, 1979).

BRAZILIAN FREE-TAILED BAT. *Tadarida brasiliensis* (I. Geof. St.-Hilaire). This small dark-brown bat with narrow wings has about half of its tail extending conspicuously beyond the edge of the tail membrane. A tropical species, its range extends from South America to the southern and southwestern United States. Smith and Goodpaster (1960) reported that a single male was collected on 31 July 1958 in a barn ten miles northwest of Portsmouth in Scioto County. Another record from the Dayton area of an autumn wanderer has been published by Barbour and Davis (1974, p. 118).

SWAMP RABBIT. *Sylvilagus aquaticus* (Bachman). This species of rabbit is very similar to, but smaller than, the cottontail. Its habitat is swamps and lowland wet areas of the southern United States. Occasional reports of the species occurring in Ohio have not been substantiated.

SNOWSHOE HARE OR **VARYING HARE.** *Lepus americanus* (Erxleben). The snowshoe hare, a northern forest species, is usually about 0.4 to 0.9 kg (1 to 2 lb) larger than the cottontail, and it undergoes a change in pelage from brown in summer to white in winter. The species was first reported in northeastern Ohio by Kirtland (1838). Evidently it was present in Ashtabula County in the nineteenth century (Bole and Moulthrop, 1942, p. 174; Hall and Kelson, 1959, p. 275), but apparently it no longer survives in Ohio. Attempts by the Ohio Division of Wildlife to restock this animal have been unsuccessful.

ORD'S KANGAROO RAT. *Dipodomys ordii* Woodhouse. A subspecies, *D.o. richardsoni* (Allen), of this western Great Plains and Rocky Mountains species was introduced on the shores of Lake Erie near Fairport Harbor, Lake County, at what is now Headlands Beach State Park (Bole and Moulthrop, 1942, p. 141). It became established in an area of sand dunes and was reported to be very abundant in spite of large populations of Norway rats (*Rattus norvegicus*), E. Bruce McLean (personal communication, 1978) reported that specimens have not been taken in recent years and that the population is probably gone.

MARSH RICE RAT. *Oryzomys palustris* (Harlan). This southern species of marshy areas has not been reported from Ohio in historic times. Evidence from bones found in Indian archaeological sites near Cincinnati, Lebanon, Portsmouth, and Chillicothe indicates, however, that this species once occurred in Ohio (Goslin, 1951; Guilday and Mayer-Oakes, 1952).

GOLDEN MOUSE. *Ochrotomys nuttalli* (Harlan). This attractive reddish-brown or golden-colored mouse, about the size of a white-footed mouse (*Peromyscus leucopus*), builds its nest in vines or trees. It has never been reported from Ohio, but Barbour and Davis (1974, p. 186) record its presence in northeastern Kentucky near the Ohio River. Although the river may be a formidable barrier to migration, eventual discovery of populations in areas of similar habitat in Adams, Scioto, and/or Lawrence counties would be noteworthy but not surprising.

BLACK RAT OR **ROOF RAT.** *Rattus rattus* (Linnaeus). The black rat probably preceded the Norway rat into Ohio, as it did elsewhere in the United States, but there are no indications that it occurred in great numbers here. Kirtland (1838, p. 161) only mentions its presence in Ohio; Bole and Moulthrop (1942, p. 164) state that in the mammal collection of the Cleveland Museum of Natural History the only specimens collected in Ohio were destroyed by fire. No locality records are given, and there are no recent records in Ohio. At any rate, once the Norway rat was introduced, it quickly displaced the smaller, less aggressive black rat.

PORCUPINE. *Erethizon dorsatum* Linnaeus. The porcupine, a slow-moving brown rodent weighing up to 9 kg (20 lb), is probably best known for its barbed quills and the defense they provide. Though its range in North America is more commonly northern and western, it occurs in Pennsylvania and once was fairly common in northeastern and northwestern Ohio (Bole and Moulthrop, 1942, p. 173). Individual specimens have been reported from Cuyahoga, Lucas, Ashtabula, and Columbiana counties. There are but two remaining museum specimens, one from Ashtabula County collected in 1906 and deposited in the Ohio State University Museum of Zoology, and the other collected in 1952 from Columbiana County and deposited in the Ohio State University Department of Zoology Teaching Collection. Reports of porcupines in northern Ohio are occasionally made to the Ohio Division of Wildlife, but existence of a breeding population in Ohio is very doubtful.

NUTRIA. *Myocastor coypus* (Molina). In the wild one of these large rodents can easily be mistaken for a small beaver or a large muskrat. An adult has a body length of 50 to 62.5 cm (20 to 25 in) and weighs between 6.75 and 11.25 kg (15 and 25 lb). The fur is usually rich dark brown. The nutria differs from the muskrat and the beaver in two ways: (1) the tail is round rather than flattened either laterally (muskrat) or dorsoventrally (beaver); and (2) there is no webbing between the fourth and fifth toes on the hind foot, although the rest of the toes are fully webbed. The webbing pattern on the hind foot plus the presence of a greatly reduced first digit on each front foot makes the track of this mammal unmistakable.

Nutria are native to Central and South America. In 1899 they were first introduced into the United States in California, where they were to be farm-raised for their fur. The initial attempt to propagate the animals in captivity failed, as did a number of similar efforts throughout the country. In many instances, following the breeding failures, the nutria were released into the wild, where they were sometimes successful in establishing feral populations. At least 16 states across the United States have reported feral nutria, with the marshes of Louisiana probably supporting the largest populations.

In Ohio, according to Bednarik (1958), "Nutria were first imported into the city of Kent, Portage County, in 1937 for the purpose of establishing a fur farm. The venture was successful and shortlived (Gilfillan, personal communication in 1955). The first nutria known to have escaped into the wild was reported at a fur farm near Kent during a flood." Feral nutria have been recorded from the following 14 counties in Ohio: Allen, Ashtabula, Champaign, Clark, Cuyahoga, Defiance, Gallia, Harrison, Lucas, Medina, Miami, Pike, Portage, and Warren (Ohio Division of Wildlife, 1978).

Further indiscriminate importation and/or release of nutria in Ohio should be prohibited because (1) wild populations are difficult to control and can do great damage to native wildfowl and muskrat habitat; (2) the monetary value of nutria fur is much lower than originally estimated (prime wild pelts have been worth about $1.50–$2.00 each as compared with Argentinian furs that have sold for as much as $25.00); and (3) nutria can cause severe damage to agricultural crops. No feral populations of nutria have been reported in Ohio since the late 1950s. They are apparently restricted by the harsh winter weather.

GRAY WOLF OR TIMBER WOLF. *Canis lupus* Linnaeus. The gray wolf, a common resident of Ohio before 1800, has succumbed to the expanding activities of man. Its feeding habits and man's animal husbandry appear to be mutually exclusive. Kirtland (1838) stated that the wolf was "becoming very rare" in Ohio. Hildreth's statement (1848) is of interest in this regard: "The wolf for thirty years was a great hindrance to the raising of sheep, and for a long period the State paid a bounty for their scalps. Neighboring farmers often associated and paid an additional bounty of ten or fifteen dollars, so as to make it an object of profit for certain old hunters to employ their whole time and skill in entrapping them. At this period [1848] the race is nearly extinct in the Ohio Company's lands." Giles (1966) quotes an early historian of Tuscarawas County who said that "as late as 1842 an organized wolf hunt took place among the Zoar hills at which seven wolves were killed." Brayton (1882) discusses wolves, but gives the impression that they were no longer present, and indeed they probably had been extirpated from Ohio by then.

A curious report of wolves in Ohio in the mid-twentieth century was published by Stewart and Negus (1961); they reported that a pack of six wolves, after killing approximately 100 sheep, had been shot during the winter of 1942–43 near Augusta in Carroll County. One mounted specimen was located in 1953 and identified definitely as a wolf by Stanley P. Young, of the U.S. Fish and Wildlife Service. How these wolves got into Ohio or where they came from is not known.

The gray wolf is listed by the U.S. Fish and Wildlife Service (1973) as an endangered species.

BLACK BEAR. *Ursus americanus* Pallas. The remains of black bears have been recovered from Indian mounds and other excavations throughout Ohio. Old historical records also indicate its widespread occurrence here before 1850. It is my impression, however, that the black bear was never plentiful in the state. Kirtland (1838) says simply, "A few still exist within our limits"; and Brayton (1882, p. 64) states, "The Black Bear must have been early driven from the woods of Ohio." According to Moseley (1906) the last native bear reported from Ohio was killed in Paulding County in 1881. In his account of the mammals of Ohio, Enders (1930) refers to Moseley's comments and continues, "Recently several bears have been introduced into the game refuge near Ironton." Enders was unable to locate their place of origin from the Division of Fish and Game, which brought them into the state. Bole and Moulthrop (1942, p. 119) also reported that the black bear had been reintroduced into Scioto County, and that it was then of casual occurrence along the Ohio-Pennsylvania border. To my knowledge there is not a single Ohio black bear skin or skull preserved in any museum collection. Oral reports about bears in the eastern and southern counties continue to be heard, but it is extremely unlikely that a breeding population is resident in Ohio. An occasional bear may wander into the state from Pennsylvania or West Virginia, however.

RINGTAIL. *Bassariscus astutus* Rhoads. This unmasked relative of the raccoon is native to the southwestern United States and Mexico. There are two acceptable records from Louisiana (Lowery,

1974) but none other east of the Mississippi River. Although Bole and Moulthrop (1942, pp. 119–21) reported on several museum skins of *Bassariscus astutus,* which they believed may have been wild specimens from Ohio, and Goodpaster and Hoffmeister (1968) reported an animal that had been shot just outside Cincinnati, I do not believe that this species is or ever has been a native resident of the state. The specimens reported from Ohio most likely had escaped from captivity.

MARTEN. *Martes americana* (Turton). This larger relative of the mink has yellowish-brown fur shading to dark brown or black on the tail and legs. Its distribution is now limited primarily to the northern coniferous forests. Kirtland (1838, p. 176), however, reports a specimen taken in the vicinity of Chillicothe. Bole and Moulthrop (1942, p. 121) report two specimens from Ashtabula County, but also indicate that the species was extirpated from Ohio previous to 1850.

FISHER. *Martes pennanti* (Erxleben). A close but larger and darker-colored relative of the marten, the fisher has a similar distribution pattern in the northern coniferous forests of central Canada. Kirtland (1838, p. 176) records two specimens of fishers taken in Ashtabula County in 1837, but by 1850 this species too probably had been extirpated from Ohio (Bole and Moulthrop, 1942, p. 122).

WOLVERINE. *Gulo gulo* Linnaeus. This dark brown mustelid weighs from 10 to 18 kg (22 to 40 lb) and generally resembles a small bear. It is a very strong mammal and feeds on many animal species. A wilderness animal, the wolverine is now limited to the coniferous forests and tundra of the Far North and the high mountain areas of the West. It was probably never common in Ohio, since the early literature rarely mentions it, and it was probably one of the first to leave the state at settlement. Hall and Kelson (1959, pp. 922–25) record it from Pennsylvania, Michigan, and Indiana.

RIVER OTTER. *Lutra canadensis* (Schreber). This large, primarily aquatic member of the weasel family has wide distribution throughout North America north of Mexico. It is most often encountered along the shores of cold, clear streams or lakes, although in the southern reaches of its range it inhabits brackish streams and salt marshes. The heavy, almost cylindrical body blends smoothly into a thick, round neck that joins to the broad, flattened head. The tail is heavy, round, and flat on the bottom and tapers gradually from its base toward the tip. The legs are noticeably short, and each foot has five toes that are fully webbed. Weights range from 5 to 11 kg (11 to 24 lb.).

Although at one time the otter was probably common in Ohio, it is now very rare and has been designated an endangered species in Ohio by the Ohio Division of Wildlife (1976). Bole and Moulthrop (1942) make no mention of the river otter. However, Enders (1930) reported that otters bred on the Mohican River at the mouth of Ball Alley Run about 1902, but because of disturbance soon left the area. Chapman (1956) reported that an otter was taken from Fulton County in northwestern Ohio in 1920. He also reported that an adult male specimen had been trapped in 1951 on Captina Creek, Belmont County, and that game protectors reported sight records from Monroe, Coshocton, and Ashtabula counties during 1954. Rumors continue to persist regarding the river otter in Ohio, especially in the eastern counties. River otters may occasionally move into Ohio from Pennsylvania and/or Michigan, or individuals may escape from fur farms and establish themselves temporarily in certain lakes or streams. To my knowledge, however, there is not an established wild population in Ohio at the present time.

MOUNTAIN LION or COUGAR or PANTHER. *Felis concolor* Linnaeus. This large, tawny to gray, long-tailed cat weighs from 36 to 96 kg (80 to 200 lb). It has the largest range of any animal in the Western Hemisphere, a range that before settlement extended from southern Canada to southern South America.

In his discussion of mammals that once occupied the vicinity of Buckeye Lake, Ohio, Trautman (1939) says:

Panthers *(Felis concolor couguar)* are frequently referred to in the literature but generally it is not clear whether the name "panther" refers to this species or the bobcat. *Felis concolor couguar* was in the general vicinity of the area in 1805 and was probably in the Great and Bloody Run swamps for several years thereafter. . . . In the autumn of 1805 Jacob Wilson, who was then living within a mile of Newark, treed with his dogs and then killed a "huge panther" that had previously raided his pig pen and carried away a pig.

Hamilton (1943) says: "Repeated press accounts of the presence of these large cats in northeastern

United States occur but such stories lack authority. The Adirondacks was the last stronghold of the panther in the northeast, and the latest record of one from this wilderness was in 1903."

Although it is still present in remote regions of Florida and Louisiana and throughout the West, the cougar had probably been extirpated from Ohio by 1850 (Trautman, 1978).

LYNX or CANADA LYNX. *Felis lynx* (Kerr). This bob-tailed cat is similar in size and form to the bobcat (*Felis rufus*), but it has a much more northern distribution in the coniferous forests of Canada and Alaska. Originally, it occurred over much of northern United States (Hall and Kelson, 1959, pp. 967-68), but settlement has reduced its range considerably. Precise records are not available, but the lynx was probably extirpated from Ohio by the mid-nineteenth century. As with the bobcat, local reports about "panthers" in an area clouded the actual situation.

BOBCAT. *Felis rufus* (Schreber). Although not as widespread today as in former years, the American bobcat still ranges through most of continental United States except Alaska. Bobcats are adaptable animals; the habitat in which they occur varies from swamps in the South to low, open, partially forested mountains in the West. In Ohio the kind of habitat where bobcats are most likely to occur is in the eastern and southern portions of the Unglaciated Allegheny Plateau. This includes cliffs and steep slopes with heavy woods bordering pastures and cultivated fields.

Externally, there is no sure way to distinguish a bobcat from a bob-tailed house cat. Usually a mature bobcat weighs from 4.5 to 18 kg (10 to 40 lb) and measures in length from 56 to 127 cm (22 to 50 in). Usually there is a short tuft of hair on each ear. Almost always the end of the short tail is black above and white below, and almost always the yellowish- to reddish-brown coat is spotted with black. However, feral house cats may exhibit all or a combination of these same characteristics, and not all bobcats have them all. The skull, however, is unmistakable. The bobcat has only three teeth behind the canines in the upper jaw, whereas the house cat has four.

To my knowledge there are no known specimens of bobcats, past or present, in any museum collection in Ohio. There are authentic literature records of the bobcat in Ohio in former days, and teeth and bones have been recovered from Indian burial mounds. Enders (1930) commented: "There are frequent newspaper reports of the capture of this large cat, but a large reward posted by a group of Columbus men has failed to produce any recent record of its existence in the state." Goodpaster (1941) was of the opinion that the animal had disappeared from the state; Bole and Moulthrop (1942) did not include it in their list of Ohio mammals. Several fur dealers with whom I have talked cannot remember ever having processed a bobcat skin that was taken in Ohio. Allen Cannon, of the Ohio Division of Wildlife, has shown me photographs of three bobcats that had been shot by hunters in Hocking (1960), Guernsey (1971), and Gallia (1978) counties. However, Cannon and other knowledgeable people in the Ohio Department of Natural Resources agree that there are probably no breeding populations of bobcats in Ohio today, although infrequent "visitors" may wander into Ohio occasionally and could possibly survive in the more remote areas of southern Ohio. The species is afforded legal protection in Ohio since it has been designated an endangered species by the Ohio Division of Wildlife (1976).

WAPITI or AMERICAN ELK. *Cervus elaphus* (Erxleben). The earliest relatives of modern elk did not arrive on this continent from Asia until fairly recently in geologic time, certainly not before the Pleistocene about two million years ago. Once here, they spread rapidly. When the white man arrived on the continent, the wapiti occupied practically the entire area from the Pacific Coast almost to the Atlantic and from southern Canada almost to Florida and Mexico (Hall and Kelson, 1959, pp. 1000-1002). Ohio and surrounding states apparently had substantial populations of elk, but reports commenting on the numbers of elk in Ohio are not precise. Brayton (1882), quoting from an early history of Ohio, says: "When Circleville was first settled, the carcasses, or rather skeletons, of 50 individuals of the family of elk lay scattered about the surface." Kirtland (1838, p. 177) says that elk were frequently observed in Ashtabula County until about 1832. Bole and Moulthrop (1942, p. 175) believe that the wapiti was very abundant since the close of the Ice Age and state: "At the time of white settlement the wapiti was found in all parts of Ohio." They claim that the last animals were killed in Ashtabula County about 1840. Murie (1951) concluded that the species disappeared from Ohio sometime between 1828 and 1838.

BISON OR AMERICAN BUFFALO. *Bison bison* (Linnaeus). In 1521 Cortez, the conqueror of the Aztecs, became the first European to see the bison. The animals that he saw, however, were in a zoo in what is now Mexico City, and it was not until 1530 that another Spaniard, de Vaca, saw the vast herds of animals on their native range in southwestern Texas. Although not present in the tremendous herds of the western prairies (estimated at 60,000,000), bison were nevertheless numerous and widely distributed throughout Ohio when the first white men arrived. In those days of early settlement, however, the numbers rapidly declined, and the last bison recorded shot and killed in Ohio was in Lawrence County in 1803 (Kochman, 1977), the very year that Ohio became a state.

SELECTED REFERENCES

Bailey, T. N. 1974. Social organization in a bobcat population. J. Wildl. Manage. 38:435–46.

Barbour, R. W., and W. H. Davis. 1969. Bats of America. Univ. Press of Ky., Lexington. 286 pp.

———. 1974. Mammals of Kentucky. Univ. Press of Ky., Lexington. 322 pp.

Bednarik, K. E. 1958. Nutria in the United States with management recommendations for Ohio. Ohio Dept. Nat. Resources, Game Management Publ. 165. Pp. 1–22.

Bole, B. P., Jr., and P. N. Moulthrop. 1942. The Ohio recent mammal collection in the Cleveland Museum of Natural History. Sci. Publ. Cleveland Mus. Nat. Hist. 5:83–181.

Brayton, A. W. 1882. Report on the mammals of Ohio. Rep. Geol. Surv. Ohio 4:1–185.

Chapman, F. B. 1946. Ohio's black bears: their last stronghold is the Little Smokies. Ohio Conserv. Bull. 10(3):12–13.

———. 1956. The river otter in Ohio. J. Mammal. 37:284.

Enders, R. K. 1930. Some factors influencing the distribution of mammals in Ohio. Occ. Papers, Univ. of Mich. Mus. Zool. 212:1–27.

Giles, R. H., Jr. 1966. Early natural history of a forested area near Dover, Ohio. Ohio J. Sci. 66(5):469–73.

Goldman, E. A. 1937. The wolves of North America. J. Mammal. 18:37–45.

Goodpaster, W. W. 1941. A list of the birds and mammals of southwestern Ohio. J. Cincinnati Soc. Nat. Hist. 22:1–47.

Goodpaster, W. W., and D. F. Hoffmeister. 1968. Notes on Ohioan mammals. Ohio J. Sci. 68:116–17.

Goslin, R. M. 1951. Evidence of the occurrence of the rice rat in prehistoric Indian village sites in Ohio. Ohio Indian Relic Collectors Soc. Bull. 26:19–22.

———. 1964. The gray bat, *Myotis grisescens* Howell, from Bat Cave, Carter County, Kentucky. Ohio J. Sci. 64(1):63.

Guilday, J. E., and W. J. Mayer-Oakes. 1952. An occurrence of the rice rat (*Oryzomys*) in West Virginia. J. Mammal. 33:253–55.

Hamilton, W. J., Jr. 1953. The mammals of eastern United States: an account of recent land mammals occurring east of the Mississippi. Comstock Publ. Co., Ithaca, N. Y. 438 pp.

Hildreth, S. P. 1848. Pioneer history: being an account of the first examinations of the Ohio valley, and the early settlement of the Northwest Territory. Hist. Soc. Cincinnati, H. W. Derby and Co., Cincinnati, Ohio. 525 pp.

Jones, H. 1881. American otter and Virginia deer. Cincinnati Soc. Nat. Hist. 4:336.

Jones, J. K., Jr., D. C. Carter, and H. H. Genoways. 1975. Revised checklist of North American mammals north of Mexico. Occ. Papers, The Museum, Tex. Tech. Univ., Lubbock. 24:1–14.

Kirtland, J. P. 1838. A catalogue of the Mammalia, birds, reptiles, fishes, Testacea, and Crustacea in Ohio. Ohio Geol. Surv. Second Annu. Rep. pp. 157–200.

Kochman, A. 1977. Vast bison herds once roamed Ohio. Ohio Hist. Soc. Echoes 16(5):3–5.

LaVal, R. K., R. L. Clawson, M. L. LaVal, and W. Caire. 1977. Foraging behavior and nocturnal activity patterns of Missouri bats, with emphasis on the endangered species *Myotis grisescens* and *Myotis sodalis*. J. Mammal. 58:592–99.

Leedy, D. L. 1948. Another nutria record. Ohio Conserv. Bull. 12(7):27.

Linsey, D. W., and R. L. Packard. 1977. *Ochrotomys nuttalli*. Amer. Soc. Mammal., Mammalian Species No. 75:1–6.

Lowery, G. H., Jr. 1974. The mammals of Louisiana and its adjacent waters. La. State Univ. Press, Baton Rouge. 565 pp.

Mech, L. D. 1974. *Canis lupus*. Amer. Soc. Mammal., Mammalian Species No. 37:1–6.

Moseley, E. L. 1906. Notes on the former occurrence of certain mammals in northern Ohio. Ohio Nat. 6:504–5.

Murie, O. J. 1951. The elk of North America. Stackpole Co., Harrisburg, Pa., and Wildlife Management Inst., Washington, D. C. 376 pp.

Ohio Division of Wildlife. 1976. Endangered wild animals in Ohio. Ohio Dept. Nat. Resources Publ. 316 (R675). 3 pp.

———. 1978. Nutria in Ohio: life history notes. Ohio Dept. Nat. Resources Publ. 94 (R478). 2 pp.

Petrides, G. A. 1947. The nutria (coypu) in Ohio. Ohio Conserv. Bull. 11(4):29.

Petrides, G. A., and D. L. Leedy. 1948. The nutria in Ohio. J. Mammal. 29:182–83.

Richardson, W. B. 1942. Ring-tailed cats *Bassariscus astutus)*: their growth and development. J. Mammal. 23:17-26.

Smith, E., and W. Goodpaster. 1960. A free-tailed bat found in Ohio. J. Mammal. 41:117.

Stewart, P. A., and N. C. Negus. 1961. Recent record of wolf in Ohio. J. Mammal. 42:420-21.

Trautman, M. B. 1939. The numerical status of some mammals throughout historic time in the vicinity of Buckeye Lake, Ohio. Ohio J. Sci. 39:133-43.

———. 1978. The Ohio country from 1750 to 1977—a naturalist's view. Ohio Biol. Surv. Biol. Notes No. 10:6.

Tuttle, M. D. 1979. Gray bat, *Myotis grisescens* (Howell, 1909), *in* Endangered species, Great Lakes Region Red Book. Fish and Wildlife Service, U.S. Dept. of Interior, Fort Snelling, Twin Cities, Minn.

U.S. Fish and Wildlife Service. 1973. Threatened wildlife of the United States. U.S. Dept. Interior, Bureau of Sport Fisheries and Wildlife. Resource Publ. 114. 289 pp., esp. 237.

Woods, C. A. 1973. *Erethizon dorsatum*. Amer. Soc. Mammal., Mammalian Species No. 29:1-6.

Young, S. P. 1958. The bobcat of North America: its history, life habits, economic status, and control. Stackpole Co., Harrisburg, Pa., and Amer. Wildl. Inst., Washington, D.C. 193 pp.

Young, S. P., and E. A. Goldman. 1944. The wolves of North America. Amer. Wildl. Inst., Washington, D.C. 636 pp.

KEY TO THE SKULLS OF OHIO MAMMALS

1. Diastema between incisors and cheek teeth (fig. 5) 2
1'. No diastema between incisors and cheek teeth (fig. 5) 28
2. More than one pair of incisors in lower jaw; incisors not chisel-like 3
2'. One pair of chisel-like incisors in lower jaw . 4
3. No incisors in upper jaw; bones in front of orbit fenestrated (porous). Order Artiodactyla, Even-toed Hoofed Ungulates; Family Cervidae, Deer . .*Odocoileus virginianus,* White-tailed Deer, p. 143
(See also "Key to the Skulls of Ungulates in Ohio," p.143
3'. Incisors in upper jaw; bones in front of orbit not fenestrated. Order Perissodactyla, Odd-Toed Hoofed Ungulates ....
.................Horse, Mule, Ass, etc.
4. Incisors in upper jaw with a pair of short, inconspicuous peg-like teeth immediately behind them (figs. 5 and 12). Order Lagomorpha, Rabbits, etc.; Family Leporidae .. 5
4'. Incisors in upper jaw without a pair of peg-like teeth behind them (figs. 4 and 5). Order Rodentia, Rodents; Families Sciuridae, Castoridae, Cricetidae, Muridae, and Zapodidae 7
5. Interparietal bones not closely fused with parietals, with obvious borders and usually easily distinguished 6
5'. Interparietal bones closely fused with parietals and without obvious borders, not easily distinguished.........................
..... *Lepus americanus,* Snowshoe Hare, p. 152
6. Postorbital processes usually fusing posteriorly with the frontal bone or so close to cranium that a slit-like opening is formed above ....
Sylvilagus floridanus, Eastern Cottontail, p. 53
6'. Postorbital processes completely fused to cranium with only a small round or oval hole formed ....
.... *Sylvilagus aquaticus,* Swamp Rabbit, p. 151
7. Zygomatic plate horizontal and narrower than infraorbital foramen. Family Zapodidae, Jumping Mice 8
7'. Zygomatic plate not horizontal, broader than infraorbital foramen 9
8. Four cheek teeth each side of upper jaw; total number of teeth 18 ....*Zapus hudsonius,* Meadow Jumping Mouse, p. 112
8'. Three cheek teeth each side of upper jaw; total number of teeth 16 ..... *Napaeozapus insignis,* Woodland Jumping Mouse, p. 115
9. Upper jaw with 4 or 5 cheek teeth on each side; total number of teeth 20 or 22. Family Sciuridae and Castoridae 10
9'. Upper jaw with 3 or 4 cheek teeth on each side; total number of teeth 16 or 18. Families Cricetidae and Muridae 17
10. Total length of skull greater than 70 mm ... 11
10'. Total length of skull less than 70 mm. Family Sciuridae (in part), Squirrels, Chipmunks, etc. 12
11. Front of incisors white; postorbital processes prominent, sharp-pointed and at right angles to the rest of the skull; 5 cheek teeth each side of upper jaw; total number of teeth 22. Family Sciuridae (in part).............
.... *Marmota monax,* Woodchuck, p. 63
11'. Front of incisors orange or yellow; postorbital processes small or missing; 4 cheek teeth each side of upper jaw; total number of teeth 20. Family Castoridae, Beaver ....
.... *Castor canadensis,* Beaver, p. 80
12. Length of skull more than 50 mm 13
12'. Length of skull less than 50 mm 14
13. Four cheek teeth each side of upper jaw (fig. 15a) .... *Sciurus niger,* Fox Squirrel, p. 73
13'. Five cheek teeth each side of upper jaw; first small and inconspicuous (rarely missing) (fig. 15b) .... *Sciurus carolinensis,* Gray Squirrel, p. 69
14. Four cheek teeth each side of upper jaw; total teeth 20 15
14'. Five cheek teeth each side of upper jaw; total teeth 22 16
15. Infraorbital foramen round, piercing zygomatic plate; rostrum with tapering sides giving a streamlined appearance ....
.... *Tamias striatus,* Eastern Chipmunk, p. 60
15'. Infraorbital foramen a vertical slit passing between the zygomatic plate and the rostrum; rostrum rounded, not streamlined in appearance ..... *Tamiasciurus hudsonicus,* Red Squirrel, p. 74
16. Interorbital region deeply notched ....
.... *Glaucomys volans,* Southern Flying Squirrel, p. 77
16'. Interorbital region not deeply notched .....
. . *Spermophilus tridecemlineatus,* Thirteen-lined Ground Squirrel, p. 66
17. Upper molars with cusps in three longitudinal rows (figs. 5 and 14). Family Muridae, Old World Rats and Mice 18
17'. Upper molars with cusps in two longitudinal rows or crowns with loops and triangles (figs. 5 and 14). Family Cricetidae; Native Rats and Mice, Voles, and Lemmings 19

18. Upper incisors with subapical notch; supraorbital and temporal ridges absent; first upper molar larger than second and third combined; length of skull less than 25 mm (figs. 20 and 21). *Mus musculus,* House Mouse, p. 110
18'. Upper incisors without subapical notch; supraorbital and temporal ridges well developed; first upper molar same size as second and third molars; length of skull more than 25 mm .. *Rattus norvegicus,* Norway Rat, p. 108
19. Molars with cusps in two longitudinal rows (figs. 5 and 14) 20
19'. Molars with enamel patterns of loops and triangles (figs. 5 and 14) 22
20. Upper incisors without deep groove down center 21
20'. Upper incisors with deep groove down center ....
 .... *Reithrodontomys humulis,* Eastern Harvest Mouse, p. 84
21. Skull length exceeds 25 mm; anterior end of palatine foramen narrow and pointed; skull appears flat in profile; interorbital width 4 mm or more (fig. 17) . *Peromyscus leucopus,* White-footed Mouse, p. 88
21'. Skull length greater than 25 mm; anterior end of palatine foramen rounded and blunt; skull appears arched in profile; interorbital width 4 mm or less (fig. 17) ....
 .. *Peromyscus maniculatus,* Deer Mouse, p. 86
22. Total length of skull less than 50 mm; first lower molar with fewer than six enamel triangles on lingual side 23
22'. Total length of skull greater than 50 mm; first lower molar with six enamel triangles on lingual side ....
 .... *Ondatra zibethicus,* Muskrat, p. 102
23. Some molars with more than two or three dentine areas surrounded by enamel (figs. 5 and 14) 24
23'. Each molar with two or three dentine areas surrounded by enamel (two or three loops) (figs. 5 and 14) ....
 .. *Neotoma floridana,* Eastern Woodrat, p. 91
24. No shallow groove down anterior-lateral surface of upper incisors 25
24'. Shallow groove down anterior-lateral surface of upper incisors (close examination required) .... *Synaptomys cooperi,* Southern Bog Lemming, p. 105
25. Crown of second upper molar with four irregular loops (fig. 5) 26
25'. Crown of second upper molar with five irregular loops (four well defined followed by a fifth smaller one) (figs. 5 and 18) ....
 Microtus pennsylvanicus, Meadow Vole, p. 95
26. Posterior border of palate with a median projection not ending in thin bony shelf 27
26'. Posterior border of palate straight, ending in thin bony shelf ....
 .... *Clethrionomys gapperi,* Red-backed Vole, p. 94
27. Crown of third upper molar with three irregular closed loops; anterior palatine foramen more than 4.5 mm (figs. 5 and 18) ....
 ..... *Microtus ochrogaster,* Prairie Vole, p. 98
27'. Crown of third upper molar with four irregular closed loops; anterior palatine foramen less than 4.5 mm (fig. 5) ....
 ... *Microtus pinetorum,* Woodland Vole, p. 99
28. Fewer than five pair of incisors in upper jaw; total number of teeth in adult 44 or less 29
28'. Five pairs of incisors in upper jaw; total number of teeth in adult: 50. Order Marsupials; Family Didelphidae, Opossums .....
 . *Didelphis virginiana,* Virginia Opossum, p. 13
29. Front of skull blunt and rounded with a U-shaped notch. Order Chiroptera, Bats ... 30
29'. Front of skull tapering and pointed without a U-shaped notch 40
30. Total number of teeth 38. *Myotis* spp. 31
30'. Total number of teeth 36 or less 34
31. Least interorbital constriction less than 4 mm in adult 32
31'. Least interorbital constriction more than 4 mm in adult ....
 .. *Myotis lucifugus,* Little Brown Myotis, p. 38
32. Total length of skull more than 14.3 mm in adult 33
32'. Total length of skull less than 14.3 mm in adult ....
 ..... *Myotis leibii,* Small-footed Myotis, p. 43
33. Median saggital crest usually present; if the skin is available use it for identification ....
 .... *Myotis sodalis,* Indiana Bat, p. 41
33'. No median saggital crest; if the skin is available, use it for identification ....
 .... *Myotis keenii,* Keen's Myotis, p. 41
34. Total number of teeth 36 35
34'. Total number of teeth less than 36 36
35. Rostrum strongly concave just behind nasal openings; skull appears flat in profile ......
 .... *Lasionycteris noctivagans,* Silver-haired Bat, p. 43
35'. Rostrum convex behind nasal openings; skull highly arched in profile ....
 .... *Plecotus rafinesquii,* Rafinesque's Big-eared Bat, p. 51
36. Total number of teeth less than 34 37
36'. Total number of teeth 34 ....

KEY TO THE SKULLS OF OHIO MAMMALS

Pipistrellus subflavus, Eastern Pipistrelle, p. 44

37. Total number of teeth 32 38
37'. Total number of teeth 30 ....
..... *Nycticeius humeralis,* Evening Bat, p. 50
38. One incisor on each side of upper jaw; first premolar minute and visible only on palatal view. *Lasiurus* spp. 39
38'. Two incisors on each side of upper jaw; first premolar not minute ....
.... *Eptesicus fuscus,* Big Brown Bat, p. 46
39. Total length of skull less than 15.5 mm .....
.... *Lasiurus borealis,* Red Bat, p. 48
39'. Total length of skull more than 15.5 mm ...
.... *Lasiurus cinereus,* Hoary Bat, p. 49
40. Molar teeth with a distinct W-pattern on crowns; canines not noticeably larger than other teeth (figs. 4 and 5). Order Insectivora, Moles and Shrews 41
40'. Molar teeth without W-pattern on crowns; canines extremely well developed and longer than other teeth (figs. 4 and 5). Order Carnivora, Flesh Eaters 48
41. Zygomatic arch present; teeth white; first pair of upper incisors not notched, and not or only slightly extending forward (fig. 7). Family Talpidae, Moles 42
41'. Zygomatic arch absent; teeth usually with red or brown coloration; first pair of upper incisors notched and projecting forward (fig. 7). Family Soricidae, Shrews 44
42. Total number of teeth 44 (22 upper jaw, 22 lower; second upper incisor minute and easily overlooked) 43
42'. Total number of teeth 36 (20 upper jaw, 16 lower) *Scalopus aquaticus,* Eastern Mole, p. 30
43. Hard palate does not extend beyond plane of last molars; first upper incisors projecting forward and easily seen from above; skull tapering evenly toward the snout ....
. *Condylura cristata,* Star-nosed Mole, p. 32
43'. Hard palate extending beyond plane of last molars; first upper incisors pointing down, not visible from above; skull not tapering evenly toward the snout ....
. *Parascalops breweri,* Hairy-tailed Mole, p. 28
44. Total number of teeth 32 (20 upper jaw, 12 lower) 45
44'. Total number of teeth 30 (18 upper jaw, 12 lower) .... *Cryptotis parva,* Least Shrew, p. 26
45. Four or five unicuspids in upper jaw visible from lateral view 46
45'. Three unicuspids in upper jaw visible from lateral view (third unicuspid small and only visible from palatal view) (fig. 7) ....

.... *Microsorex hoyi,* Pygmy Shrew, p. 22
46. Five unicuspids in upper jaw visible from lateral view (fig. 7). *Sorex* spp............. 47
46'. Four unicuspids in upper jaw visible from lateral view (fifth present but small and hidden, only visible from palatal view) ....
. *Blarina brevicauda,* Short-tailed Shrew, p. 23
47. First three unicuspids equal in size, fourth unicuspid smaller than third..............
.... *Sorex fumeus,* Smoky Shrew, p. 21
47'. First two unicuspids equal in size and larger than third and fourth which are equal in size.
.... *Sorex cinereus,* Masked Shrew, p. 18
48. Total number of teeth 30 or less; molars with cutting surfaces. Family Felidae, Cats 49
48'. Total number of teeth 34 or more; some molars with crushing surfaces 50
49. Four cheek teeth each side of upper jaw (two premolars, two molars) ....
.... *Felis domesticus,* Domestic Cat, p. 155
49'. Three cheek teeth each side of upper jaw (two premolars, one molar ....
.... *Felis rufus,* Bobcat, p. 155
50. Total number of teeth 42 51
50'. Total number of teeth fewer than 42 55
51. Last molar in upper jaw not 1.5-2 times as long as wide, not noticeably larger than molar immediately in front of it; molars with both crushing and cutting crowns. Family Canidae, Dogs, Foxes, and Wolves 52
51'. Last molar in upper jaw 1.5-2 times as long as wide, noticeably larger than molar immediately in front of it; molars with crushing surfaces on crowns and cusps not modified for cutting. Family Ursidae, Bears ....
.... *Ursus americanus,* Black Bear, p. 153
52. Frontal bones with concave depression along postorbital processes 53
52'. Frontal bones convex along postorbital processes 54
53. Temporal ridges widely separated forming a U-shaped ridge, notch in lower jaw just below ramus; front of incisors not lobed (fig. 22). *Urocyon cinereoargenteus,* Gray Fox, p. 126
53'. Temporal ridges converge forming a V-shaped ridge; no notch in lower jaw below ramus; front of incisors lobed (fig. 22) .....
.... *Vulpes vulpes,* Red Fox, p. 123
54. Distance from the front of the first premolars to back of socket of last molar at least 3.1 times the distance between the inner borders of the sockets of the first premolars; upper canines extend below the two anterior mental foramina; upper incisors set close to-

gether and usually in contact (figs. 23 and 24) .... *Canis latrans,* Coyote, p. 120

54'. Distance from the front of the first premolars to back of socket of last molar 2.8 times or less the distance between the inner borders of the sockets of the first premolars; upper canines do not extend below the two anterior mental foramina; upper incisors usually with spaces between them (fig. 23) .....
.... *Canis familiaris,* Domestic Dog, p. 122

55. Total number of teeth less than 40. Family Mustelidae, Weasels, Skunk, Mink, Otter, and Badger.......................... 56
55'. Total number of teeth 40. Family Procyonidae, Raccoons and Allies ....
.... *Procyon lotor,* Raccoon, p. 128

56. Total number of teeth 34 57
56'. Total number of teeth 36 ....
.... *Lutra canadensis,* River Otter, p. 154

57. Last upper molar not square; hard palate extends well beyond posterior border of last molar 58
57'. Last upper molar squarish; hard palate extends to or only slightly past posterior border of last molar ....
...... *Mephitis mephitis,* Striped Skunk, p. 140

58. Skull length less than 70 mm; last upper molar dumbbell-shaped; brain case not triangular 59
58'. Skull length over 100 mm; last upper molar more or less triangular; brain case triangular
.... *Taxidea taxus,* Badger, p. 138

59. Skull length less than 55 mm 60
59'. Skull length greater than 55 mm ....
.... *Mustella vison,* Mink, p. 137

60. Total length of skull less than 33 mm. (In very rare instances, a small female specimen of *M. frenata* or *M. erminea* might be included here. See note below.) ....
.... *Mustella nivalis,* Least Weasel, p. 132
60'. Total length of skull more than 33 mm either
...*Mustella frenata,* Long-tailed Weasel, p. 134
or *Mustella erminea,* Ermine, p. 131

Note: Because of size discrepancies between males and females in both *M. frenata* and *M. erminea* (and even *M. nivalis* in rare instances), skull measurements may overlap. In most cases it is necessary to have the skin with the skull to make positive identification. *M. erminea,* however, is most uncommon and will be found only in the northern-most counties; *M. frenata* and *M. nivalis* occur throughout the state.

APPENDIX

Appendix Table 1. Selected English-metric conversions.

Symbol	When you know	Multiply by	To find	Symbol
in	inches	2.54	centimeters	cm
ft	feet	30.0	centimeters	cm
mi	miles	1.6	kilometers	km
a	acres	0.4	hectares	ha
mi^2	square miles	2.6	square kilometers	km^2
oz	ounces	28.0	grams	g
lb	pounds	0.45	kilograms	kg
qt	quarts	0.95	liters	l
°F	Fahrenheit temperature	5/9 (after subtracting 32)	Celsius temperature	°C
mm	millimeters	0.04	inches	in
cm	centimeters	0.4	inches	in
m	meters	3.3	feet	ft
km	kilometers	0.6	miles	mi
ha	hectares (10,000 m^2)	2.5	acres	
km^2	square kilometers	0.4	square miles	mi^2
g	grams	0.035	ounces	oz
kg	kilograms	2.2	pounds	lb
l	liters	1.06	quarts	qt
°C	Celsius temperature	9/5 (then add 32)	Fahrenheit temperature	°F

APPENDIX

Appendix Table 2. Total pelts reported by Ohio fur buyers, 1951-52 through 1977-78 and monetary value of Ohio fur harvests, 1965-66 through 1977-78. (Adapted from Bednarik, 1976[1]; Mountz & Bonsel, 1978[2]).

Season	Mink	Muskrat	Raccoon	Oppossum	Red Fox	Gray Fox	Weasel	Skunk	Badger	Beaver	Total
Total pelts											
1951-52	7,111	429,958	150,381	27,306	1,984	1,410	2,692	4,650	...	...	625,492
1952-53	8,803	510,770	146,927	28,508	1,671	1,211	2,754	6,235	...	...	706,879
1953-54	4,128	260,306	75,630	9,732	515	835	1,553	2,221	...	...	354,920
1954-55	3,666	215,239	103,756	5,675	268	553	1,151	1,319	...	...	331,627
1955-56	6,557	416,345	142,866	6,893	788	613	7,882	2,380	...	...	584,324
1956-57	6,802	550,052	185,803	6,298	...	...	2,075	2,864	...	...	753,894
1957-58	6,644	842,437	165,955	6,739	828	527	2,295	4,043	...	...	1,029,468
1958-59	5,931	680,942	121,582	4,151	541	313	1,274	2,250	2	...	816,986
1959-60	7,798	419,777	154,222	11,066	1,976	1,119	1,972	2,474	3	...	672,407
1960-61	8,888	493,436	159,904	20,420	2,456	1,585	1,976	2,930	4	140	691,739
1961-62	8,717	447,833	159,399	31,816	4,678	1,850	1,873	2,648	2	182	658,998
1962-63	8,743	429,276	133,788	21,748	4,811	1,981	1,239	1,097	1	148	602,832
1963-64	10,327	519,573	145,714	11,793	4,169	1,809	1,357	1,128	4	183	696,057
1964-65	7,375	418,601	143,985	9,756	3,515	1,656	1,265	688	6	245	587,092
1965-66	6,461	417,365	156,784	15,931	6,685	3,444	1,326	814	6	176	608,992
1966-67	5,399	408,219	165,985	15,381	8,093	4,104	858	726	7	375	609,147
1967-68	4,717	435,593	146,131	11,885	8,035	3,980	484	370	1	384	611,580
1968-69	5,179	565,672	176,430	14,960	9,787	5,446	518	556	2	631	779,181
1969-70	7,727	741,085	174,535	17,535	12,980	6,931	449	490	6	532	962,270
1970-71	8,786	568,788	165,527	10,824	10,834	6,357	257	370	5	238	771,986
1971-72	8,990	586,921	233,685	15,265	9,313	6,873	188	486	4	532	862,257
1972-73	10,279	526,770	244,890	29,945	10,953	12,350	321	1,014	5	1,747	838,274
1973-74	15,107	872,275	299,843	62,735	15,499	18,972	795	3,154	9	1,551	1,289,940
1974-75	14,525	641,161	334,261	51,279	16,824	19,174	612	1,701	13	1,208	1,088,758
1975-76	12,339	592,009	410,785	47,938	19,374	23,220	381	1,726	14	1.095	1,108,881
1976-77	13,824	596,840	294,079	56,867	21,521	24,448	734	1,490	38	977	1,010,818
1977-78	10,170	550,352	310,843	34,081	22,716	24,038	521	1,000	12	918	954,651
Total money paid for pelts											
1965-66	$41,750	$705,100	$439,301	$7,820	$40,302	$6,985	$1,100	$669	$11	$2,123	$1,245,161
1966-67	30,994	413,312	318,725	5,657	23,060	5,594	461	476	14	4,332	802,625
1967-68	27,261	365,239	343,595	5,501	31,129	7,272	253	270	2	4,887	785,409
1968-69	36,944	581,981	787,444	8,527	76,368	17,673	244	461	0	11,332	1,520,974
1969-70	39,889	887,124	495,418	9,375	61,235	22,483	186	354	18	6,681	1,522,763
1970-71	36,312	631,756	304,008	4,758	57,334	15,780	90	247	15	2,424	1,052,724
1971-72	59,617	942,488	1,001,002	6,496	83,238	25,100	60	395	6	7,826	2,126,228
1972-73	111,426	1,187,065	1,661,610	27,924	181,917	106,264	156	1,239	18	30,426	3,308,045
1973-74	164,877	2,096,990	2,667,335	96,925	382,102	273,840	489	5,108	52	22,858	5,710,576
1974-75	93,547	1,610,683	2,752,748	83,444	308,219	210,236	284	2,414	128	14,135	5,075,838
1975-76	94,677	2,045,677	5,273,895	57,962	717,697	528,014	218	1,823	147	12,981	8,733,091
1976-77	205,072	2,848,308	5,110,296	111,519	914,881	706,466	458	3,226	456	16,784	9,917,466
1977-78	121,162	2,547,940	4,858,913	72,444	979,835	678,773	292	1,758	127	10,889	9,272,133
Average price paid per pelt											
1965-66	$6.46	$1,69	$2.80	$0.49	$6.03	$2.03	$0.83	$0.82	$1.83	$12.06	
1966-67	5.74	1.01	1.92	.37	2.85	1.36	.54	.66	2.00	11.55	
1967-68	5.78	.84	2.35	.46	3.87	1.83	.52	.73	2.00	12.73	
1968-69	7.13	1.03	4.46	.57	7.80	3.25	.47	.83	...	17.96	
1969-70	5.16	1.20	2.84	.53	4.72	3.24	.41	.72	3.00	12.56	
1970-71	4.13	1.11	1.84	.44	5.29	2.48	.35	.67	3.00	10.18	
1971-72	6.63	1.61	4.28	.43	8.94	3.65	.32	.81	1.50	14.71	
1972-73	10.84	2.25	6.79	.93	16.61	8.60	.49	1.22	3.60	17.42	
1973-74	10.91	2.40	8.90	1.54	24.65	14.43	.62	1.62	5.78	14.74	
1974-75	6.44	2.51	8.24	1.41	18.32	10.96	.46	1.42	9.85	11.70	
1975-76	7.67	3.46	12.84	1.21	37.04	22.74	.57	1.06	10.50	11.86	
1976-77	14.83	4.77	17.38	1.96	42.51	28.90	.62	2.17	12.00	17.18	
1977-78	11.91	4.63	15.63	2.13	43.13	28.24	.56	1.76	10.58	11.86	

[1]Bednarik, Karl E. 1976. Furbearers. Pp. 3-1 to 3-6 in Reasons for seasons, hunting and trapping. Ohio Dept. Nat. Resources, Div. Wildlife Publ. 68 (R 1276). 58 pp.

[2]Mountz, G.L., and R.E. Bonsel. 1978. Fur harvest in Ohio, 1968-1977. Ohio Dept. Nat. Resources, Div. Wildlife Publ. 178 (R 778).

LONG-TAILED WEASEL
Mustela frenata

Color transparency by Karl H. Maslowski

MINK
Mustela vison

BADGER
Taxidea taxus

Appendix Table 3. Summary of Ohio deer hunting seasons 1943–1978. Adapted and updated from Donohoe and Stoll (1976)[1] with data from 1976–78 supplied by M.W. McClain of the Ohio Division of Wildlife (1978, personal communication) and Ohio Division of Wildlife Publ. No. 166 (R379).

	No. days[2]		No. open gun counties	Season type[3]	No. permits sold	No. deer harvested[4]	No. permits sold per deer bagged
	Gun	Bow					
1943	12	12	3	B	8,500	168	51
1944	6	6	2	B	9,200	117	79
1945	6	6	3	B	7,700	62	124
1947	3	2	8	A	9,669	1,000	10
1948	5	5	13	A	23,098	1,600	14
1950	4	4	19	A	22,728	3,500	6
1952	3	3	27	B	14,081	450	31
1953	3	14	40	A	30,033	4,000	8
1955	3	19	42	A	36,419	4,200	9
1956	4	27	88	A	48,334	3,911	12
1957	4	67	88	A	46,435	4,784	10
1958	4	67	88	A	42,777	4,415	10
1959	3	67	88	A	37,976	2,960	13
1960	2	67	88	A	27,430	2,584	11
1962	2	67	88	A	22,806	2,114	11
1963	2	67	88	A	33,298	2,074	16
1964	2	67	88	A	32,400	1,326	24
1965	3,4	64,67	36	A,B	12,808	406	32
1966	3,5	61	48	A,B	24,031	1,073	22
1967	3,5	62	50	A,B	28,108	1,437	20
1968	5,6	68	50	A,B	35,266	1,396	25
1969	5,6	80	56	A,B	45,014	2,105	21
1970	3,5,6	74	51	A,B	53,537	2,387	22
1971	3,5,9	80	63	A,B	74,709	3,831	20
1972	3,5,9	74	63	A,B	89,216	5,074	18
1973	3,5,6	74	59	A,B,P	108,130	7,594	14
1974	3,5,6	92	65	A,B,P	106,107	10,747	10
1975	3,5,6	92	68	A,B,P	113,000	13,480	8
1976	5,3,5,6	86	68	A,B,P	139,352	23,431	6
1977	5,5,6	86	73	A,B,P	155,423	22,319	7
1978	5,5,6	80	74	A,B,P	161,212	22,967	7

[1]Donohoe, R.W., and R. Stoll, Jr. 1976. Forest game. Pp. 2-1 to 2-17 in Ohio Division of Wildlife. Reasons for seasons, hunting and trapping. Ohio Dept. Nat. Resources Publ. No. 68 (R1276).

[2]More than one number in a year indicates a difference between management zones.

[3]A = any deer, B = buck only, P = county anterless permits issued. More than one letter in a year indicates differences between management zones.

[4]Variations in weather, type of season, and season length have affected the annual hunting presure, success rate, and total harvest.

GLOSSARY

Aestivate: to pass the summer in a dormant state; cf. *hibernate*
Agouti: a grizzled fur coloration of alternating light and dark color banding
Albinism: a congenital condition of lack of pigment; thus pinkish or white in appearance
Annulation: formation of rings
Anterior: front; head end; cf. *posterior*
Anterior mental foramen: opening in the front of the lower jaw bone (see fig. 23, p. 122)
Anterior palatine foramina: paired elongate openings forward on the palate just behind the incisors
Anus (anal): posterior opening of the digestive tract
Apiculturist: bee-keeper
Arboreal: living in trees
Articulated: connected by joints; jointed
Auditory bulla: rounded, thin-walled, bony capsule enclosing the bones of the inner ear (see fig. 4, p. 9)
Auditory meatus: ear opening
Baculum: bone in the penis of certain mammals
Bicuspid: tooth with two major cusps; cf. *unicuspid*
Blastocyst (blastula): hollow-ball stage of development of an embryo
Boreal: northern
Bulla: see *auditory bulla*
Calcaneum: heel bone
Calcar: a cartilaginous process arising from the ankle joint of a bat and extending toward the tail
Canid: any member of the family Canidae (dog-like carnivores)
Canine: one of the four kinds of teeth found in mammals; located singly on each side immediately behind incisors and just in front of premolars in upper and lower jaws; often elongated, unicuspid, and single-rooted (see fig. 4, p. 9)
Carnassial: the last (fourth) premolar tooth in the upper jaw and first molar in the lower jaw in the Carnivora; a tooth enlarged and modified for crushing, cutting, and shearing
Carnivorous: preying and feeding on animals; flesh-eating

Castor glands: scent glands located on either side of, and just within, the anal opening of beavers
Castoreum: strong-smelling, oily substance from the castor glands of the beaver
Chitin: hard material in the outer covering of insects
Cloaca: common collecting chamber of reproductive, excretory, and digestive systems; literally, a sewer
Coition (coitus): reproductive act; sexual intercourse
Commensal: (1) in the usual sense, an organism living symbiotically with a host, where the host neither benefits nor suffers from the relationship; (2) in the specific sense, rats and mice living at man's expense, eating his foods, living in his houses, and sharing some of his diseases without contributing anything beneficial to the relationship
Convoluted: coiled, twisted
Coprophagy: eating of dung; e.g., rabbits reingest their soft pellets
Copulation: see *coition*
Cranium: the skull; or more specifically, the portion of the skull enclosing the brain
Crepuscular: active at twilight; e.g., bats move about primarily during late evening and early morning hours
Cricetine: a cricetid rodent of the subfamily Cricetinae (see p. 83); cf. *microtine*
Cursorial: adapted to running; e.g., deer
Cusp: raised area (point) on the crown or biting surface of a tooth
Cyclic: the repetitive variation of mammal population numbers from year to year and over long periods of time—sometimes high, sometimes low—in more or less predictable patterns
Deciduous: shedding at a certain season or stage of growth, as deer antlers
Dentary bone: the bone making up half the lower jaw on each side
Dentition: number and kinds of teeth and their arrangement in the mouth
Diastema: a space (gap) between teeth in the jaw; in rodents, the large space between the incisors and the cheek teeth

GLOSSARY

Dichotomous: branching into two parts; e.g., a dichotomous key provides two choices at each point of decision.

Digitigrade: walking on the toes of the foot, the heel not touching the ground, e.g., fox, cat

Dimorphism: two forms of individuals in the same species; e.g., red bats show sexual dimorphism—the males have red fur, the females yellow fur

Diurnal: active in the daytime; cf. *nocturnal*

Dorsum (dorsal): back or upper surface of the body; cf. *venter*

Endangered: having survival and reproduction in immediate jeopardy

Epididymis: first convoluted portion of the excretory duct of the testis, storing sperm and connecting testis to vas deferens

Epipubic bones: in the marsupials, a pair of bones extending forward from the pelvic girdle in both sexes, supporting the pouch in the female

Estrus (estrous): a state of sexual stimulation in which the female accepts mating with the male and is capable of conceiving

Extinct: eliminated from all areas of its range

Extirpated: eliminated from portions of its range but still surviving elsewhere

Feral: untamed; previously domesticated, but returning to the wild

Fibula: long, thin, outer, and smaller bone of the two in the lower portion of the hind limb

Follicle: a small cavity or pit; e.g., individual hairs grow out of hair follicles; Graafin follicles in the female's ovaries contain developing eggs

Foot tubercles: small fleshy pads on the soles of the feet

Foramen (foramina): an opening or perforation, as through a bone

Fossorial: adapted to digging; living underground

Friable: referring to soil easily crumbled or dug through

Frontal bones: in the skull, the anterior pair of bones on the roof of the braincase between the orbits (see fig. 4, p. 9)

Fulvous: dull reddish- or brownish-yellow

Fur: short, soft thick hairs lying next to the body; cf. *guard hairs*

Gestation period: period of embryonic development between fertilization of the egg and birth of the young

Glans penis: distal end of the penis

Gregarious: living in herds; sociable; cf. *solitary*

Guard hairs: longer, coarser hairs that lie over the fur; cf. *fur*

Herbivorous: feeding on plant material (vegetation)

Heterodont: having several different kinds of teeth in the jaws; cf. *homodont*

Hispid: covered with stiff hairs or small spines

Histogenesis: tissue development and differentiation

Homodont: having all teeth the same; cf. *heterodont*

Hormonal: regulated by hormones, as the estrous cycle

Hypogonodal: deficient in development of secondary sexual characteristics

Implantation: condition of being embedded or fixed firmly in place; e.g., fertilized eggs are implanted in the uterine wall

Incisor: the most anterior of the four kinds of teeth found in mammals (see fig. 4, p. 9); usually chisel-shaped, especially in the rodents

Infraorbital foramen: in the skull, an opening through the zygomatic process of the maxilla (see fig. 4, p. 9)

Inguinal: in the region of the groin

Insectivorous: feeding chiefly on insects

Interfemoral membrane: membrane stretched between the hind legs of bats, enclosing the tail

Keel: a ridge or expanded part of the calcar providing extra attachment surface for the interfemoral membrane

Labyrinth: intricate network of winding passages; a maze

Lachrymal (lacrimal) bone: in the skull, a small bone in the anterior wall of each orbit

Lactation: secretion of milk

Lingual: of the tongue or side toward the tongue

Loops and whorls: enamel ridges bordered by dentine on grinding surface of teeth of rodents

Lophodont: a molar tooth with a pattern of lophs (ridges) on the grinding surface; or a mammal with such teeth (e.g., *Microtus,* the field mouse)

Mammae: mammary glands

Mandible: the lower jaw, in mammals consisting of two dentary bones

Marsupium: an abdominal pouch enclosing the mammary glands of most marsupials; e.g., the opossum

Mast: nuts of forest trees such as beechnuts and acorns used as animal food

Maxilla: the upper jaw

Melanistic: having unusual dark or black coloration

Mesenteric: contained within the mesenteries, membranes that support internal organs in the abdominal cavity

Mesophytic: adapted to living under medium conditions of moisture

Metabolic rate: use of energy by an organism; e.g., shrews burn many calories in a short time, and thus have a high metabolic rate; opossums have a low metabolic rate

Microtine: a cricetid rodent of the subfamily Microtinae (see p. 83); cf. *cricetine*

Midden: refuse heap

Migratory: roving; wandering; moving from one area to another

Molar: one of four kinds of teeth found in mammals; cheek teeth behind (posterior to) the premolars, with no deciduous precursors (see fig. 4, p. 9)

Monestrous: having a single estrus cycle per year; cf. *polyestrous*

Multiparous: producing more than one young at birth

Multituberculate: having several tubercles, or cusps, on the surface of a molar tooth

Musk: secretion from any of the scent glands of a mammal

Mustelid: a member of the weasel family, Mustelidae

Nape: back of the neck

Necrosis: death or decay of a particular body tissue

Nocturnal: active at night; cf. *diurnal*

Nonretractile: unable to be retracted or drawn back; e.g., the claws of the Canidae (dogs, foxes) cannot be drawn back into protective sheaths; cf. *retractile*

Occiput: back part of the skull or head

Occlude: to close the upper and lower jaws with the cusps of the teeth fitting together

Omnivorous: eating both plant and animal matter

Opposable: capable of being placed opposite something else; e.g., the big toe of an opossum can be placed opposite each of the other toes on the foot

Orbit: in the skull, the bony socket that houses the eye

Ovaries: in the female, the primary reproductive structures in which eggs are produced

Ovulation: release of an egg or eggs from the ovary

Palate (palatal, palatine): the roof of the mouth

Palatine foramen: opening in the palate (see fig. 17, p. 85)

Papilla: small nipple-like projection or process

Parturition: the process of giving birth

Pectoral: located in or on the chest area; pectoral limbs are the forelimbs

Pelage: hair covering the body of a mammal

Pencil: tuft of hair extending beyond end of tail

Penis: male organ of copulation

Pinna: fleshy, external ear

Placenta: the organ uniting the fetus to the maternal uterus in most animals

Plantar tubercles: raised areas on the soles of the feet of mammals; friction pads

Plantigrade: walking on the soles of the feet with the heel touching the ground, e.g., bear

Polyestrous: having several estrous cycles during a single year; cf. *monestrous*

Posterior: at or toward the hind end of the body; cf. *anterior*

Postorbital process: in the skull, the projection of the frontal bone that marks the posterior boundary of the orbit (see fig. 4, p. 9)

Prehensile: adapted to grasping

Prehibernal: before winter; or before going into hibernation

Premolar: one of the four kinds of teeth found in mammals; located anterior to the molars and posterior to the canines (see fig. 4, p. 9)

Prolific: producing many young

Promiscuous: in reference to breeding, mating with several different individuals

Ramus: the vertical posterior part of the lower jawbone

Retractile: able to be retracted or drawn back, as cats' claws into their sheaths; cf. *nonretractile*

Rostrum: facial region of the skull, anterior to the orbits

Rufous: brownish-red; rust-colored

Runway: paths or runs on the surface of the ground used repeatedly by mammals when moving about

Rutting season: annually recurring season of sexual activity when mating occurs, especially in reference to male deer

Sagittal crest: in the skull, the median ridge or crest on top of the brain case resulting from the fusion of the temporal ridges (see fig. 22, p. 121)

Saltatorial: adapted to jumping

Sciurid: member of the family Sciuridae (squirrels)

Scrotum: the pouch of skin outside the body cavity that contains the testes

Sculling: rowing; an action produced by the tail of beavers and muskrats

Sectorial: referring to teeth modified for cutting and tearing

Semiplantigrade: walking partially, but not completely, on the soles of the feet, e.g., raccoon

Solitary: living alone or in pairs; not colonial; cf. *gregarious*

Sternum: breastbone; flat bone or cartilage to which most of the ribs are attached

Subapical notch: notch on upper incisor (see fig. 21, p. 111)

Subcutaneous: beneath the skin

Subnivean: beneath the snow

Subspecies: A subdivision of a given species made on minor characteristic differences such as color, size, and weight

Subterranean: underground

Supraorbital process: in the skull, the bones occurring above the orbits of the eye

Tactile: relating to the sense of touch

Temporal ridges: in the skull, two usually curved, raised lines or ridges, one on each side of the brain case; if they meet posteriorly, they give rise to the sagittal crest (see fig. 22, p. 121)

Testes: the male reproductive glands

Tibia: innermost and larger of the two bones in the lower hind leg; the shin bone

Tragus: prominent extension of skin immediately in front of, and at the bottom of, the external ear opening, especially in bats

Tubercle: a small knob or protrusion

Tularemia: an infectious bacterial disease of rodents and rabbits, transmissible to man by handling the flesh of infected animals or by the bite of certain ticks and insects

Tympanic bulla: see *auditory bulla*

Unguligrade: walking on hoofs, e.g., deer

Unicuspid: tooth with a single cusp, e.g., the canine tooth; cf. *bicuspid*

Uterine tract: the portion of the female reproductive tract in which the embryo develops

Vaginal orifice: external opening of the female reproductive tract

Vas deferens: long, convoluted duct carrying sperm from the epididymus to the penis

Vector: an organism or carrier that transmits a disease-producing organism from one host to another

Venter (ventral): the under or lower surface of the body; the belly; cf. *dorsum*

Vibrissae: long, stiff hairs around the nose, on the lips, or on other areas of the head of many mammals; e.g., the "whiskers" of a mouse

Vole: a short-tailed mouse, with tail equal to, shorter than, or, at most, slightly longer than the hind foot

Zygomatic arch: in the skull, an arch of bone that encloses the orbit (see fig. 4, p. 9)

Zygomatic plate: the flattened, expanded part of the maxillary in front of the eye socket from which the anterior part of the zygomatic arch arises

GENERAL REFERENCES

This is not intended to be a complete list of all general mammalogy works. It includes selected publications of historic interest or of a more general nature than many of the references listed after each species account.

Allen, G. M. 1939. Bats. Harvard Univ. Press, Cambridge, Mass. 368 pp.

———. 1942. Extinct and vanishing mammals of the western hemisphere. Spec. Publ. Comm. Intnatl. Wildl. Protect. Assoc., Lancaster, Pa. 620 pp.

Anderson, S., and J. K. Jones, Jr., eds. 1967. Recent mammals of the world: a synopsis of families. Ronald Press Co., New York. 453 pp.

Anthony, H. E. 1928. Field book of North American mammals. G. P. Putnam & Sons, New York. 625 pp.

Asdell, S. A. 1964. Patterns of mammalian reproduction. 2d ed. Comstock Publ. Assoc., Ithaca, N.Y. 670 pp.

Audubon, J. J., and J. Bachman. 1846. The viviparous quadrupeds of North America. V. G. Audubon, New York. 334 pp. Reprinted 1974 as: The quadrupeds of North America. Arno Press, New York. 3 vols.

Banfield, A. W. F. 1974. The mammals of Canada. Publ. for Natl. Mus. Nat. Sci. (Canada) by Univ. of Toronto Press, Toronto, Ont. 438 pp.

Barbour, R. W., and W. H. Davis. 1969. Bats of America. Univ. Press of Ky., Lexington. 286 pp.

———. 1974. Mammals of Kentucky. Univ. Press of Ky., Lexington. 322 pp.

Blair, W. F. 1957. Mammals. Part 6. Pp. 615-774 *in* Blair, Blair, Brodkorb, Cagle, and Moore. The vertebrates of the United States. McGraw-Hill, New York. 819 pp.

Bole, B. P., Jr. 1932. An addition to the known list of Ohio mammals. Ohio J. Sci. 32:402.

———. 1939. The quadrat method of studying small mammal populations. Sci. Publ. Cleveland Mus. Nat. Hist. 5:15-77.

Bole, B. P., Jr., and P. N. Moulthrop. 1942. The Ohio recent mammal collection in the Cleveland Museum of Natural History. Sci. Publ. Cleveland Mus. Nat. Hist. 5:83-181.

Borror, D. J. 1960. Dictionary of wordroots and combining forms. N-P Publ., Palo Alto, Calif. 134 pp.

Bouliere, F. 1955. Mammals of the world, their life and habits. Knopf Press, New York. 223 pp.

———. 1956. The natural history of mammals. 2d ed. Knopf Press, New York. 364 pp.

Brayton, A. W. 1882. Report on the mammals of Ohio. Rep. Geol. Surv. Ohio 4:1-185.

Burt, W. H. 1946. The mammals of Michigan. Univ. of Mich. Press, Ann Arbor. 288 pp.

———. 1957. Mammals of the Great Lakes region. Univ. of Mich. Press, Ann Arbor. 246 pp.

Burt, W. H., and R. P. Grossenheider. 1976. A field guide to the mammals. 3d ed. Houghton Mifflin, Boston, Mass. 289 pp.

Cahalane, V. H. 1947. Mammals of North America. MacMillan Co., New York. 682 pp.

Chapman, F. B. 1938. The development and utilization of the wildlife resources of unglaciated Ohio. 2 vols. Ph. D. dissertation, Ohio State Univ., Columbus. 791 pp.

Cochran, R. W., Jr. 1960. Ohio's wildlife resources. Ohio Dept. Nat. Resources, Div. Wildlife, Columbus. 210 pp.

Cockrum, E. L. 1962. Introduction to mammalogy. Ronald Press, New York. 455 pp.

Colbert, E. H. 1955. Evolution of the vertebrates: a history of the backboned animals through time. John Wiley and Sons, New York. 479 pp. 2d ed., 1969. 535 pp.

Collins, H. H. 1959. Complete field guide to American wildlife. Harper and Row Publ. Co., New York. 683 pp.

Condit, J. M. 1966. The classification of recent mammals of Ohio. Ohio Hist. Soc., Ohio State Mus. Info. Series 1(4). 8 pp. Mimeo.

DeCapita, M. E. 1975. Evaluation of strip-mine reclamation for terrestrial wildlife restoration. M. S. thesis, Ohio State Univ., Columbus. 134 pp.

DeCapita, M. E., and T. A. Bookhout. 1975. Small mammal populations, vegetational cover, and hunting use of an Ohio strip-mined area. Ohio J. Sci. 75:305-13.

DeVos, A. 1964. Range changes of mammals in the Great Lakes region. Amer. Midl. Nat. 71:210-31.

Donachy, N. D. 1958. A survey of intestinal helminthes of small mammals from Vinton County, Ohio. M. S. thesis, Ohio State Univ., Columbus. 34 pp.

Doutt, J. K., C. A. Heppenstall, and J. E. Guilday. 1977. Mammals of Pennsylvania. 4th ed. Pa. Game Commission, Harrisburg, in coop. with Carnegie Museum, Pittsburgh. 288 pp.

Doutt, J. K., A. B. Howell, and W. B. Davis. 1945. The mammal collections of North America. J. Mammal. 26:231-72.

Eadie, W. R. 1954. Animal control in field, farm, and forest. MacMillan Co., New York. 257 pp.

Ellerman, J. R. 1940. The families and genera of living rodents. Vol. 1. Publ. Brit. Mus. Nat. Hist., London. 689 pp.

Elton, C. S. 1942. Voles, mice, and lemmings; problems in population dynamics. Clarendon Press, Oxford. 496 pp.

Enders, R. K. 1930. Some factors influencing the distribution of mammals in Ohio. Occ. Papers, Univ. of Mich. Mus. Zool. 212:1-27.

Giles, R. H., Jr. 1966. Early natural history of a forested area near Dover, Ohio. Ohio J. Sci. 66:469-73.

Glass, B. P. 1951. A key to the skulls of North American mammals. Dept. Zool., Okla. State Univ., Stillwater. 54 pp.

Golley, F. B. 1962. Mammals of Georgia: a study of their distribution and functional role in the ecosystem. Univ. of Ga. Press, Athens. 218 pp.

Good, E. E. 1963. Some reports concerning interesting mammals in Ohio. Wheaton Club (Columbus, Ohio) Bull. 8:25-27.

Goodpaster, W. W. 1941. Mammals of southwestern Ohio. J. Cincinnati Soc. Nat. Hist. 22:41-47.

Goodpaster, W. W., and D. F. Hoffmeister. 1968. Notes on Ohioan mammals. Ohio J. Sci. 68:116-17.

Gottschang, J. L. 1965. Winter populations of small mammals in old-fields in southwestern Ohio. J. Mammal. 46:44-52.

Griffin, D. R. 1958. Listening in the dark. Yale Univ. Press, New Haven, Conn. 413 pp.

Gunderson, H. L., and J. R. Beer. 1953. The mammals of Minnesota. Univ. of Minn. Press, Minneapolis. 196 pp.

Hahn, W. L. 1909. Mammals of Indiana. 33d Annu. Rep., Ind. Dept. Geol. and Nat. Resources. Pp. 419-654.

Hall, E. R., and K. R. Kelson. 1959. The mammals of North America. Ronald Press Co., New York. 2 vols., 1162 pp.

Hamilton, W. J., Jr. 1939. American mammals: their lives, habits, and economic relations. McGraw-Hill Book Co., New York. 434 pp.

———. 1943. The mammals of eastern United States: an account of recent land mammals occurring east of the Mississippi. Comstock Publ. Co., Ithaca, N. Y. 438 pp.

Handley, C. O. 1971. Appalachian mammalian geography—Recent Epoch. In Perry C. Holt, ed. The distribution history of the biota of the southern Appalachians. Part 3, Vertebrates. Research Div. Monograph 4, Va. Polytech. Inst. and State Univ., Blacksburg, Va.

Handley, D. E. 1962. Wetlands game. Game Research Ohio 1:60-79.

Harper, F. 1929. Notes on the mammals of the Adirondacks. Handbook, N. Y. State Museum 8:51-118.

Hart, E. D. 1949. A study of the food habits and population of mammals in central Ohio crop fields. M. S. thesis, Ohio State Univ., Columbus. 60 pp.

Hibbard, C. W., D. E. Ray, D. E. Savage, D. W. Taylor, and J. E. Guilday. 1965. Quaternary mammals of North America. Pp. 509-25 in H. E. Wright, Jr., and D. G. Frey, eds. The Quaternary of the United States. Princeton Univ. Press, Princeton, N. J. 922 pp.

Hicks, L. E. 1938. The status of game mammals in Ohio. Trans. Third North Amer. Wildl. Conf. 3:415-20.

Hine, J. S. 1906. Notes on some Ohio mammals. Ohio Nat. 7:550-51.

———. 1929. Distribution of Ohio mammals. Annu. Rept. Proc. Ohio Acad. Sci. 8(6):260-68.

Hoffmeister, D. F., and C. O. Mohr. 1957. Fieldbook of Illinois mammals. Manual No. 4, Ill. Nat. Hist. Surv. Reprinted 1972, Dover Publ., New York. 233 pp.

Jackson, H. H. T. 1961. Mammals of Wisconsin. Univ. of Wisc. Press, Madison. 504 pp.

Kellogg, R. 1937. Annotated list of West Virginia mammals. Proc. U. S. Natl. Mus. 84(3022):443-79.

Kirtland, J. P. 1838. Report on the zoology of Ohio. Second Annu. Rep., Geol. Surv. Ohio. 2:157-200.

Laub, K. W. 1975. Wildlife conservation in Ohio: the role of hunting and trapping. Ohio Dept. Nat. Resources, Div. Wildl. Publ. 273. 81 pp.

Long, R. J. 1951. The Sciuridae of Ohio: a study of their ecology, anatomy of their digestive tract, and brain. Ph. D. dissertation, Ohio State Univ., Columbus. 152 pp.

Lyon, M. W., Jr. 1936. Mammals of Indiana. Univ. of N. D. Press, Notre Dame, Ind. 384 pp. Or Amer.

Midl. Nat. 17:1-384.

———. 1942. Additions to the "Mammals of Indiana." Amer. Midl. Nat. 27:790-91.

Matthews, L. H. 1970. The life of mammals. Vol. 1. Universe Books, New York. 340 pp.

Metzger, B. 1955. Notes on the mammals of Perry County, Ohio. J. Mammal. 36:101-5.

Miller, G. S., and R. Kellogg. 1955. List of North American recent mammals. Bull. U. S. Natl. Mus. 205. 924 pp.

Moseley, E. L. 1906. Notes on the former occurrence of certain mammals in northern Ohio. Ohio Nat. 6:504-5.

Mountz, G. L., and R. E. Bonsel. 1978. Fur harvest in Ohio, 1968-1977. Ohio Dept. Nat. Resources, Div. Wildl. Publ. 178 (R 778).

Mumford, R. E. 1969. Distribution of the mammals of Indiana. Monograph No. 1, Ind. Acad. Sci., Indianapolis, Ind. 114 pp.

Murie, O. J. 1975. A field guide to animal tracks. 2d. ed. Houghton Mifflin Co., Boston, Mass. 375 pp.

Necker, W. L., and D. M. Hatfield. 1941. Mammals of Illinois. Chicago Acad. Sci. Bull. 6:17-60.

Novick, A. 1969. The world of bats. Holt, Rinehart & Winston, New York. 171 pp.

Ohio Division of Wildlife. 1976a. Endangered wild animals in Ohio. Ohio Dept. Nat. Resources Publ. 316 (R 576). 3 pp.

———. 1976b. Reasons for seasons, hunting and trapping. Ohio Dept. Nat. Resources Publ. No. 68 (R 1276). 57 pp.

———. 1978. Trapping in Ohio. Ohio Dept. Nat. Resources Publ. 115. 58 pp.

Palmer, R. S. 1954. The mammal guide. Doubleday, Garden City, N. Y. 384 pp.

Peterson, R. L. 1966. The mammals of eastern Canada. Oxford Univ. Press, Toronto, Ont. 465 pp.

Petrides, G. 1948. Sex and age ratios: their determination and uses in game and fur animals of the eastern United States. Ph. D. dissertation, Ohio State Univ., Columbus. 229 pp.

Phillips, R. S. 1947. Notes on some mammals of Hancock County, Ohio. J. Mammal. 28:189-90.

Preble, N. A. 1942. Notes on the mammals of Morrow County, Ohio. J. Mammal. 23:82-86.

Reynolds, L. P. 1939. Ecological distribution of certain small mammals in southeastern Ohio. M. S. thesis, Ohio Univ., Athens. 63 pp.

Sanderson, I. T. 1955. Living mammals of the world. Hanover House, Garden City, N. J. 303 pp.

Schwartz, C. W., and E. R. Schwartz. 1959. The wild mammals of Missouri. Univ. of Mo. Press and Mo. Conserv. Comm. 341 pp.

Scott, W. B. 1913. A history of land mammals in the western hemisphere. MacMillan Co., New York, N. Y. 786 pp. Rev. ed., 1937, 786 pp.

Seton, E. T. 1925-28. Lives of game animals. 4 vols. Doubleday, Page & Co., Garden City, N. Y.

Simpson, G. G. 1945. The principles of classification and a classification of mammals. Bull. Amer. Mus. Nat. Hist. 85. 350 pp.

Smith, C. E. 1940. A study of small mammals of central Ohio. M. S. thesis, Ohio State Univ., Columbus. 66 pp.

Smith, H. G., R. K. Burnard, E. E. Good, and J. M. Keener. 1973. Rare and endangered vertebrates of Ohio. Ohio J. Sci. 73:257-71.

Taylor, W. P., ed. 1956. The deer of North America: the white-tailed, mule, and black-tailed deer, genus *Odocoileus*, their history and management. Stackpole Co., Harrisburg, Pa., and Wildl. Manage. Inst., Washington, D. C. 668 pp.

Thomas, E. S. 1951. Distribution of Ohio animals. Ohio J. Sci. 51:153-67.

Thomas, E. S., and B. P. Bole, Jr. 1938. List of the mammals of Ohio. Bull. Ohio State Mus. 1:1-2.

Thomas, E. S., S. Kress, and C. Triplehorn. 1967. Neotoma vertebrates, exclusive of birds. Wheaton Club (Columbus, Ohio) Bull. 12(1):25-26.

Trautman, M. B. 1931. List of the mammals of Ohio. Bur. Sci. Research, Dept. Conserv., Ohio Dept. Agric. Bull. 54. 2 pp.

———. 1939. The numerical status of some mammals throughout historic time in the vicinity of Buckeye Lake, Ohio. Ohio J. Sci. 39:133-43.

———. 1957. The fishes of Ohio. Ohio State Univ. Press, Columbus. 683 pp.

———. 1977. The Ohio country from 1750 to 1977—a naturalist's view. Ohio Biol. Surv. Biol. Notes No. 10. 25 pp.

Vaughan, T. A. 1972. (2d ed., 1978). Mammalogy. W. B. Saunders Co., Philadelphia, Pa. 522 pp.

Walker, E. P., ed. 1975. Mammals of the world. 3d ed. 2 vols. Johns Hopkins Univ. Press, Baltimore.

Wilson, L. W., and J. E. Friedel. 1942. A list of mammals collected in West Virginia. Proc. W. Va. Acad. Sci. for 1941, Vol. 15. (W. Va. Univ. Bull. Ser. 42, Nos. 8-11.) pp. 85-92.

Young, J. Z. 1962. The life of vertebrates. 2d ed. Oxford Univ. Press, New York. 820 pp.

Zim, H. S., and D. F. Hoffmeister. 1955. Mammals. Simon and Schuster, New York. 160 pp.

INDEX

Species are indexed under both their common and scientific names.
For technical terms, see Glossary, pp. 167-70.
For physiographic regions of Ohio, see fig. 1, p. 5, and discussion, p. 4.
For Ohio counties, see fig. 2, p. 6.
For common body measurements, see fig. 3, p. 8.
For names of skull bones and teeth, see figs. 4-5, pp. 9-10.

Allegheny woodrat. *See* Eastern woodrat
American beaver. *See* Beaver
American buffalo. *See* Bison
American elk. *See* Wapiti
Antlers, 143, 147
Artiodactyla, order, 143-49, 155-56

Badger, 138-39
Bassariscus astutus, 153
Bat, 38-52
   big brown, 46-47
   eastern pipistrelle, 44-45
   evening, 50-51
   gray, 151
   hoary, 49-50
   Indiana, 41-42
   Keen's, 41
   little brown, 38-40
   Rafinesque's big-eared, 51-52
   red, 48-49
   silver-haired, 43-44
   small-footed, 43
Bear, black, 153
Beaver, 80-83
Bison, 156
Bison bison, 156
Blarina brevicauda, 23-25
Bobcat, 155

Canidae, family, 119-28, 153
   key to species, 120
Canis latrans, 120-23
Canis lupus, 153
Carnivora, order, 119-42, 153-55
   key to families, 119
Castor canadensis, 80-83

Castoridae, family, 80
Cervidae, family, 143
Cervus elaphus, 155
Checklist, mammals of Ohio, 11-12
Chickaree. *See* Squirrel, red
Chiroptera, order, 35-52
   echo-location, 36
   feeding, 36
   metabolism, 36
   parasites, 36
   rabies, 36
   reproduction, 36
   walking, 36
   wings, 36
Clethrionomys gapperi, 94-95
Condylura cristata, 17, 32-33
Coprophagy, 53, 55
Corynorhinus rafinesquii.
   See Plecotus rafinesquii
Cottontail rabbit. *See* Eastern cottont
Cougar, 154
Coyote, 120-23
Cricetidae, family, 83
   key to species, 83-84
Cricetinae, 84-94
Cryptotis parva, 17, 26-28
   C. p. elasson, 28
   C. p. harlani, 28
   C. p. parva, 28

Deer hunting seasons 1943-78,
   summary of, 165
Deer, white-tailed, 143-47
Didelphidae, family, 13
Didelphis virginiana, 13-16
Dipodomys ordii, 152
Distribution maps, explanation of, 5

Eastern chipmunk, 60-62
Eastern cottontail, 53-57
Eastern woodrat, 91-93
Eptesicus fuscus, 46-47
Ermine, 131-32
Erethizon dorsatum, 152
Extirpated species, 151-57

Felis concolor, 154
Felis lynx, 155
Felis rufus, 155
Fisher, 154
Foods and feeding habits.
   See individual species.
Fossil record, 1
Fossil Ohio mammals, 1
Fox
   gray, 126-27
   red, 123-25
Furs, numbers sold and prices paid
   1952-78, 164

General information, Ohio mammals, 3-4
Glaucomys volans, 77-79
Glossary, 167-70
Groundhog, 63-65
Gulo gulo, 154

Hare, snowshoe or varying, 152

Insectivora, order, 17-34
   key to families and species, 17

Key to orders of mammals in Ohio, 9-10
   key to skulls, 159-62
Keys, how to use, 7

175

INDEX

Lagomorpha, order, 53–57
Lasionycteris noctivagans, 43–44
Lasiurus borealis, 48–49
Lasiurus cinereus, 49–50
Lemming, southern bog, 105–7
Leporidae, family, 53, 151, 152
Lion, mountain, 154
Lutra canadensis, 154
Lynx, Canada, 155

Mammals, general
  hair, 3
  molt, 3
  reproduction, 3
  home range, 4
  territoriality, 4
  identification, 7
  common body measurements, fig. 3, p. 8
  skull bones and teeth, figs. 4–5, pp. 9–10
  collecting, 9
  habitat preference, 4
Marmota monax, 63–65
Marsupialia, order, 13
Marten, 154
Martes americana, 154
Martes pennanti, 154
Mephitis mephitis, 140–42
Metric conversion table, 163
Microsorex hoyi, 22
Microtinae, 95–107
Microtus ochrogaster, 98–99
Microtus pennsylvanicus, 95–97
Microtus pinetorum, 99–102
Mink, 137–38
Mole, 28–34
  eastern, 30–32
  hairy-tailed, 28–29
  star-nosed, 32–33
  Townsend's, 31
Mouse
  Cooper's, 105–7
  deer, 86–88
  Eastern harvest, 84–86
  field, 95–97
  golden, 152
  house, 110–12
  meadow jumping, 112–15
  pine, 99–102
  white-footed, 88–91
  woodland jumping, 115–17
Muridae, family, 107
  key to species, 107
Mus musculus, 110–12
Muskrat, 102–4
Mustela erminea, 131–32
Mustela frenata, 134–36
Mustela nivalis, 132–34
Mustela vison, 137–38
Mustelidae, family, 131–42, 154
  key to species, 131

Myocastor coypus, 152
Myotis grisescens, 151
Myotis keenii, 41
Myotis leibii, 43
Myotis Lucifugus, 38–40
Myotis sodalis, 41–42

Napaeozapus insignis, 115–17
Natural environment, Ohio, 4
Neotoma floridana, 91–94
Nutria, 152
Nycticeius humeralis, 50–51

Ochrotomys nuttalli, 152
Odocoileus virginianus, 143–47
Ondatra zibethicus, 102–4
Orders, key to Ohio, 9–10
Oryctolagus sp., 55
Oryzomys palustris, 152
Otter, river, 154

Pack rat. *See* Eastern woodrat
Panther, 154
Parascalops breweri, 17, 28–29
Peromyscus leucopus, 88–91
Peromyscus maniculatus, 86–88
Physiographic sections of Ohio, fig. 1, p. 5
Pipistrellus subflavus, 44–46
Plecotus rafinesquii, 51–52
Poison bite, shrew, 24
Porcupine, 152
Possum. *See* Virginia opussum
Procyon lotor, 128–31
Procyonidae, family, 128

Rabbit
  domestic, 143
  swamp, 151
  snowshoe, 152
Raccoon, 128–31
Rat
  eastern wood, 91–93
  black or roof, 152
  marsh rice, 152
  Ord's kangaroo, 152
  Norway, common, or brown, 108–9
Rattus, norvegicus, 108–9
Rattus rattus, 152
Reithrodontomys humulis, 84–86
Reproduction and reproductive habits. *See* individual species
Ringtail, 153
Rodentia, order, 59–117
  key to families, 59

Scalopus aquaticus, 17, 30–32
Scapanus townsendii, 31
Sciuridae, family, 60
  key to species, 60
Sciurus carolinensis, 69–72
Sciurus niger, 73–74

Shrew
  least, 26–28
  masked, 18–21
  pygmy, 22–23
  short-tailed, 23–26
  smoky, 21–22
Skunk, striped, 140–42
Snowshoe hare, 152
Sorex cinereus, 18–21
  S. c. lesurii, 20
  S. c. ohionensis, 20
Sorex femeus, 21–22
Soricidae, family, 17–28
  key to species, 17
Spermophilus tridecemlineatus, 66–69
Squirrel
  fox, 73–74
  gray, 69–71
  red, 74–77
  southern flying, 77–79
  thirteen-lined ground, 66–69
Striped gopher. *See* Squirrel, thirteen-lined ground
Swamp rabbit, 151
Sylvilagus floridanus, 53–57
Synaptomys cooperi, 105–7

Tadarida brasiliensis, 151
Talpidae, family, 28–34
Tamias striatus, 60–63
Tamiasciurus hudsonicus, 74–77
Taxidea taxus, 138–39
Tularemia, 56

Urocyon cinereoargenteus, 126–27
Ursus americanus, 153

Vespertilionidae, family, 36–52
  key to species, 37–38
  migratory vs. nonmigratory, 37
  populations, 36–37
Virginia oppossum, 13–16
Vole
  meadow, 95–98
  prairie, 98–99
  woodland (or pine), 99–102
  southern red-backed, 94–95
Vulpes vulpes, 123–25

Wapiti, 155
Weasel
  least, 132–34
  long-tailed, 134–36
  short-tailed, 131–32
Wolf, gray or timber, 153
Wolverine, 154
Woodchuck, 63–65

Zapodidae, family, 112
  key to species, 112
Zapus hudsonius, 112–5

DATE DUE